630.973 LOV 2006

Love of the land :
essential farm and

Love of the Land

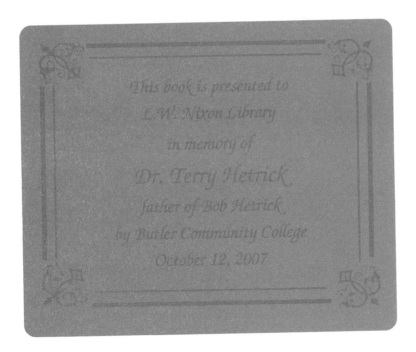

LOVE OF THE LAND

Essential Farm and Conservation
Readings from an American
Golden Age, 1880–1920

Zachary Michael Jack

CAMBRIA
PRESS

YOUNGSTOWN, NEW YORK

Copyright 2006 Zachary Jack

All rights reserved
Printed in the United States of America

No part of this publication may be reproduced, stored in or introduced into a retrieval system, or transmitted, in any form, or by any means (electronic, mechanical, photocopying, recording, or otherwise), without the prior permission of the publisher. Requests for permission should be directed to permissions@cambriapress.com, or mailed to Permissions, Cambria Press, PO Box 350, Youngstown, New York 14174-0350.

Library of Congress Cataloging-in-Publication Data
   Love of the land : essential farm and conservation readings from an American Golden Age, 1880-1920 / edited by Zachary Michael Jack.
      p. cm.
   Includes bibliographical references.
   ISBN-13: 978-1-934043-33-2 (alk. paper)
   ISBN-10: 1-934043-33-8
   1. Agricultural conservation—United States—History—19th century. 2. Agricultural conservation—United States—History—20th century. I. Jack, Zachary Michael, 1979- II. Title.

S604.6.L68 2006
630.973'09034--dc22

2006101727

630.973 LOV 2006

Love of the land :
essential farm and

*For Mule Team I, Mule Team II,
and the rest of the headstrong crew*

TABLE OF CONTENTS

Preface xiii

Acknowledgments xvii

America's Love of the Land: An Introduction 1

Editor's Notes 17

Forerunners: Readings 1770–1880

    Introducing the Forerunners 19

        "Letter II, On the Situation, Feelings,
            and Pleasures of an American Farmer"
            Jean de Crevecoeur 21

        "Notes on the State of Virginia, Query XIX"
            Thomas Jefferson 31

        "How Democratic Institutions
            and Manners Tend to Raise Rents ..."
            Alexis de Tocqueville 34

        "Nature"
            Ralph Waldo Emerson 38

        "Address Delivered Before the Agricultural
            Society of Rutland County, 1847"
            George Perkins Marsh 41

"The Ploughboy"
    John Muir 48

"Walking"
    Henry David Thoreau 53

"Birds and the Services They Render"
    Edmund Morris.. 62

"Preliminary Report Upon Yosemite
    and Big Tree Grove"
    Frederick Law Olmsted 69

"Phases of Farm Life"
    John Burroughs... 73

Part I: The Gilded Age, Readings 1880–1899

Introducing the Gilded Age 81

Farm Pages: The Gilded Age 94

"Footpaths"
    John Burroughs... 101

"Selections from Specimen Days
    and Notes Left Over"
    Walt Whitman 107

"Ranching the Badlands"
    Theodore Roosevelt.. 113

"The Home Garden"
    Tuisco Greiner... 122

Table of Contents ix

"On Sectionalism: Selections from
   The Farmers' Alliance History"
   by Leonidas K. Polk
   and Benjamin Hutchinson Clover 130

"On Roles for Women: Selections from
   The Farmers' Alliance History"
   by Bettie Gay and Jennie E. Dunning 140

"History of the Colored Farmers'
   National Alliance and Cooperative Union"
   R. M. Humphrey 152

"Physical Development"
   Helen Gilbert Ecob 158

"Destruction of the Forest"
   Edwin J. Houston 168

"The Prevalence of Weeds"
   Thomas Shaw 172

"The Significance of the Frontier in
   American History"
   Frederick Jackson Turner 178

"A Windstorm in the Forest"
   John Muir 182

"The Advantages of Irrigation"
   Lucius "Lute" Wilcox 192

"The Fertility of the Land:
   A Chat with a Young Farmer"
   Isaac Phillips Roberts 198

"Women as Farmers"
Frances Elizabeth Willard 204

"The Agricultural Colleges"
Edward Francis Adams 211

"The Martyr and the Pioneer"
Charlotte Perkins Gilman 232

"Conspicuous Consumption"
Thorstein Veblen 241

Part II: The Golden Age, Readings 1900–1920

Introducing the Golden Age 249

Farm Pages: The Golden Age 267

"Renovating the Deserted Homestead"
Edward Payson Powell 276

"The Land of Little Rain"
Mary Hunter Austin 283

Introducing the Boone and Crockett Club 289

"The Mountain Sheep and Its Range"
George Bird Grinnell 293

"Life of the Sequoia and the History of Thought"
Henry Fairfield Osborn 296

"Wilderness Reserves"
Theodore Roosevelt 299

Table of Contents xi

"Earth and Man"
   Nathaniel Southgate Shaler 302

"Local Degeneracy"
   Wilbert Lee Anderson 311

"The Nature Movement"
   Dallas Lore Sharp 322

"The Garden Yard"
   Bolton Hall 330

Introducing the Country Life Commission 333

"To the Senate and the House of Representatives"
   Theodore Roosevelt 335

"Woman's Work on the Farm"
   Liberty Hyde Bailey et al. 343

"Nature Study and School Grounds"
   H. W. Foght.. 350

"The Moral Issue"
   Gifford Pinchot... 359

"The Argument for Cooperation"
   John Lee Coulter 364

"Birds"
   Mary Huston Gregory 372

"Protection of Water Supplies as
    a Conservation of Natural Resources"
    Ellen H. Richards 378

"Church and Community"
    Warren H. Wilson.. 383

"Selections from Farm Boys and Girls"
    William A. McKeever.. 388

"Hetch Hetchy Valley"
    John Muir 403

"The Former Abundance of Wildlife"
    William Temple Hornaday 409

"It Is Kindly"
    Liberty Hyde Bailey 419

"The Country Girl, Where Is She?"
    Martha Foote Crow.. 423

"The Health of Farm Folk"
    Henry Wallace 431

"Country Fetes"
    Charles Josiah Galpin... 437

"The Farming That Is Not Farming"
    Kenyon L. Butterfield... 447

"At a World Table"
    Marion Florence Lansing 459

About the Editor 465

Preface

Mine is a family that loves the land and has lived from it as far back as any of us can remember. My great-grandfather, Walter Thomas Jack, author of the soil conservation classic *The Furrow and Us*, wrote perhaps the best known chapter in our love letter to the land in 1946, leaving me to plumb the depths, three generations later, of such agrarian koans as *the soil can never be better than the tiller* or this equally provocative aphorism breathed by Walter's Quaker brethren in the gloaming of 1890s West Branch, Iowa: *if thee tries to beat nature, nature will beat thee.*

Soil can never be better than the tiller. If thee tries to beat nature, nature will beat thee. For better or for worse, I grew up on such stuff, translated, as it was, through intervening generations. And while my great-grandfather is indeed the deepest seed from which this work grows, my grandfather, Edward, and my father, Michael—both award-winning conservation tillers—are equally important, as are their partners, my grandmother and mother.

The writers anthologized in *Love of the Land: Essential Farm and Conservation Readings from an American Golden Age, 1880–1920*— this first-of-its-kind interdisciplinary anthology—represent a who's who of farm and conservation writers fresh from a golden age. The table of contents represents preeminent agrarian literary voices—Mary Hunter Austin, Charlotte Perkins Gilman, Dallas Lore Sharp, Henry David Thoreau, and Walt Whitman prominent among them; groundbreaking social scientists—Wilbert Lee Anderson, Kenyon L. Butterfield, John Lee Coulter, H. W. Foght, Charles Josiah Galpin, Frederick Jackson Turner, Thorstein Veblen, and others; and the household names of conservation studies—Liberty Hyde Bailey, John Burroughs, John Muir, Gifford Pinchot, Ellen H. Richards, Theodore Roosevelt, and Henry Wallace. Made by referral and endorsement of top scholars in the field, the selections in *Love of the Land* reflect thorough peer review. Their breadth suggests the true diversity of those who love the land and its people. Their presentation as primary works means the giants of conservation speak for themselves. Their nature, concise and highly readable, suggests a careful and voluminous survey of the literature for consumption and agriculture by nature enthusiast and credentialed historian alike. Finally, their extensive introductions, author biographies, and special sections—such as the supplementary anthology-within-an-anthology of agrarian and conservation forerunners, 1770–1880—offer context, as rich and full as space allows.

The result is a book to which I plan to return often in my own teaching of rural studies and environmental narrative and in my own life lived on the land. For me, the voices behind the more than fifty readings in *Love of the Land*, sampled from a golden age of farming and conservation writing, echo eerily across the generations, just as surely as my great-grandfather's whisper (*Because a man owns a farm does not mean that he owns the land to do with as he pleases*) resonates for me. I hope the reader finds here writing to be quoted often and well, writing that is engaging and enlightening, and writing that seems uncannily contemporary in its substance. And I sincerely hope the work anthologized here not only informs a love of the land, but sustains it.

I have worked as a groundskeeper, a librarian, a journalist, and a professor, among other callings, and yet my most enduring role is that of a conservation farmer's great-grandson, grandson, and son. Our own family farm—still operating and still in the family—received its due recognition at the Iowa State Fair this year, earning the Heritage Award for more than one hundred and fifty years of continuous family ownership. One hundred and fifty years takes us well beyond the golden age featured in this collection—well south of 1880—to the middle life of Ralph Waldo Emerson and Henry David Thoreau.

The land endures.

Zachary Michael Jack
Clinton County, Iowa
August 2006

ACKNOWLEDGMENTS

The preparation of this volume has been greatly aided by several useful research libraries and their archives, exhibits, and special and circulating collections, in particular those of Cornell University, the Library of Congress, the University of Northern Illinois, and the University of Iowa. Faculty Enhancement release time from North Central College provided time for the editing of this volume. As always, I am most grateful to my family for instilling in me the dialogue between agriculturist, conservationist, and naturalist that informs this work. My most profound appreciation goes to the pioneering farm and wilderness writers whose tireless work and inspired advocacy vivify this collection.

Love of the Land

AMERICA'S LOVE OF THE LAND: AN INTRODUCTION

The need for a volume of historical writing that presents, side by side, the worlds of agriculture and conservation has long been unmet. What efforts have been towards this worthwhile goal, including, most notably, Randall S. Beeman's and James A. Pritchard's recent *A Green and Permanent Land: Ecology and Agriculture in the Twentieth Century*, have made an invaluable, though belated, contribution. And yet even a book as useful as that has its limitations—bound principally to the years 1930 to 1990 and unable to give extended, interrupted space to writers of the era. In search of a detailed and truly comprehensive history of land use in America, one has, up until now, been forced to take an exclusively disciplinary route, reading well and widely in environmental history, on the one hand, and agricultural history, on the other—hashing out the differences on one's own. In changing this status quo, *Love of the Land: Essential Farm and Conservation Readings from an American Golden Age, 1880–1920* aims to be of service.

Prior to *Love of the Land*, to read even the most comprehensive historical monographs on the history of environmentalism through

1920 has been to read much about George Perkins Marsh, David Thoreau, John Muir, Teddy Roosevelt, and Gifford Pinchot but to read little about prominent rural scholars and writers such as John Lee Coulter, the eldest Henry Wallace, and Isaac Phillips Roberts. In fact, a quick survey of the indices of half a dozen of the best and most comprehensive environmental histories finds no entries for Isaac Phillips Roberts—the Cornell University professor Liberty Hyde Bailey called "the wisest farmer I have ever known." Moreover, repeating the search for arguably the most influential agrarian writer of his day, Liberty Hyde Bailey, is to find *no entries* under his name in nearly half of the recent treatments of U.S. environmental histories, including Ted Steinberg's fine book *Down to Earth: Nature's Role in American History* (2002) and the second edition of *First Along the River: A Brief History of the U.S. Environmental Movement* (2000). Conversely, to read the definitive studies of agricultural history, including R. Douglas Hurt's *American Agriculture, A Brief History* and David B. Danbom's *Born in the Country: A History of Rural America* is to find no entries for John Muir—the most prominent naturalist of his day, a farmer's son by birth, and a practicing farmer well into his early adulthood—and a single entry between the two books for one Henry David Thoreau. The problem here is not poor or incomplete scholarship—these are the most excellent histories available, and this book owes much to them. The problem is essentially disciplinary, which is to say agricultural history and environmental history, as emerging fields of graduate study and the stuff of professional conference- and association-making, have varied interests and, in any case, much already on their disciplinary plates. Each is involved, as in older, more established degree-granting academic programs, in defining and redefining its canon. This simple fact remains: a rural historian will find different historical significance and attach different historical weight to, for example, the Homestead Act than will an environmental historian, as the Homestead Act affected farmers differently than wilderness advocates. That is, perhaps, as it should be. And yet the few topics on which agricultural history and environmental history emphatically and equitably overlap often assume disproportionate importance at the expense of other equally crucial areas of mutual concern, including, for example, agricultural wetlands and farmland wildlife biology. General readers and historians searching for a truly interdisciplinary approach are, as a consequence, left to construct their own metaphoric Venn diagrams or to feel otherwise verboten among even the new generation of welcoming interdisciplinary associations and annual conferences, such as the well-subscribed Association for the Study of Literature and the Environment (ASLE). Truth be known, to be a student of both rural history and environmental history equally, or purposefully betwixt and between, often feels to the devoted scholar and lover of the land to be a kind of new Unitarianism—or perhaps an old-fashioned limbo.

The American Progressive Era, roughly inclusive of the years 1890 to 1918, presents the historian with men and women so eclectic as to be utterly singular. Notable among them are Frederick Law Olmsted, John Muir, the elder Henry Wallace, Theodore Roosevelt, Liberty Hyde Bailey, and Mary Hunter Austin, to name a few, who farmed or ranched whether as children or adults and who also played a prominent role in the conservation and preservation movements. Indeed, Liberty Hyde Bailey, whose book *The Holy Earth* is considered a conservation classic, doubled as the editor of *Country Life* magazine and the chair of the Country Life Commission. Likewise, the eldest Henry Wallace served as a Country Life Commissioner while soon after presiding over the Third National Conservation Congress in 1910. In a Renaissance fashion, these characters in the farming and conservation drama resist categorization, appearing over and over in progressively more nuanced and conflicted guises. To compile an anthology of works wide enough to showcase their eclectic interests and expertise is a daunting task indeed.

Thus William Butler Yeats, himself a combination of poet, painter, philosopher, dramatist, mystic, politician and ruralist, found cause to say, "Out of the quarrel with others we make rhetoric; out of the quarrel with ourselves we make poetry." It should not surprise, then, that John Muir once described himself to friend and conservationist co-conspirator Robert Underwood Johnson as "poeticio-trampo-geologist-bot, and ornith-natural, etc.!—!—!"[1] To wit: Each of the Progressive Era's prominent farming and conservation writers had to come to terms as much with their own divided loyalties as with others'. Roderick Frazier Nash's *Wilderness and the American Mind,* perhaps the best historical treatment of the enlarged, land-use theme, traces the roots of Muir's ambivalence as follows:

> The prime of John Muir's life coincided with the advent of national concern over conservation. At first, and superficially, the problem seemed simple: "exploiters" of natural resources had to be checked by those determined to protect them. Initially, anxiety over the rapid depletion of raw materials, particularly forests, was broad enough to embrace many points of view. A common enemy united the early conservationists. But soon they realized that as wide differences existed within their own house as between it and the exploiters. (ibid)

Nash further considers how, "[in] juxtaposing the needs of civilization with the spiritual and aesthetic value of wilderness, the conservation issue extended the old dialogue between pioneers and romantics" (ibid). Clearly, capturing the diversity of a thoroughly complicated man such as John Muir requires a number of snapshots varying in width and depth of field, documenting the individual at various stages in their evolution. Hence, the figures who most nearly represent a hybrid of the conservationist and the agriculturalist—John Burroughs, John Muir, Teddy Roosevelt, and Liberty Hyde Bailey—are represented in *Love of the Land* by several essays. For the same reason, a cast of forerunners and fountainheads, including the likes of Thomas Jefferson, Jean Crevecoeur, and Frederick Law Olmsted, merit inclusion as emblems of the shared roots from which the agrarian and conservation movements sprang.

---

[1] Roderick Frazier Nash, *Wilderness and the American Mind,* 4th Edition (New Haven: Yale University Press, 2001), 122.

Up and through the Gilded Age of American farm and conservation writing, half of American workers depended on the farm for their income and were, quite literally, tuned in—by newspaper, magazine, journal, book, and late in the Golden Age, by radio—to farming and to the land in general. Of the remaining urbanized half, the vast majority were, by consequence, no more than a generation or two removed from the land.

Importantly, the varied, interdisciplinary readings in this historic volume come from both sides of the proverbial fence: from college-trained, urban intellectuals studying then-emerging disciplines such as sociology, anthropology, hydrology, and forestry—the way today's scholars have made urban and suburban studies and environmental studies suddenly relevant as academic disciplines—from the farmers and foresters who made a living with their hands, and from the wilderness advocates who made a living by their transcendent ramblings, as did the likes of John Burroughs, Henry Thoreau, John Muir, and Walt Whitman. Each of the latter found, in the run-up to the Golden Age of farm and conservation writing—and often in unexpected places—a reading public hungry for their particular brand of hands-on, hard-won knowledge. In fact, the many farm-dedicated, nature-invested presses and publishers of the day soon learned that writing hard-hewn from the land proved popular among prosperous urban- and suburbanites. In fact, the moral culture of the day, threatening to the children of the nouveau riche, linked wealth and ease with "arrested development."[2] In such a climate, Henry Ward Beecher preached this unlikely, fire-and-brimstone sermon at his wealthy congregation: "How blessed, then, is the stroke of disaster which sets the children free, and gives them over to the hard but kind bosom of Poverty, who says to them 'Work!' and, working, makes them men" (ibid 7). Little wonder that the farmer and his farm and the naturalist and his nature became the locus of a class-fascinated dialogue between those who worked with their hands and those who studied those who worked with their

---

[2] Samuel Haber, *Efficiency and Uplift: Scientific Management in the Progressive Era 1890–1929* (Chicago: Chicago UP, 1973), 6.

hands—a debate played out in the copious, illuminating, and now mostly forgotten agrarian and conservation treatises of the day. The readings anthologized here aim to capture that love of the land in its myriad manifestations and across disciplines.

Of course, a so-called "Golden Age" of any profession, indeed of anything or anyone, proves largely academic—more useful, perhaps, as a retrospective, rhetorical lens than for its actual historical coinage. Some agricultural historians, David Danbom among them, have settled on the first twenty years of the twentieth century as agriculture's Golden Age. In his definitive history of rural America, *Born in the Country*, Danbom cites the "rare prosperity" of 1900 to 1920 as good reason for the golden appellation. Gross farm incomes, Danbom writes, doubled while farm values tripled.[3] Others, including the Library of Congress, occasionally widen the window for archival purposes to 1880 to 1920. Those inclined to take the broader view of the era do so on largely economic or socio-historical grounds. Economically, the uneven but still dramatic agricultural growth occurring between 1870 and 1900—when more land, according to scholar William Serrin, came under cultivation than in the previous two hundred and fifty years and when the ranching business boomed in the West—argues for a wider golden era.[4] Similarly, agricultural nostalgics and Luddites have been inclined to define agriculture's halcyon days more expansively, perhaps even colloquially, to include the last two decades of the nineteenth century—days when farmers formed communal alliances, fielded Populist presidential candidates, and controlled statehouses across the Great Plains. There, they claim, existed an era when overall farm prosperity did not mean rampant consumerism and conspicuous consumption on the farm as it would for some rural residents after 1900, with the arrival of rural mail delivery, the internal combustion engine, and

---

[3] David Danbom, *Born in the Country: A History of Rural America* (Baltimore: Johns Hopkins, 1995), 161.

[4] William Serrin. "American Farm Policy Helps/Hurts The American Farm" *APF Reporter* 1, no. 4, (1978).

ultimately, electricity and radio. Still others, including agriculture extension specialist Don Hofstrand prefer to speak not in terms of gilded epochs but what Hofstrand calls "waves of farm prosperity" to include not only the late 1800s but also the period of high production and high prices during the mid to late 1940s.[5]

The debate over the chronological bookends of farming's Golden Age is not, however, the thrust of this volume. Instead, the real import of the moniker *Golden Age* is in the coin of the realm it suggests—in the sudden interest in everything the farmer and the naturalist had to say, and perhaps even more so, what was said and written about the farmer and the naturalist.

The twenty-first century reader will find it difficult, no doubt, to imagine the political, literary clout made possible by the 50 percent of Americans daily earning their living from the land: whether by farming, ranching, lumbering, or from a host of related occupations. Our times offer no ready analog for the popularity of farm and conservation writings between 1880 and 1920. These days periodic surveys administered to youth—polls such as those sponsored by Junior Achievement—list doctor, businessperson, athlete, and teacher to be the most sought-after professions, with no single profession garnering more than 10 percent of the youth vote. Thus, contemporary America is occupationally diverse by comparison with the Golden Age; its reading tastes and political priorities are likewise more diffuse.

In this light, today's reader, rightly, comes to appreciate the singularity of the age that produced the farm and conservation readings collected in *Love of the Land*. One reads in them an unprecedented national absorption in a constellation of agrarian vocations and avocations, a nexus encompassing the related fields of conservation, forestry, horticulture, botany, hydrology, geology, forestry, and agriculture together with the educational, governmental, recreational, ecclesiastical, and entrepreneurial institutions formed in service to agrarian communities.

---

[5] Don Hofstrand, "The Changing Face of Agriculture," *AgDM Newsletter*, March 1999. http:www.extension.iastate.edu/agdm/articles/hof/HofMar99.htm.

During the years featured in *Love of the Land*, rural America grew into a particularly contentious relationship with the institutions pledged to serve it, changing the political landscape forever. Towards the end of the Golden Age, the American farmer was increasingly advised to view himself as a businessman rather than the animal husband, land steward, craftsperson, and all-around yeoman he had theretofore labeled himself. A pattern, conspicuous in the readings that follow, suggests itself—a pattern by which cyclical agrarian boom times beget interventionist policies that either burst the bubble outright or deflate it by virtue of sudden and suffocating attentions. Some thirty years after the wave of agricultural prosperity that began in the 1940s, for example, the American farmer would once again draw attention to himself via his prosperity; over the succeeding two decades he would once more—and yet again, it seemed, because of earlier economic and cultural gains—be told to "get big or get out."[6] And once more, shortly after he heeded those words, he would find himself in agricultural crisis.

Thus the Progressive Era of the late nineteenth and early twentieth century begets, both for the American farmer and conservationist, a curious and enduring paradox, wherein experts promise to make life better, safer, and cleaner for the rural American, but do so by methods often injurious to agrarian and wilderness lifestyles. In such an environment, the agrarian and the preservationist, historically isolationist, further deepen their trademark separation. By analogy, the sick patient, or a patient thought to be sick, comes to distrust the doctor consistently able to identify the malady but unable or unwilling to provide a cure. Conversely, the doctor's capacity for do-gooding seems predicated on the patient's (rural and wild America in this analogy) illness to test his or her diagnosis. Wilbert Lee Anderson, in his sociologically-minded *The Country Town: A Study of Rural Evolution*, captures the need for both accurate diagnosis and compassionate cure thusly: "If this book had the

---

[6] Secretary of Agriculture Earl Butz quoted in Judy Alieson, "Embattled Farmers: 1776–2003." www.commondreams.org (accessed December 2000).

gift of prophecy and knew all mysteries and all knowledge, if it had all faith so as to remove mountains, and did not prompt the deeds of love, it would be nothing." Thus, the back and forth nature of a doctor-patient exchange is captured in the primary readings that follow in ways not possible in an exclusively interpretive history or conventional secondary textbook.

In a modern era when surveys document a growing public distrust of government officials and newspaper journalists as well as doctors and policymakers, the Golden Age of agriculture and conservation reasserts its relevance. Moreover, in an energy-starved twenty-first century economy turning increasingly to renewable bio-fuels produced by those living close to the land, the farmer and the conservationist become once more hot commodities. Whenever our society urgently considers the double bind of limitless war and finite natural resources, as it did during the Golden Age and as it does now, the rural and the remote are looked to as litmus tests. Meanwhile, the pundits—by which we mean the entire class of analysts, consultants, academics, and policymakers mobilized in times of war and shortage—come to the fore along with, but not necessarily alongside, the farmer and the conservationist. The pundit and the pundit's think tank, the bureaucrat and bureaucrat's government—draw their substance from the Progressive Era portrayed here, when they were mobilized not only to win wars, build roads, bust trusts, and educate citizens, but to harness—which is to say study, apprehend—the engine that is, or was, the American landscape.

The historic agrarian and conservation writings of *Love of the Land* echo, sometimes uncannily, the political present. Perhaps even more than his father, George W. Bush, the record shows, has modeled his presidency after Teddy Roosevelt. Covering his campaign stop for the *Augusta Chronicle* in June of 2000, for example, journalist Justin Martin reported Bush's admiration of Teddy Roosevelt's "boundless energy" and "great vision."[7] Four years later, when Bush the Younger was locked

---

[7] Justin Martin, "Presidential Hopeful Pushes Positive Vision" *Augusta Chronicle*, http://quest.cjonline.com/stories/060700/gen_vision.shtml (accessed June 7, 2000).

in a close race with John Kerry, conservative columnist George Will opened his pro-Bush column with a particularly suggestive epigraph from the *New York Sun's* 1904 endorsement of Teddy Roosevelt's reelection bid: *THEODORE! With all thy faults.*

By now critics of the Iraq War have made the comparison between the Progressive Era and the Bush administration ubiquitous. A case in point is John B. Judis' book *The Folly of Empire: What George W. Bush Could Learn from Theodore Roosevelt and Woodrow Wilson*, which compares Roosevelt's unilateral war in the Philippines to modern day Iraq. Citing Roosevelt's ultimate realization that the Philippines proved a "heel of Achilles" for his countrymen, Judis likewise predicts a latter-day Bush reversal regarding the administration's alleged imperialist agenda. A senior editor at the *New Republic*, Judis points out Bush's Rooseveltian willingness to use the proverbial "big stick" internationally and his Woodrow Wilson-esque calls for democracy overseas.

The similarities do not end there. Like Roosevelt, Bush makes political hay of his personal brand of agrarianism, taking advantage of lengthy, well-covered working vacations at his Crawford, Texas ranch to instill in his political base some sense of a grounded executive. TR, of course, did much the same in the Bighorn Mountains, cutting his naturalist's teeth on something larger, and grander, than himself in full view of the nation. Roosevelt, in fact, viewed the countryside as the nation's savior, praising the farmer's "best service in governing himself in time of peace and also in fighting in time of war."[8] And like Roosevelt's day, our current political and economic climate features a hotly contested battle between natural resource pragmatists—utilitarians—who opportunistically define conservation much as Roosevelt's chief forester Gifford Pinchot did ("the art of producing from the forest whatever it can yield for the service of man") and environmentalists and preservationists allied in their opposition to

---

[8] George Mowry, *The Era of Theodore Roosevelt and the Birth of Modern America* (New York: Harper, 1958), 92.

aggressive logging in the national forests and proposed drilling in the Artic National Wildlife Refuge (ANWR). Thus despite the fact that farmers now make up less than 2 percent of the nation's workforce, the Golden Era of agricultural and conservation is perhaps more relevant today than at any time in our nation's history. Just how *country* we are as a nation and what exactly *country* means shed light on our actions at home and abroad.

The Progressive Era reasserts itself, too, in contemporary pop culture, most especially in our popular concern for, and delight in, nature. Richard's Louv's surprise 2005 bestseller, *Last Child in the Woods: Saving our Children from Nature-Deficit Disorder* resurrects, Lazarus-like, the nature study agendas long-ago championed by the contributors to this volume: Liberty Hyde Bailey, John Burroughs, Kenyon L. Butterfield, H. W. Foght, John Muir, Edward Payson Powell, Theodore Roosevelt, and Dallas Lore Sharp, among others. And just as Progressive Era social scientist Kenyon L. Butterfield proposed heading off delinquency and depravity among boys and girls via garden clubs and city farming, so our era relentlessly catalogues the deficiencies of children by measuring them against cotemporary notions of naturalness. Cultural observers decry the estimated 16 percent of U.S. children that are obese[9] and the thirty-nine hours per week the Kaiser Family Foundation reports Americans aged eight to eighteen spend with television, computers, and video games.[10] Studies undertaken with new zeal find that children who play on natural playgrounds are more likely to make up their own games and be more cooperative than those who play on man-made equipment. And recent work from the University of Illinois shows that exposure to nature may effectively reduce attention deficit disorders in children.[11] Moreover,

---

[9] Colleen Long, "Journalist Says Kids Suffer from 'Nature-Deficit Disorder,'" Associated Press State and Local Wire, July 13, 2005, *LexisNexis*.
[10] Leslie Brody, "Take a Hike; Reintroducing Kids to Nature Becomes a Crusade," *The Record* (Bergen County, NJ), A01, June 8, 2006, *LexisNexis*.
[11] Julie Deardorff, "Nature Deficit Sends Kids Down a Desolate Path," *Herald News* (Passaid County, NJ), B04, February 28, 2006, Knight Ridder News Service.

the "new day" Butterfield envisioned in 1919, coinciding with the back-to-the-country movement, has been realized in our own century. A 2003 study at Cornell University suggests that children who merely grow up with a "green view" from their window are better equipped to handle stress.[12] Nature education, reports the California Department of Education, resulted in 27 percent better science scores for weeklong campers than for those trapped in traditional classrooms (ibid). And yet for all their newness, these studies confirm what those who love the land have long known. Golden Age nature writer Dallas Lore Sharp said this in his book *The Lay of the Land* in 1908: "The best good, the deep healing, comes when one, no longer a stranger, breaks away from his getting and spending, from his thinking with men, and camps under the open sky, where he knows without thinking, and worships without priest or chant or prayer."

Record-high national park attendance likewise affirms the American public's subliminal understanding of nature's balm and succor, the very affinity between humans and the natural world biologist Edward O. Wilson calls "biophilia."[13] We love nature, or the idea of nature, to death, it seems, and to such a degree that media coverage of national parks in the last decade has focused on degradation caused by overuse, over-love. In the face of record attendance in the early and mid 1990s—figures that have remained steady in the decade since[14]—Americans increasingly find in the popular press not coverage of the parks' natural wonders but advice on how to avoid peak times at the five most visited national parks—Great Smoky Mountains, Grand Canyon, Yosemite, Olympic, and Yellowstone—and, unbelievably, how to stay safe. To wit: the Associated Press's Matthew Daly details in "Forest Service Reports Surge in Violence Against Rangers" an all-time high in "attacks, threats, and lesser fights" in the parks in 2005. In the same article, Agriculture Undersecretary Mark Rey, charged

---

[12] "How Heading to the Woods Can Heal a Child," *Toronto Star*, A05, June 11, 2006.
[13] Deardorff, "Nature Deficit Sends Kids …"
[14] Richard Louv, "The End of Nature—Parcel by Parcel," *The San Diego Union-Tribune*, G-3, June 1, 2003, *LexisNexis*.

with overseeing the Forest Service, describes the motley crew of criminal perpetrators as a mix of "drunks, drugs users, or deranged environmental protestors."[15] And Rey's revelation seems only the latest to document societal incivility and overcrowding. Here again, comparisons between our own age and a golden age prove our loving nature to death, or to danger, is nothing new. Facing the prospects of thousands of new tourists in Hetch Hetchy Valley, an exasperated John Muir registered a similarly cynical truth in 1912, writing, "Ever since the establishment of the Yosemite National Park, strife has been going on around its borders, and I suppose this will go on as part of the universal battle between right and wrong, however much its boundaries may be shorn, or its wild beauty destroyed."

Indeed, history shows the parks were created expressly to keep such rowdiness at bay. The creation of Yellowstone, Nathaniel P. Langford argued in 1871, would keep speculators and "squatters"—a term understood to mean poor whites and Indians—away from a region made precious by its "beautiful decorations."[16] Meanwhile, tourists from the East became the anointed visitors, the desirables, the cash cows that brought the railroad interest into the pro-Yellowstone camp. *Scribner's* and other periodicals in the East salivated at the prospects of "Yankee enterprise dot[ting] the new park with hostelries and ... lines of travel" (ibid 113). Hearing such braggadocio, those who loved the parks best were torn between promotion and preservation, whether to let all the world know or to keep the secret to themselves. Just as the one-armed Civil War veteran Major John Wesley Powell turned his 1869 expedition to Utah, Colorado, and Arizona into guidebooks, and guidebooks into cash, so the current debate among guidebook publishers is whether, or how much, to spread the word of the nation's natural wonders. Apropos to the outdoor recreation industry, which has been recording dramatic annual sales increases

---

[15] Daly, Matthew, "Park Service Reports Surge in Violence Against Rangers," *Washington Post*, June 27, 2006, Associated Press, www.washingtonpost.com.
[16] Nash, *Wilderness and the American Mind*, 112.

for more than a decade, outdoor photographer Scott Smith poses the following question in *High Country News*, "Are we exploiting the environment by telling and showing all—the way that some magazines exploit women?"[17]

In a golden age of farm and conservation writing, stories of encounters with nature—its unforgiving cruelty and its unrivaled beauty—enthralled, as they do in our own age of reality television. *Survivor* has brought the Golden Age's obsession with Social Darwinism back to American television screens to the tune of thirty Emmy nominations and a tally of 23 million weekly viewers at its peak.[18] Similarly, readers in the Progressive Era devoured the man versus nature novels of Jack London and hunted headlines for news of "wild man" Joseph Knowles. Knowles, whose heroics were celebrated by a 1913 *Boston Post* headline reading "Naked He Plunges into the Woods," trekked into the back country to live alone for two months[19] in his own version of *Survivor: Maine*. For the next two months, Knowles titillated readers with missives written with charcoal and birchbark, regaling them with tales of clothes made from woven bark and meals of hand-caught trout and partridge. By the time the *Post* reported that Knowles had trapped a bear and was wearing its hide as a coat, newspapers throughout the country had picked up the serial story.

Perhaps the most enduring feature of the Progressive Era to resurface in the new millennium is the spirit of land-centered volunteerism, particularly among youth. Many of the youth organizations popularized during the era—most notably the Boy Scouts of America, begun in 1910 by popular nature writer Ernest Thompson Seton, and the Young Men's Christian Association (YMCA)—were based on physical fitness and life in the outdoors as an antidote to urban blight. As present day iterations of the Progressive impulse, AmeriCorps VISTA,

---

[17] Christopher Smith, "I Came, I Saw, I Wrote A Guidebook," *High Country News*, www.hcn.org (accessed September 4, 1995).
[18] Steve Rogers, "Episode 4 of 'Survivor: The Amazon' Tops Ratings …," *Reality TV World*, www.realitytvworld.com (accessed September 4, 1995).
[19] Nash, *Wilderness and the American Mind*, 141.

Teach for America, and the Peace Corps well document today's youth commitment to service learning and social service. All three of the programs have recorded staggering increases since 2000. Applications for AmeriCorps, in particular, jumped 50 percent since 2004.[20] Significantly, AmeriCorps, where volunteers renovate parks, build paths, help with disaster relief, and otherwise lend a hand where needed, is tied most closely to domestic conservationist concerns. Among its sister programs, it has recorded, not coincidentally, the largest increases.

As the 2000 U.S. Census makes clear, Americans are choosing to live ever closer to the land, reversing the conspicuous urbanization of an earlier age. Here again, the Golden Age offers a compelling analog, as Kenyon L. Butterfield noted the very same trend in his 1919 book *The Farmer and the New Day*. Writing of the proliferation of country estates, Butterfield concludes, "The practice of living in the country for at least half the year is rapidly growing. It is a healthy, normal, educative movement. It leads to outdoor life, to a new understanding of country things, and occasionally helps to educate a community to better farming." And yet Butterfield worried, as land-use scholars do today, that such palatial places gobbled up otherwise productive farmland.

In their report "The Rural Rebound: Recent Nonmetropolitan Demographic Trends in the United States," Kenneth Johnson of Loyola University, Chicago, and Calvin L. Beale of the U.S. Department of Agriculture document the dramatic repopulation of rural America for in a new millennium. Their work describes a demographic shift whereby nonmetropolitan areas now contain 56.1 million residents and where a staggering 1702 of 2303 nonmetropolitan counties grew between 1990 and 2000.[21] Interestingly, this population phenomenon—"selective decentralization"—occurs, in part, because the rural environs have made great strides in quality of life. Johnson and Beale attribute the dra-

---

[20] Beth Walton, "Volunteer Rates Hit Record Numbers; Peace Corps Others See Surge," *USA Today*, 1A, July 7, 2006, *LexisNexis*.
[21] Kenneth M. Johnson and Calvin L. Beale, "The Rural Rebound: Recent Nonmetropolitan Demographic Trends in the United States," http:www.luc.edu/depts./sociology/Johnson/p99webn.html.

matic rural repopulation to improvements in transportation and communications infrastructure—representing a further evolution of similar technological and infrastructural concerns (electrification, mail delivery, road improvement) identified by Roosevelt's Country Life Commission in 1909.

In the final analysis, *Love of the Land* is not intended merely as a collection of agricultural and environmental history, but as a cultural primer for the many now returning, without context, to a life on the land. As the twenty-first century citizen debates a move back to the country, perhaps even to occupy a long-abandoned family homestead, they will find herein timeless advice from, for example, Edward Payson Powell. As they consider how to raise their son or daughter in rural America, they will find still-relevant, gender-specific counsel from Martha Foote Crow, William A. McKeever, Isaac Phillips Roberts, and Henry Wallace. And as those same parents consider educational options for their children in a rural school district, they will no doubt find traces of both the obstacles and the pleasures of nature study as outlined by Mary Hunter Austin, Liberty Hyde Bailey, H. W. Foght, John Muir, Theodore Roosevelt, Dallas Lore Sharp, Henry David Thoreau, Walt Whitman, and others who believed nature the best teacher.

Love of the Land: Essential Farm and Conservation Readings from an American Golden Age, 1880–1920 aims to be the best kind of historical environmental reader, anthologizing farm and conservation readings that seem as near as today's news. As Mark Twain said, "History does not repeat itself, but history rhymes." And nowhere is that rhyme more apparent than in these classic agrarian and conservation writings from a golden age.

Editor's Notes

The editing of historic writing presents unique challenges. As a guiding editorial principle, the source text has been modified minimally. Spelling and usage have been modernized and regularized as has, to a lesser extent, punctuation. Throughout, earnest efforts have been made to maintain the original feel and flavor of the text, as evident, for example, in the anomalous frequency of semi-colons and capitalization for emphasis of concepts such as "Nature" or "Art." Original paragraphing has been preserved in nearly every instance where a readily verifiable original text exists. Occasionally, and as noted, related sections or chapters of a given monograph have been combined for the sake of concision and readability. In such instances where organizational clarity made beneficial an umbrella title or grouping, mention has been made either in the introduction to the work(s) or in the footnotes. Elsewhere, and in the interest of making room for a truly diverse digest, chapter excerpts begin, as noted, a specified number of paragraphs from the beginning of the original chapter or section as it appeared in the source edition.

Where possible, transcription has been made directly from the page, though, in some instances, especially the early nineteenth-century writings of the supplementary anthology-within-an-anthology, electronic transcriptions from reputable sources have been employed. For purposes of historical transcription, first editions were preferred. For nineteenth-century work in particular the earliest, readily available edition for which verifiable page numbers existed was used. Unless otherwise noted, work has been anthologized in ascending chronological order by publication date of the edition used as a source; where publication years matched for any two or more selections, works have been arranged in alphabetical order by the author's last name. Original page numbers, as assigned in the source edition, have been footnoted for scholarly use. In typesetting and keying these texts, every attempt has been made to avoid errors in transcription, though some error is inevitable.

No attempt has been made to alter these historic readings to include gender-inclusive or politically-correct language. Instead, underrepresented groups have purposefully been given voice in the selection of these diverse readings.

In consultation with the U.S. Copyright Office, the historical work contained in this anthology is believed to be out of copyright and in the public domain. Sources for individual, anthologized works are fully acknowledged in footnotes listing bibliographic data.

FORERUNNERS: READINGS 1770–1880

INTRODUCING THE FORERUNNERS: Jean de Crevecoeur, Thomas Jefferson, Alexis de Tocqueville, Ralph Waldo Emerson, George Perkins Marsh, John Muir, Henry David Thoreau, Edmund Morris, Frederick Law Olmsted, and John Burroughs.

The following ten readings represent early engagements between farming and conservation that would blossom into a complete but sometimes fraught matrimony by the early twentieth century. Significantly, these selections portray salad days for the American republic when farmers and conservationists (though the term *conservation* itself would not enjoy popular usage until the days of the Theodore Roosevelt administration) were not yet fully aware of their differences nor polarized by labels concerning belief and practice. The earliest presidents—Washington, Jefferson, and Madison—considered themselves cultivators, habitually blending politics, war-making, and surveying with farming. Farmers, as the writings of Jean Crevecoeur and Edmund Morris here show, were "conservationists" by necessity, isolated, as they were, from distant markets. Many if not most American farmers kept livestock and poultry, farmed a crop, provided for their subsistence, and sold what little, if any, remained. As the writings of Crevecoeur, Morris, and Burroughs indicate, many ventured into the "wild" to collect honey or fruit for home use or for market. At the turn of the nineteenth century, in fact, perhaps 80 percent of the population was engaged in agriculture, a percentage that would remain nearly steady in the South, at least, until the beginning of the Civil War, while, in the country as a whole, the percentage would drop to around 50 percent.

Still, "conservation" as a theory and practice in the years 1770–1880 applied more in the East, where land claims were longer settled than west of the Appalachians, where ignorant squatters and fly-by-night cultivator-speculators often overfarmed and overgrazed a plot of ground before leaving for "virgin" lands to the west—a practice condemned as "land-skinning." The courtship of farming and conservation—a hybrid later known by such terms as agro-ecology, sustainable farming, and permanent agriculture—began in the relatively more prosperous, more urbane East, which could afford circumspection of the kind practiced by transcendentalists Thoreau and Emerson, whose studies in philosophy and religion formed the basis of a spiritual land ethic. Thomas Jefferson and Vermont Congressman George Perkins Marsh were well positioned on the eastern seaboard to read the latest agricultural news and reviews from Europe. Fellow northeasterners John Burroughs and Frederick Law Olmsted came by their conservationist tendencies via American farm upbringings leavened by European travel. Alone in this grouping, John Muir came from the hinterlands, Wisconsin territory, and it is perhaps no surprise that his writings highlight especially hard-won wisdoms.

The supplementary anthology-within-an-anthology that follows shows the diversity and energy of agricultural and ecological dialogues underway in the Antebellum period and shortly thereafter. With the possible exception of John Muir, these forerunners explicitly or implicitly endorse the Jeffersonian notion that "cultivators are the most virtuous citizens," though, in doing, they are forced to more closely examine the virtues of their own agricultural practice. Given the "land-skinning" and shameless speculating taking place west of the Appalachians, each of these forerunners was left to consider, with sudden gravity, Jefferson's claim that "corruption of morals in the mass of cultivators is a phenomenon of which no age nor nation has furnished an example."

The concise readings that follow were selected specifically for the agrarian birthright or practice they embody (Burroughs, Crevecoeur, Jefferson, Morris, Muir, Olmsted); for their comparison of farm to wilderness (Thoreau, Emerson, Marsh), and, in one case, (de Tocqueville) for the objectivity available only to a non-citizen. The essays are presented chronologically according to publication date or, in the case of several authors publishing widely in periodicals of the time—Burroughs, Emerson, and Thoreau, in particular—by the date at which the featured work was made available to the reading public. Significantly, John Muir's memoir is here placed according to the era it recollects rather than the date of its publication. Frederick Law Olmsted's "Yosemite and the Mariposa Grove" is reprinted here from a typed transcription of the Library of Congress manuscript collection, as the large-scale publishing of Olmsted's controversial report was allegedly suppressed by the Yosemite Commission.

JEAN DE CREVECOEUR

Representing one of the earliest accounts of conservationist farming in America, *Letters from an American Farmer*, first published in 1782, seems remarkably contemporary in its intimacies. Frenchman Michel-Guillaume Jean de Crevecoeur first came to North America as an officer under Montcalm in the French-Indian War. After being wounded and ultimately captured with the defeat of the French in Quebec, he negotiated for his release, changed his name to J. Hector St. John de Crevecoeur, and sailed for New York to take up surveying and trading on the frontier. Ultimately wearied by the transient life, the Frenchman became a naturalized citizen of Orange County, New York in 1765, marrying into a prominent British-American family four years later. Happily betrothed and financially stable, Crevecoeur and his bride settled into a 120-acre farm, "Pine Hill," from whence he wrote *Letters from an American Farmer* as well as agricultural articles, under the pseudo *Agricola*, for American newspapers.

Crevecoeur lived a fundamentally ambivalent existence. He was considered a Tory, perhaps explaining why his books met with significantly more acclaim in Europe than they did in America. His "American" letters, meanwhile, struck some as disingenuous, in nationality and in genre, as they were alternately considered sketches, essays, stories, and epistles. Crevecoeur may have intended this genre multiplicity, as he clearly hoped to court literary as well as agricultural audiences with his missives, as indicated by the many revisions he made to his "letters" in later editions. Like John Burroughs and Henry Thoreau, his identity as an "American farmer" seems largely self-constructed, though his popular accounts of his adopted country, America, and his new profession, farming, are considered the first literary success by an American author in Europe.

Though the often anthologized "Letter III" has much to say about the American Dream, "Letter II," excerpted here, better testifies to the unique blending of Crevecoeur the naturalist with Crevecoeur the farmer. Blending precise notation (Crevecoeur was a surveyor and a mapmaker) with a frontier farmer's appreciation for the preciousness of life in all its forms (Crevecoeur would lose two loves in his life, including his first love and, later, his wife, who was killed in an Indian raid while Crevecoeur was abroad), Crevecoeur's writing is at once methodical and vulnerable. His observations of the nurturing instincts of local birds intimate his own long-suffering caretaking as a land steward, animal husband, and father. Also evident in his natural observations is an interest in nationality and governance, as when he describes the "republic" of bees and the "government" of their society. Despite self-representations as a humble farmer, Crevecoeur was a well-educated, well-traveled French noblemen who, in public at least, opposed the American revolution. And despite farming Pine Hill for less than a decade, he was a devoted and productive agrarian, credited with introducing the culture of European crops, especially alfalfa, to America and, conversely, transplanting the American potato into regions of France, where he would eventually return to popular literary acclaim. In France he was distrusted by the French, the American, and the British governments, all of whom reportedly considered Crevecoeur, at various times, a spy, while nevertheless placing him in occasional governmental posts.

Letter II, On the Situation, Feelings, and Pleasures of an American Farmer[22]

I do not know an instance in which the singular barbarity of man is so strongly delineated, as in the catching and murdering those harmless birds, at that cruel season of the year.[23] Mr. _____, one of the most famous and extraordinary farmers that has ever done honor to the province of Connecticut, by his timely and humane assistance in a hard winter, saved this species from being entirely destroyed. They perished all over the country, none of their delightful whistlings were heard the next spring, but upon this gentleman's farm, and to his humanity we owe the continuation of their music. When the severities of that season have dispirited all my cattle, no farmer ever attends them with more pleasure than I do; it is one of those duties which is sweetened with the most rational satisfaction. I amuse myself in beholding their different tempers, actions, and the various effects of their instinct now powerfully impelled by the force of hunger. I trace their various inclinations, and the different effects of their passions, which are exactly the same as among men; the law is to us precisely what I am in my barnyard, a bridle and check to prevent the strong and greedy from oppressing the timid and weak. Conscious of superiority, they always strive to encroach on their neighbors; unsatisfied with their portion, they eagerly swallow it in order to have an opportunity

---

[22] Hector St. John de Crevecoeur, *Letters From an American Farmer* (New York: Fox, Duffield, 1904), 32–47.
[23] For the sake of concision and readability, the excerpt begins with paragraph seven of Letter II.

of taking what is given to others, except they are prevented. Some I chide, others, unmindful of my admonitions, receive some blows. Could victuals thus be given to men without the assistance of any language, I am sure they would not behave better to one another, nor more philosophically than my cattle do.

The same spirit prevails in the stable; but there I have to do with more generous animals, there my well-known voice has immediate influence, and soon restores peace and tranquility. Thus by superior knowledge I govern all my cattle as wise men are obliged to govern fools and the ignorant. A variety of other thoughts crowd on my mind at that peculiar instant, but they all vanish by the time I return home. If in a cold night I swiftly travel in my sledge, carried along at the rate of twelve miles an hour, many are the reflections excited by surrounding circumstances. I ask myself what sort of an agent is that which we call frost? Our minister compares it to needles, the points of which enter our pores. What is become of the heat of the summer; in what part of the world is it that the N. W. keeps these grand magazines of nitre? When I see in the morning a river over which I can travel, that in the evening before was liquid, I am astonished indeed! What is become of those millions of insects which played in our summer fields, and in our evening meadows; they were so puny and so delicate, the period of their existence was so short, that one cannot help wondering how they could learn, in that short space, the sublime art to hide themselves and their offspring in so perfect a manner as to baffle the rigor of the season, and preserve that precious embryo of life, that small portion of ethereal heat, which if once destroyed would destroy the species! Whence that irresistible propensity to sleep so common in all those who are severely attacked by the frost. Dreary as this season appears, yet it has like all others its miracles, it presents to man a variety of problems which he can never resolve; among the rest, we have here a set of small birds which never appear until the snow falls; contrary to all others, they dwell and appear to delight in that element.

It is my bees, however, which afford me the most pleasing and extensive themes; let me look at them when I will, their government, their industry, their quarrels, their passions, always present me with something new; for which reason, when weary with labor, my common place of rest is under my locust tree, close by my bee house. By their movements I can predict the weather, and can tell the day of their swarming; but the most difficult point is, when on the wing, to know whether they want to go to the woods or not. If they have previously pitched in some hollow trees, it is not the allurements of salt and water, of fennel, hickory leaves, et cetera, nor the finest box, that can induce them to stay; they will prefer those rude, rough habitations to the best polished mahogany hive. When that is the case with mine, I seldom thwart their inclinations; it is in freedom that they work: were I to confine them, they would dwindle away and quit their labor. In such excursions we only part for a while; I am generally sure to find them again the following fall. This elopement of theirs only adds to my recreations; I know how to deceive even their superlative instinct; nor do I fear losing them, though eighteen miles from my house, and lodged in the most lofty trees, in the most impervious of our forests. I once took you along with me in one of these rambles, and yet you insist on my repeating the detail of our operations: it brings back into my mind many of the useful and entertaining reflections with which you so happily beguiled our tedious hours.

After I have done sowing, by way of recreation, I prepare for a week's jaunt in the woods, not to hunt either the deer or the bears, as my neighbors do, but to catch the more harmless bees. I cannot boast that this chase is so noble, or so famous among men, but I find it less fatiguing, and full as profitable; and the last consideration is the only one that moves me. I take with me my dog, as a companion, for he is useless as to this game; my gun, for no man you know ought to enter the woods without one; my blanket, some provisions, some wax, vermilion, honey, and a small pocket compass. With these implements I proceed to such woods as are at a considerable distance from any settlements. I carefully examine whether they abound with large trees; if so, I make a small fire on some flat stones, in a convenient place; on the fire I put some wax; close by this fire, on another stone, I drop honey in distinct drops, which I surround with small quantities of vermilion, laid on the stone; and then I retire carefully to watch whether any bees appear. If there are any in that neighborhood, I rest assured that the smell of the burnt wax will unavoidably attract them; they will soon find out the honey, for they are fond of preying on that which is not their own, and in their approach they will necessarily tinge themselves with some particles of vermilion, which will adhere long to their bodies. I next fix my compass, to find out their course, which they keep invariably straight, when they are returning home loaded. By the assistance of my watch, I observe how long those are returning which are marked with vermilion. Thus possessed of the course, and, in some measure, of the distance, which I can easily guess at, I follow the first, and seldom fail of coming to the tree where those republics are lodged. I then mark it, and thus, with patience, I have found out sometimes eleven swarms in a season, and it is inconceivable what a quantity of honey these trees will sometimes afford. It entirely depends on the size of the hollow, as the bees never rest nor swarm till it is all replenished, for like men, it is only the want of room that induces them to quit the maternal hive. Next I proceed to some of the nearest settlements, where I procure proper assistance to cut down the trees, get all my prey secured, and then return home with my prize. The first bees I ever procured were thus found in the woods, by mere accident; for at that time I had no kind of skill in this method of tracing them. The body of the tree being perfectly sound, they had lodged themselves in the hollow of one of its principal limbs, which I carefully sawed off and with a good deal of labor and industry brought it home, where I fixed it up again in the same position in which I found it growing. This was in April; I had five swarms that year, and they have been ever since very prosperous. This business generally takes up a week of my time every fall, and to me it is a week of solitary ease and relaxation.

The seed is by that time committed to the ground; there is nothing very material to do at home, and this additional quantity of honey enables me to be more generous to my home bees, and my wife to make a due quantity of mead. The reason, Sir, that you found mine better than that of others is that she puts two gallons of brandy in each barrel, which ripens it, and takes off that sweet, luscious taste, which it is apt to retain a long time. If we find anywhere in the woods (no matter on whose land) what is called a bee-tree, we must mark it; in the fall of the year when we propose to cut it down, our duty is to inform the proprietor of the land, who is entitled to half the contents; if this is not complied with we are exposed to an action of trespass, as well as he who should go and cut down a bee-tree which he had neither found out nor marked.

We have twice a year the pleasure of catching pigeons, whose numbers are sometimes so astonishing as to obscure the sun in their flight. Where is it that they hatch? For such multitudes must require an immense quantity of food. I fancy they breed toward the plains of Ohio, and those about Lake Michigan, which abound in wild oats; though I have never killed any that had that grain in their craws. In one of them, last year, I found some undigested rice. Now the nearest rice fields from where I live must be at least five hundred and sixty miles, and either their digestion must be suspended while they are flying, or else they must fly with the celerity of the wind. We catch them with a net extended on the ground, to which they are allured by what we call tame wild pigeons, made blind, and fastened to a long string; his short flights, and his repeated calls, never fail to bring them down. The greatest number I ever caught was fourteen dozen, though much larger quantities have often been trapped. I have frequently seen them at the market so cheap that for a penny you might have as many as you could carry away, and yet from the extreme cheapness you must not conclude that they are but an ordinary food; on the contrary, I think they are excellent. Every farmer has a tame wild pigeon in a cage at his door all the year round, in order to be ready whenever the season comes for catching them.

The pleasure I receive from the warblings of the birds in the spring is superior to my poor description, as the continual succession of their tuneful notes is forever new to me. I generally rise from bed about that indistinct interval, which, properly speaking, is neither night or day, for this is the moment of the most universal vocal choir. Who can listen unmoved to the sweet love tales of our robins, told from tree to tree? Or to the shrill catbirds? The sublime accents of the thrush from on high always retard my steps that I may listen to the delicious music. The variegated appearances of the dewdrops, as they hang to the different objects, must present even to a clownish imagination, the most voluptuous ideas. The astonishing art which all birds display in the construction of their nests, ill provided as we may suppose them with proper tools, their neatness, their convenience, always make me ashamed of the slovenliness of our houses; their love to their dame, their incessant careful attention, and the peculiar songs they address to her while she tediously incubates their eggs, remind me of my duty could I ever forget it. Their affection to their helpless little ones is a lively precept, and, in short, the whole economy of what we proudly call the brute creation, is admirable in every circumstance, and vain man, though adorned with the additional gift of reason, might learn from the perfection of instinct, how to regulate the follies, and how to temper the errors which this second gift often makes him commit. This is a subject on which I have often bestowed the most serious thoughts; I have often blushed within myself, and been greatly astonished, when I have compared the unerring path they all follow, all just, all proper, all wise, up to the necessary degree of perfection, with the coarse, the imperfect systems of men, not merely as governors and kings, but as masters, as husbands, as fathers, as citizens. But this is a sanctuary in which an ignorant farmer must not presume to enter.

If ever man was permitted to receive and enjoy some blessings that might alleviate the many sorrows to which he is exposed, it is certainly in the country, when he attentively considers those ravishing scenes with which he is everywhere surrounded. This is the only time of the year in which I am avaricious of every moment, I therefore lose none that can add to this simple and inoffensive happiness. I roam early throughout all my fields; not the least operation do I perform which is not accompanied with the most pleasing observations; were I to extend them as far as I have carried them, I should become tedious; you would think me guilty of affectation, and I should perhaps represent many things as pleasurable from which you might not perhaps receive the least agreeable emotions. But, believe me, what I write is all true and real.

Some time ago, as I sat smoking a contemplative pipe in my piazza, I saw with amazement a remarkable instance of selfishness displayed in a very small bird, which I had hitherto respected for its inoffensiveness. Three nests were placed almost contiguous to each other in my piazza: that of a swallow was affixed in the corner next to the house, that of a phoebe in the other, a wren possessed a little box which I had made on purpose, and hung between. Be not surprised at their tameness, all my family had long been taught to respect them as well as myself. The wren had shown before signs of dislike to the box which I had given it, but I knew not on what account; at last it resolved, small as it was, to drive the swallow from its own habitation, and to my very great surprise it succeeded. Impudence often gets the better of modesty, and this exploit was no sooner performed, than it removed every material to its own box with the most admirable dexterity; the signs of triumph appeared very visible, it fluttered its wings with uncommon velocity, an universal joy was perceivable in all its movements. Where did this little bird learn that spirit of injustice? It was not endowed with what we term reason! Here then is a proof that both those gifts border very near on one another; for we see the perfection of the one mixing with the errors of the other! The peaceable swallow, like the passive Quaker, meekly sat at a small distance and never offered the least resistance, but no sooner was the plunder carried away, than the injured bird went to work with unabated ardor, and in a few days the depredations were repaired. To prevent, however, a repetition of the same violence, I removed the wren's box to another part of the house.

In the middle of my new parlor I have, you may remember, a curious republic of industrious hornets; their nest hangs to the ceiling, by the same twig on which it was so admirably built and contrived in the woods. Its removal did not displease them, for they find in my house plenty of food, and I have left a hole open in one of the panes of the window, which answers all their purposes. By this kind usage they are become quite harmless; they live on the flies, which are very troublesome to us throughout the summer; they are constantly busy in catching them, even on the eyelids of my children. It is surprising how quickly they smear them with a sort of glue, lest they might escape, and when thus prepared, they carry them to their nests, as food for their young ones. These globular nests are most ingeniously divided into many stories, all provided with cells, and proper communications. The materials with which this fabric is built, they procure from the cottony furze, with which our oak rails are covered; this substance tempered with glue, produces a sort of pasteboard, which is very strong, and resists all the inclemency of weather. By their assistance, I am but little troubled with flies. All my family are so accustomed to their strong buzzing, that no one takes any notice of them, and though they are fierce and vindictive, yet kindness and hospitality has made them useful and harmless.

We have a great variety of wasps; most of them build their nests in mud, which they fix against the shingles of our roofs, as nigh the pitch as they can. These aggregates represent nothing, at first view, but coarse and irregular lumps, but if you break them, you will observe, that the inside of them contains a great number of oblong cells, in which they deposit their eggs, and in which they bury themselves in the fall of the year. Thus immured they securely pass through the severity of that season, and on the return of the sun are enabled to perforate their cells, and to open themselves a passage from these recesses into the sunshine. The yellow wasps, which build underground, in our meadows, are much more to be dreaded, for when the mower unwittingly passes his scythe over their holes they immediately sally forth with a fury and velocity superior even to the strength of man. They make the boldest fly, and the only remedy is to lie down and cover our heads with hay, for it is only at the head they aim their blows; nor is there any possibility of finishing that part of the work until, by means of fire and brimstone, they are all silenced. But though I have been obliged to execute this dreadful sentence in my own defense, I have often thought it a great pity, for the sake of a little hay, to lay waste so ingenious a subterranean town, furnished with every convenience, and built with a most surprising mechanism.

I never should have done were I to recount the many objects which involuntarily strike my imagination in the midst of my work, and spontaneously afford me the most pleasing relief. These appear insignificant trifles to a person who has traveled through Europe and America, and is acquainted with books and with many sciences; but such simple objects of contemplation suffice me, who have no time to bestow on more extensive observations. Happily, these require no study, they are obvious, they gild the moments I dedicate to them, and enliven the severe labors which I perform. At home my happiness springs from very different objects; the gradual unfolding of my children's reason, the study of their dawning tempers attract all my paternal attention. I have to contrive little punishments for their little faults, small encouragements for their good actions, and a variety of other expedients dictated by various occasions. But these are themes unworthy your perusal, and which ought not to be carried beyond the walls of my house, being domestic mysteries adapted only to the locality of the small sanctuary wherein my family resides. Sometimes I delight in inventing and executing machines, which simplify my wife's labor. I have been tolerably successful that way, and these, Sir, are the narrow circles within which I constantly revolve, and what can I wish for beyond them? I bless God for all the good he has given me; I envy no man's prosperity, and with no other portion of happiness than that I may live to teach the same philosophy to my children, and give each of them a farm, show them how to cultivate it, and be like their father, good substantial independent American farmers—an appellation which will be the most fortunate one a man of my class can possess, so long as our civil government continues to shed blessings on our husbandry. Adieu.

THOMAS JEFFERSON

Thomas Jefferson's *Notes on the State of Virginia*, first published privately in 1784 and released in a first English edition in 1787 in London, has become, almost in spite of its original, somewhat mundane purpose, the defining document of American agrarianism. Jefferson wrote these short pensees at the request of Francois Marbois of the French legation at Philadelphia. And yet Jefferson used Marbois' practical request for information about the new country and one of its most influential states, Virginia, to press his own agenda. While Jefferson described his responses to Marbois as an almost haphazard collection of "loose papers, bundled without order," scholars such as George Alan Davy have examined the ways in which *Notes*, Jefferson's only book, provides a venue in which the Virginian gentleman farmer-legislator-statesman moves freely from statistical calculation, to philosophic history, to geographical inventory. It is Jefferson's argumentative passages, such as Query XIX, that have caused historians to challenge the notion of the document as a mere "informal handbook."

Query XIX articulates, in the fluid, eloquent prose style of the Declaration of Independence, Jefferson's agrarian bias as a Virginia planter and gentleman farmer—an underlying rationale for the Louisiana Purchase in 1803 and the expansion of the agrarian republic it enabled. While the War of 1812 caused Jefferson to concede the need for domestic manufacturing, his vision of an America feed and governed by virtuous cultivators lasted well into the nineteenth century, and in certain regional and academic pockets, persists to the present day. Contemporary agrarian Gene Logsdon, for example, paraphrases Jeffersonian logic in his book *At Nature's Pace*, wherein he writes, "sustainable farms are to today's headlong rush toward the earth's destruction what the monasteries were to the Dark Ages." In any case, Jefferson's insistence that "those who labor in the earth are the chosen people of God," and the keepers of the country's "sacred fire" have been rallying cries for agrarians for more than two centuries.

Notes on the State of Virginia, Query XIX[24]

The present state of manufactures, commerce, interior and exterior trade? We never had an interior trade of any importance. Our exterior commerce has suffered very much from the beginning of the present

---

[24] Thomas Jefferson, *Notes on the State of Virginia* (Boston: Lilly and Wait, 1832), 171–173.

contest. During this time we have manufactured within our families the most necessary articles of clothing. Those of cotton will bear some comparison with the same kinds of manufacture in Europe; but those of wool, flax and hemp are very coarse, unsightly, and unpleasant: and such is our attachment to agriculture, and such our preference for foreign manufactures, that be it wise or unwise, our people will certainly return as soon as they can to the raising raw materials, and exchanging them for finer manufactures than they are able to execute themselves. The political economists of Europe have established it as a principle that every state should endeavor to manufacture for itself: and this principle, like many others, we transfer to America, without calculating the difference of circumstance which should often produce a difference of result. In Europe the lands are either cultivated, or locked up against the cultivator. Manufacture must therefore be resorted to of necessity not of choice, to support the surplus of their people. But we have an immensity of land courting the industry of the husbandman. Is it best then that all our citizens should be employed in its improvement, or that one-half should be called off from that to exercise manufactures and handicraft arts for the other? Those who labor in the earth are the chosen people of God, if ever he had a chosen people, whose breasts he has made his peculiar deposit for substantial and genuine virtue. It is the focus in which he keeps alive that sacred fire, which otherwise might escape from the face of the earth. Corruption of morals in the mass of cultivators is a phenomenon of which no age nor nation has furnished an example. It is the mark set on those, who not looking up to heaven, to their own soil and industry, as does the husbandman, for their subsistence, depend for it on the casualties and caprice of customers. Dependence begets subservience and venality, suffocates the germ of virtue, and prepares fit tools for the designs of ambition. This, the natural progress and consequence of the arts, has sometimes perhaps been retarded by accidental circumstances: but, generally speaking, the proportion which the aggregate of the other classes of citizens bears in any state to that of its husbandmen, is the proportion of its unsound to its healthy parts, and is a good enough barometer whereby to measure its degree of corruption. While we have land to labor then, let us never wish to see our citizens occupied at a workbench, or twirling a distaff. Carpenters, masons, smiths, are wanting in husbandry: but, for the general operations of manufacture, let our workshops remain in Europe. It is better to carry provisions and materials to workmen there, than bring them to the provisions and materials, and with them their manners and principles. The loss by the transportation of commodities across the Atlantic will be made up in happiness and permanence of government. The mobs of great cities add just so much to the support of pure government, as sores do to the strength of the human body. It is the manners and spirit of a people which preserve a republic in vigor. A degeneracy in these is a canker which soon eats to the heart of its laws and constitution.

ALEXIS DE TOCQUEVILLE

Alexis de Tocqueville's *Democracy in America* is required reading for many American students, though its contribution to agrarianism has consistently been overlooked. *Democracy in America*, first published in large edition in 1835, earned its author the Montyon Prize from the French Academy and catapulted him into a career in politics as a deputy and, for a brief time, as the Minister of Foreign Affairs under the presidency of Louis Napoleon. The passage below, from the Henry Reeve translation of 1899, shows the degree to which de Tocqueville's vision of American democracy is founded on the availability of land for the yeoman. "In America," de Tocqueville writes, "there are, properly speaking, no farming tenants." While an enduring system of tenant farming, particularly in the American South, renders this statement substantively inaccurate, the limitations of de Tocqueville and Gustave de Beaumont's travels in America in 1831 partially explain the overgeneralization. The Frenchmen, who had come to the United States to assess the new country's prison system, spent most of their time in urban Boston, New York, and Philadelphia, though they did travel as far south as New Orleans and as far west as Michigan, albeit briefly.

A Parisian born of a noble family, de Tocqueville is considered primarily a political philosopher, politician, and historian rather than an agrarian, though his distance from the yeoman life lends objectivity to his observations. Tocqueville's methodology, borrowed in part from his historical studies and consisting of personal interview and on-site visits, appears to compensate for whatever intrinsic knowledge of farming he may have lacked. In the passage that follows, de Tocqueville represents a dynamic, often antagonistic relationship between the American tenant farmer and his landlord, if any, and the chronic instability of such a system, particularly for the American landlord. The land, in de Tocqueville's estimation, was a safety valve for the tension of a capitalistic society as well as a constant source of hope. In positing an American democracy where every son could expect a homestead, he does overlook systems of progenitor imported from Europe and widely transplanted in America. De Tocqueville's emphasis on land and localism—"the strength of free people resides in local community"—lays important agrarian foundations.

How Democratic Institutions and Manners Tend to Raise Rents and Shorten the Terms of Leases[25]

What has been said of servants and masters is applicable to a certain extent to landowners and farming tenants, but this subject deserves to be considered by itself.

In America there are, properly speaking, no farming tenants; every man owns the ground he tills. It must be admitted that democratic laws tend greatly to increase the number of landowners and to diminish that of farming tenants. Yet what takes place in the United States is much less attributable to the institutions of the country than to the country itself. In America land is cheap and anyone may easily become a landowner; its returns are small and its produce cannot well be divided between a landowner and a farmer. America therefore stands alone in this respect, as well as in many others, and it would be a mistake to take it as an example.

---

[25] Alexis de Tocqueville, *Democracy in America*, vol. 2, ed. Francis Bowen, trans. Henry Reeve (Cambridge: Sever and Francis, 1862, 1899), 226–229.

I believe that in democratic as well as in aristocratic countries there will be landowners and tenants, but the connection existing between them will be of a different kind. In aristocracies the hire of a farm is paid to the landlord, not only in rent, but in respect, regard, and duty; in democracies the whole is paid in cash. When estates are divided and passed from hand to hand, and the permanent connection that existed between families and the soil is dissolved, the landowner and the tenant are only casually brought into contact. They meet for a moment to settle the conditions of the agreement and then lose sight of each other; they are two strangers brought together by a common interest, who keenly talk over a matter of business, the sole object of which is to make money.

In proportion as property is subdivided and wealth distributed over the country, the community is filled with people whose former opulence is declining, and with others whose fortunes are of recent growth and whose wants increase more rapidly than their resources. For all such persons the smallest pecuniary profit is a matter of importance, and none of them feel disposed to waive any of their claims or to lose any portion of their income.

As ranks are intermingled, and as very large as well as very scanty fortunes become more rare, every day brings the social condition of the landowner nearer to that of the farmer: the one has not naturally any uncontested superiority over the other; between two men who are equal and not at ease in their circumstances, the contract of hire is exclusively an affair of money. A man whose estate extends over a whole district and who owns a hundred farms is well aware of the importance of gaining at the same time the affections of some thousands of men. This object appears to call for his exertions, and to attain it he will readily make considerable sacrifices. But he who owns a hundred acres is insensible to similar considerations, and cares but little to win the private regard of his tenant.

An aristocracy does not expire, like a man, in a single day; the aristocratic principle is slowly undermined in men's opinion before it is attacked in their laws. Long before open war is declared against it, the tie that had hitherto united the higher classes to the lower may be seen to be gradually relaxed. Indifference and contempt are betrayed by one class, jealousy and hatred by the others. The intercourse between rich and poor becomes less frequent and less kind, and rents are raised. This is not the consequence of a democratic revolution, but its certain harbinger; for an aristocracy that has lost the affections of the people once and forever is like a tree dead at the root, which is the more easily torn up by the winds the higher its branches have spread. In the course of the last fifty years the rents of farms have amazingly increased, not only in France, but throughout the greater part of Europe. The remarkable improvements that have taken place in agriculture and manufactures within the same period do not suffice, in my opinion, to explain this fact; recourse must be had to another cause, more powerful and more concealed. I believe that cause is to be found in the democratic institutions which several European nations have adopted and in the democratic passions which more or less agitate all the rest.

I have frequently heard great English landowners congratulate themselves that at the present day they derive a much larger income from their estates than their fathers did. They have perhaps good reason to be glad, but most assuredly they do not know what they are glad of. They think they are making a clear gain when it is in reality only an exchange; their influence is what they are parting with for cash, and what they gain in money will before long be lost in power.

There is yet another sign by which it is easy to know that a great democratic revolution is going on or approaching. In the Middle Ages almost all lands were leased for lives or for very long terms; the domestic economy of that period shows that leases for ninety-nine years were more frequent then than leases for twelve years are now. Men then believed that families were immortal; men's conditions seemed settled forever, and the whole of society appeared to be so fixed that it was not supposed anything would ever be stirred or shaken in its structure. In ages of equality the human mind takes a different bent: the prevailing notion is that nothing abides, and man is haunted by the thought of mutability. Under this impression the landowner and the tenant himself are instinctively averse to protracted terms of obligation; they are afraid of being tied up tomorrow by the contract that benefits them today. They do not trust themselves; they are afraid that, their standards changing, they may have trouble in ridding themselves of the thing which had been the object of their longing. And they are right to fear this, for in democratic times what is most unstable, in the midst of the instability of everything, is the heart of man.

RALPH WALDO EMERSON

Unlike many of the other forerunners of the American conservation and agrarian movements, Ralph Waldo Emerson, the son of a Unitarian minister, had little direct experience with farming, though he endured a life of loss equal to that of most frontier farmers. Born in Boston, Emerson lost his father when he was eight, his first wife when he was twenty, and his son Waldo in 1842, six years after the publication of *Nature* in 1836.

Prompted by the loss of his first wife to tuberculosis, Emerson's soul-searching journey to England deepened him, steeping him in the Romantic philosophies of Carlyle, Coleridge, and Wordsworth. Emerson's lifelong vision problems, which shortened his studies at Harvard Divinity School, seem, ironically, to have sharpened the poet-preacher-philosopher's appreciation for nature, in which he found divinity. An ordained minister of the Second Church in Boston circa 1829, Emerson's Unitarianism invited his reading of Persian and Indic mystical texts as well as sharpened his commitment to self-reliance and spiritual and intellectual independence—qualities he would bring to bear in his opposition to slavery and to the mistreatment of Native Americans.

In the classic essay below, Emerson contrasts the poet and the farmer, arguing that the poet's very lack of property imbues him with a purist's view unavailable to the plowman, for all the latter's deeds. The healthful man, Emerson suggests, views the divine world with the eyes of a child, casting off his years, by balm of the woods, "as the snake his slough." Though Emerson was viewed by some, then as now, as a hopeless romantic prone to sentimentalist claims ("Nature never wears a mean appearance") he achieves dynamic equipoise in

"Nature." Emerson's assertion "In the woods, we return to reason and faith" profoundly influenced Thoreau—a lifelong friend of Emerson's who lived on his property for a time—and Whitman, whose *Leaves of Grass* Emerson called "the most extraordinary piece of wit and wisdom that America has yet contributed." The essential harmony of man and nature, as set forth in this essay, would be amplified in Emerson's developing transcendentalism, which comes to full flower in his 1841 essay, "The Over-Soul," where he writes, "We see the world piece by piece, as the sun, the moon, the animal, the tree: but the whole, of which these are shining parts, is the soul."

Nature[26]

To go into solitude, a man needs to retire as much from his chamber as from society. I am not solitary whilst I read and write, though nobody is with me. But if a man would be alone, let him look at the stars. The rays that come from those heavenly worlds, will separate between him and what he touches. One might think the atmosphere was made transparent with this design, to give man, in the heavenly bodies, the perpetual presence of the sublime. Seen in the streets of cities, how great they are! If the stars should appear one night in a thousand years, how would men believe and adore; and preserve for many generations the remembrance of the city of God which had been shown! But every night come out these envoys of beauty, and light the universe with their admonishing smile.

The stars awaken a certain reverence, because though always present, they are inaccessible; but all natural objects make a kindred impression, when the mind is open to their influence. Nature never wears a mean appearance. Neither does the wisest man extort her secret, and lose his curiosity by finding out all her perfection. Nature never became a toy to a wise spirit. The flowers, the animals, the mountains, reflected the wisdom of his best hour, as much as they had delighted the simplicity of his childhood.

---

[26] Ralph Waldo Emerson, *The Prose Works of Ralph Waldo Emerson. In Two Volumes* (Boston: Fields, Osgood, & Co., 1870), 6–9.

When we speak of nature in this manner, we have a distinct but most poetical sense in the mind. We mean the integrity of impression made by manifold natural objects. It is this which distinguishes the stick of timber of the woodcutter from the tree of the poet. The charming landscape which I saw this morning is indubitably made up of some twenty or thirty farms. Miller owns this field, Locke that, and Manning the woodland beyond. But none of them owns the landscape. There is a property in the horizon which no man has but he whose eye can integrate all the parts, that is, the poet. This is the best part of these men's farms, yet to this their warranty deeds give no title.

To speak truly, few adult persons can see nature. Most persons do not see the sun. At least they have a very superficial seeing. The sun illuminates only the eye of the man, but shines into the eye and the heart of the child. The lover of nature is he whose inward and outward senses are still truly adjusted to each other; who has retained the spirit of infancy even into the era of manhood. His intercourse with heaven and earth becomes part of his daily food. In the presence of nature, a wild delight runs through the man, in spite of real sorrows. Nature says, he is my creature, and maugre all his impertinent griefs, he shall be glad with me. Not the sun or the summer alone, but every hour and season yields its tribute of delight; for every hour and change corresponds to and authorizes a different state of the mind, from breathless noon to grimmest midnight. Nature is a setting that fits equally well a comic or a mourning piece. In good health, the air is a cordial of incredible virtue. Crossing a bare common, in snow puddles, at twilight, under a clouded sky, without having in my thoughts any occurrence of special good fortune, I have enjoyed a perfect exhilaration. I am glad to the brink of fear. In the woods, too, a man casts off his years, as the snake his slough, and at what period soever of life, is always a child. In the woods, is perpetual youth. Within these plantations of God, a decorum and sanctity reign, a perennial festival is dressed, and the guest sees not how he should tire of them in a thousand years. In the woods, we return to reason and faith. There I feel that nothing can befall me in life—no disgrace, no calamity, (leaving me my eyes,) which nature cannot repair. Standing on the bare ground, my head bathed by the blithe air, and uplifted into infinite space, all mean egotism vanishes. I become a transparent eyeball; I am nothing; I see all; the currents of the Universal Being circulate through me; I am part or particle of God. The name of the nearest friend sounds then foreign and accidental: to be brothers, to be acquaintances, master or servant, is then a trifle and a disturbance. I am the lover of uncontained and immortal beauty. In the wilderness, I find something more dear and connate than in streets or villages. In the tranquil landscape, and especially in the distant line of the horizon, man beholds somewhat as beautiful as his own nature.

The greatest delight which the fields and woods minister, is the suggestion of an occult relation between man and the vegetable. I am not alone and unacknowledged. They nod to me, and I to them. The waving of the boughs in the storm is new to me and old. It takes me by surprise, and yet is not unknown. Its effect is like that of a higher thought or a better emotion coming over me, when I deemed I was thinking justly or doing right.

Yet it is certain that the power to produce this delight, does not reside in nature, but in man, or in a harmony of both. It is necessary to use these pleasures with great temperance. For, nature is not always tricked in holiday attire, but the same scene which yesterday breathed perfume and glittered as for the frolic of the nymphs is overspread with melancholy today. Nature always wears the colors of the spirit. To a man laboring under calamity, the heat of his own fire hath sadness in it. Then, there is a kind of contempt of the landscape felt by him who has just lost by death a dear friend. The sky is less grand as it shuts down over less worth in the population.

GEORGE PERKINS MARSH

In the following reading, George Perkins Marsh, then Congressman from Vermont and later the American minister to Turkey and to Italy, both encourages and chastises farmers for their traditional practices—most especially their inattention to predatory forestry. A wonderfully rich record of its times made by a proud New Englander who was a lawyer and scholar rather than a farmer, "Address Delivered

Before the Agricultural Society" reveals its age in the tensions inherent in Marsh's cosmopolitan knowledge balanced against his local, Vermont-centered boosterism. At times humorous—as when Marsh cites Vermont's "three winters"—sober truth-telling best characterizes the tone of this address given to an expressly agrarian audience. Startling is Marsh's cautionary if not somewhat exaggerated claim that the natural landscape of Vermont had changed so much in one generation as to be unrecognizable. Congressman Marsh reveals both his own legislative leanings as well as an astute read of the future of American conservation when he writes that the concerned citizen can no longer rely on "enlightened self-interest" to introduce reforms or "check abuses." These notions would be refined and developed in Marsh's more voluminous, less readable magnum opus *Man and Nature* (1864). "None had a more powerful impact on the subsequent history of the conservation movement than George Perkins Marsh," write the curators of the Library of Congress exhibit *The Evolution of the Conservation Movement*. An abolitionist and an opponent of the Mexican American War, Marsh's blend of environmental, economic, and political thinking closely parallels that of fellow New Englanders Thoreau and Emerson.

Address Delivered Before the Agricultural Society of Rutland County, 1847[27]

It is little to the credit of our agriculturists that the greatest progress in these and other modern improvements should have been made by persons not bred to agricultural pursuits, and it has often been said that mechanics, merchants and professional men make in the end the best farmers. If there is any truth in this opinion, it is probably because these persons, commencing their new calling at a period of life when judgment is mature, tied down by habit to no blind routine of antiquated practice, and ridden by no nightmare of hereditary prejudice

---

[27] George Perkins Marsh, "Address Delivered Before ..." (Rutland, VT: Herald, 1848), 17–21.

in regard to particular modes of cultivation, are conscious of the necessity of observation and reflection, in an occupation, the successful pursuit of which requires so much of both, and feel themselves at liberty to select such processes as are commended by the results of actual experience, or accord with the known laws of vegetable physiology. Under such circumstances, a judicious man, encouraged by the stimulus of novelty, would be likely to study the subject with earnestness, and to profit by his own errors, as well as by the experience of others.

There are certain other improvements connected with agriculture, to which I desire to draw your special attention. One of these is the introduction of a better economy in the management of our forest lands. The increasing value of timber and fuel ought to teach us that trees are no longer what they were in our fathers' time, an encumbrance. We have undoubtedly already a larger proportion of cleared land in Vermont than would be required, with proper culture, for the support of a much greater population than we now possess, and every additional acre both lessens our means for thorough husbandry, by disproportionately extending its area, and deprives succeeding generations of what, though comparatively worthless to us, would be of great value to them. The functions of the forest, besides supplying timber and fuel, are very various. The conducting powers of trees render them highly useful in restoring the disturbed equilibrium of the electric fluid; they are of great value in sheltering and protecting more tender vegetables against the destructive effects of bleak or parching winds, and the annual deposit of the foliage of deciduous trees, and the decomposition of their decaying trunks, form an accumulation of vegetable mould, which gives the greatest fertility to the often originally barren soils on which they grow, and enriches lower grounds by the wash from rains and the melting snows. The inconveniences resulting from a want of foresight in the economy of the forest are already severely felt in many parts of New England, and even in some of the older towns in Vermont. Steep hillsides and rocky ledges are well suited to the permanent growth of wood, but when in the rage for improvement they are improvidently stripped of this protection, the action of sun and wind and rain soon deprives them of their thin coating of vegetable mould, and this, when exhausted, cannot be restored by ordinary husbandry. They remain, therefore, barren and unsightly blots, producing neither grain nor grass, and yielding no crop but a harvest of noxious weeds to infest with their scattered seeds the richer arable grounds below. But this is by no means the only evil resulting from the injudicious destruction of the woods. Forests serve as reservoirs and equalizers of humidity. In wet seasons, the decayed leaves and spongy soil of woodlands retain a large proportion of the falling rains, and give back the moisture in time of drought, by evaporation or through the medium of springs. They thus both check the sudden flow of water from the surface into the streams and low grounds, and prevent the droughts of summer from parching our pastures and drying up the rivulets which water them. On the other hand, where too large a proportion of the surface is bared of wood, the action of the summer sun and wind scorches the hills which are no longer shaded or sheltered by trees, the springs and rivulets that found their supply in the bibulous soil of the forest disappear, and the farmer is obliged to surrender his meadows to his cattle, which can no longer find food in his pastures, and sometime even to drive them miles for water. Again, the vernal and autumnal rains, and the melting snows of winter, no longer intercepted and absorbed by the leaves or the open soil of the woods, but falling everywhere upon a comparatively hard and even surface, flow swiftly over the smooth ground, washing away the vegetable mould as they seek their natural outlets, fill every ravine with a torrent, and convert every river into an ocean. The suddenness and violence of our freshets increases in proportion as the soil is cleared; bridges are washed away, meadows swept of their crops and fences, and covered with barren sand, or themselves abraded by the fury of the current, and there is reason to fear that the valleys of many of our streams will soon be converted from smiling meadows into broad wastes of shingle and gravel and pebbles, deserts in summer, and seas in autumn and spring. The changes, which these causes have wrought in the physical geography of Vermont within a single generation are too striking to have escaped the attention of any observing person, and every middle-aged man who revisits his birthplace after a few years of absence, looks upon another landscape than that which formed the theatre of his youthful toils and pleasures. The signs of artificial improvement are mingled with the tokens of improvident waste, and the bald and barren hills, the dry beds of the smaller streams, the ravines furrowed out by the torrents of spring, and the diminished thread of interval that skirts the widened channel of the rivers, seem sad substitutes for the pleasant groves and brooks and broad meadows of his ancient paternal domain. If the present value of timber and land will not justify the artificial replanting of grounds injudiciously cleared, at least nature ought to be allowed to reclothe them with a spontaneous growth of wood, and in our future husbandry a more careful selection should be made of land for permanent improvement. It has long been a practice in many parts of Europe, as well as in our older settlements, to cut the forests reserved for timber and fuel at stated intervals. It is quite time that this practice should be introduced among us. After the first felling of the original forest, it is indeed a long time before its place is supplied, because the roots of old and full-grown trees seldom throw up shoots, but when the second growth is once established, it may be cut with great advantage, at periods of about twenty-five years, and yields a material, in every respect but size, far superior to the wood of the primitive tree. In many European countries, the economy of the forest is regulated by law; but here, where public opinion determines, or rather in practice constitutes law, we can only appeal to an enlightened self-interest to introduce the reforms, check the abuses, and preserve us from an increase of the evils I have mentioned.

There is a branch of rural industry hitherto not much attended to among us, but to the social and economical importance of which we are beginning to be somewhat awake. I refer to the agreeable and profitable art of horticulture. The neglect of this art is probably to be ascribed to the opinion that the products of the garden and the fruit yard are to be regarded rather as condiments or garnishing than as nutritious food, as something calculated to tickle the palate, not to strengthen the system; as belonging in short to the department of ornament, not to that of utility. This is an unfortunate error. The tendency of our cold climate is to create an inordinate appetite for animal food, and we habitually consume much too large a proportion of that stimulating aliment. This, when we compare the relative cost of a given quantity of nutritive matter obtained from animals and vegetables, seems very indifferent economy, and considerations of health most clearly indicate the expediency of increasing the proportion of our fruit and vegetable diet. We cannot in this latitude expect to rival the pomona of more favored climes, but in most situations, we may, with little labor or expense, rear such a variety of fruits as to supply our tables with a succession of delicious and healthful viands throughout the entire year. It is for us a happy circumstance that most fruits attain their highest perfection near the northern limit of their growth, and though the fig and the peach cannot be naturalized among us, we may, to say nothing of the smaller fruits, successfully cultivate the finer varieties of the apple, the pear, and even the grape.

Another mode of rural improvement may be fitly mentioned in connection with this last. I refer to the introduction of a better style of domestic architecture, which shall combine convenience, warmth, and reasonable embellishment. A well-arranged and well-proportioned building costs no more than a misshapen disjointed structure, and commodity and comfort may be had at as cheap a rate as inconvenience and confusion. Neither is a little expenditure in ornament thrown away. The paint which embellishes tends also to preserve, and the shade trees not only furnish a protection against the exhausting heats of summer, but they serve, if thickly planted, to break the fury of the blasts of winter, and in the end they furnish a better material for fuel or mechanical uses than the spontaneous forest growth. The habit of domestic order, comfort, and neatness will be found to have a very favorable influence in the manner in which the outdoor operations of husbandry are connected. A farmer whose house is neatly and tastefully constructed and arranged will never be a slovenly agriculturist. The order of his dwelling and his courtyard will extend to his stables, his barns, his granaries, and his fields. His beasts will be well-lodged and cared for, his meadows free from stumps, and briars, and bushes, and the strength of his fences will secure him against the trespasses of his thriftless neighbor's unruly cattle. Another consideration, which most strongly recommends attention to order and comfort and beauty in domestic and rural arrangements is that all these tend to foster a sentiment, of which the enterprising and adventurous Yankee has in general, far too little—I mean a feeling of attachment to his home, and by a natural association, to the institutions of his native New England. To make our homes in themselves desirable is the most effectual means of compensating for that rude climate which gives us three winters each year—two southern, with a Siberian intercalated between—and of arming our children against the tempting attractions of the milder sky and less laborious life of the South, and the seductions of the boasted greatness and exaggerated fertility of the West. A son of Vermont who has enjoyed beneath the paternal roof the blessings of a comfortable and well-ordered home, and whose eye has been trained to appreciate the charms of rural beauty, which his own hands have helped perhaps to embellish, will find little to please in the slovenly husbandry, the rickety dwellings and the wasteful economy of the southern planter, little to admire in the tame monotony of a boundless prairie, and little to entice in the rude domestic arrangements, the coarse fare and the coarser manners of the western squatter. A youth will not readily abandon the orchard he has dressed, the flowering shrubs which he has aided his sisters to rear, the fruit or shade tree planted on the day of his birth, and whose thrifty growth he has regarded with as much pride as his own increase of statue; and who that has been taught to gaze with admiring eye on the unrivalled landscapes unfolded from our every hill, where lake, and island, and mountain and rock, and well-tilled field, and evergreen wood, and purling brook, and cheerful home of man are presented at due distance and in fairest proportion, would exchange such scenes as these, for the mirey sloughs, the puny groves, the slimy streams, which alone diversify the dead uniformity of Wisconsin and Illinois!

I have now shown, I hope, rather by suggestion than by argument, that the profession of agriculture in this age and land is an honorable, and in its true spirit, and elevated and an enlightened calling; I have adverted to its importance as an instrument of primary civilization, endeavored to indicate its present position as an art, and hinted at its future hopes and encouragements.

JOHN MUIR

Published one year before his death in 1914, John Muir's *The Story of My Boyhood and Youth* is presented here out of chronological order, both in deference to Muir's status as a fountainhead for the conservation and environmental movements and in recognition of the era it documents—Muir's teenage years in the 1850s. John Muir—arguably the most famous naturalist, "Father of Our National Park System," and defender of the Yosemite and Sierra Nevada—spent a good deal of his life as a farmer. Muir first labored as a farmer's son, as described in the passage below, later herded sheep upon his arrival in the Yosemite in the late 1860s, and ultimately partnered with his father-in-law in successfully managing the family fruit ranch near Martinez, California.

Like Emerson, early problems with vision—Muir was temporarily blinded via an eye injury sustained while working in a carriage parts shop in Indianapolis—made John Muir a greater, more sublime witness to nature. It was Muir's unbounded enthusiasm for the Sierra Nevada and other natural places that ultimately led Emerson and Roosevelt to his doorstep and drew readers by the thousands to his books. Long-lived like Emerson and struck likewise with wanderlust, Muir's oeuvre reflects his travels in Australia, South America, Africa, Europe, China and Japan and articulates what might rightly, and literally, be called his *natural philosophy*. Incredibly prolific, Muir published more than three hundred articles and ten major books, the most famous of which is *Our National Parks*, the book which prompted Roosevelt's visit and which served as the foundation for the President's conservationist policies. Roosevelt and Muir would later come to sharp disagreement over the Hetch Hetchy Dam that both Teddy and his chief forester, Gifford Pinchot, supported. And, though Muir would lose the Hetch Hetchy conflict (many say the emotional toll of his defeat contributed to his death), he consistently won other battles for the establishment of Yosemite National Park and the creation of Sequoia, Mount Rainier, Petrified Forest, and Grand Canyon National Parks.

The autobiographical passage below is infused with the wry circumspection of an aged Muir reflecting on his adolescence on the family farm near Kingston, Wisconsin. Muir's father, a hard-driving Scotsman who had brought the family to America in 1849, is omnipresent in these pages—an enduring emblem of his son's strength and also his vulnerability. Neither overtly reactionary nor revolutionary, Muir's understated ethos in *The Story of My Boyhood* hints at a seedbed for this farm son's revolutionary thinking regarding the white man versus Native Americans, recreation versus work, and agrarian versus industrial economies.

The Ploughboy[28]

The mixed lot of settlers around us offered a favorable field for observation of the different kinds of people of our own race.[29] We were swift to note the way they behaved, the differences in their religion and morals, and in their ways of drawing a living from the same kind of soil under the same general conditions; how they protected themselves from the weather; how they were influenced by new doctrines and old ones seen in new lights in preaching, lecturing, debating, bringing up their children, et cetera, and how they regarded the Indians, those first settlers and owners of the ground that was being made into farms.

I well remember my father's discussing with a Scotch neighbor, a Mr. George Mair, the Indian question as to the rightful ownership of the soil. Mr. Mair remarked one day that it was pitiful to see how the unfortunate Indians, children of Nature, living on the natural products of the soil, hunting, fishing, and even cultivating small cornfields on the most fertile spots, were now being robbed of their lands and pushed ruthlessly back into narrower and narrower limits by alien races who were cutting off their means of livelihood. Father replied that

---

[28] John Muir, *The Story of My Boyhood and Youth* (Boston: Houghton Mifflin, 1913), 217–227.
[29] This excerpt picks up with the twenty-second paragraph of Chapter VI, "The Ploughboy."

surely it could never have been the intention of God to allow Indians to rove and hunt over so fertile a country and hold it forever in unproductive wildness, while Scotch and Irish and English farmers could put it to so much better use. Where an Indian required thousands of acres for his family, these acres in the hands of industrious, God-fearing farmers would support ten or a hundred times more people in a far worthier manner, while at the same time helping to spread the gospel.

Mr. Mair urged that such farming as our first immigrants were practicing was in many ways rude and full of the mistakes of ignorance, yet, rude as it was, and ill-tilled as were most of our Wisconsin farms by unskillful, inexperienced settlers who had been merchants and mechanics and servants in the old countries, how should we like to have specially trained and educated farmers drive us out of our homes and farms, such as they were, making use of the same argument, that God could never have intended such ignorant, unprofitable, devastating farmers as we were to occupy land upon which scientific farmers could raise five or ten times as much on each acre as we did? And I well remember thinking that Mr. Mair had the better side of the argument. It then seemed to me that, whatever the final outcome might be, it was at this stage of the fight only an example of the rule of might with but little or no thought for the right or welfare of the other fellow if he were the weaker; that "they should take who had the power, and they should keep who can," as Wordsworth makes the marauding Scottish Highlanders say.

Many of our old neighbors toiled and sweated and grubbed themselves into their graves years before their natural dying days, in getting a living on a quarter-section of land and vaguely trying to get rich, while bread and raiment might have been serenely won on less than a fourth of this land, and time gained to get better acquainted with God. I was put to the plow at the age of twelve, when my head reached but little above the handles, and for many years I had to do the greater part of the plowing. It was hard work for so small a boy; nevertheless, as good plowing was exacted from me as if I were a man, and very soon I had to become a good plowman, or rather plowboy. None could draw a straighter furrow. For the first few years the work was particularly hard on account of the tree stumps that had to be dodged. Later the stumps were all dug and chopped out to make way for the McCormick reaper, and because I proved to be the best chopper and stump-digger, I had nearly all of it to myself. It was dull, hard work leaning over on my knees all day, chopping out those tough oak and hickory stumps, deep down below the crowns of the big roots. Some, though fortunately not many, were two feet or more in diameter.

And as I was the eldest boy, the greater part of all the other hard work of the farm quite naturally fell on me. I had to split rails for long lines of zigzag fences. The trees that were tall enough and straight enough to afford one or two logs ten feet long were used for rails, the others, too knotty or cross-grained, were disposed of in log and cordwood fences. Making rails was hard work and required no little skill. I used to cut and split a hundred a day from our short, knotty oak timber, swinging the axe and heavy mallet, often with sore hands, from early morning to night. Father was not successful as a rail-splitter. After trying the work with me a day or two, he in despair left it all to me. I rather liked it, for I was proud of my skill, and tried to believe that I was as tough as the timber I mauled, though this and other heavy jobs stopped my growth and earned for me the title "runt of the family."

In those early days, long before the great labor-saving machines came to our help, almost everything connected with wheat-raising abounded in trying work—cradling in the long, sweaty dog-days, raking and binding, stacking, thrashing—and it often seemed to me that our fierce, over-industrious way of getting the grain from the ground was too closely connected with gravedigging. The staff of life, naturally beautiful, oftentimes suggested the gravedigger's spade. Men and boys, and in those days even women and girls, were cut down while cutting the wheat. The fat folk grew lean and the lean leaner, while the rosy cheeks brought from Scotland and other cool countries across the sea faded to yellow like the wheat. We were all made slaves through the vice of over-industry. The same was in great part true in making hay to keep the cattle and horses through the long winters. We were called in the morning at four o'clock and seldom got to bed before nine, making a broiling, seething day seventeen hours long loaded with heavy work, while I was only a small stunted boy; and a few years later my brothers David and Daniel and my older sisters had to endure about as much as I did. In the harvest dog-days and dog-nights and dog-mornings, when we arose from our clammy beds, our cotton shirts clung to our backs as wet with sweat as the bathing suits of swimmers, and remained so all the long, sweltering days. In mowing and cradling, the most exhausting of all the farm work, I made matters worse by foolish ambition in keeping ahead of the hired men. Never a warning word was spoken of the dangers of over-work. On the contrary, even when sick we were held to our tasks as long as we could stand. Once in harvest time I had the mumps and was unable to swallow any food except milk, but this was not allowed to make any difference, while I staggered with weakness and sometimes fell headlong among the sheaves. Only once was I allowed to leave the harvest field—when I was stricken down with pneumonia. I lay gasping for weeks, but the Scotch are hard to kill and I pulled through. No physician was called, for father was an enthusiast, and always said and believed that God and hard work were by far the best doctors.

None of our neighbors were so excessively industrious as father; though nearly all of the Scotch, English, and Irish worked too hard, trying to make good homes and to lay up money enough for comfortable independence. Excepting small garden patches, few of them had owned land in the old country. Here their craving land-hunger was satisfied, and they were naturally proud of their farms and tried to keep them as neat and clean and well-tilled as gardens. To accomplish this without the means for hiring help was impossible. Flowers were planted about the neatly kept log or frame houses; barnyards, granaries, et cetera, were kept in about as neat order as the homes, and the fences and corn rows were rigidly straight. But every uncut weed distressed them; so also did every ungathered ear of grain, and all that was lost by birds and gophers; and this overcarefulness bred endless work and worry.

As for money, for many a year there was precious little of it in the country for anybody. Eggs sold at six cents a dozen in trade, and five-cent calico was exchanged at twenty-five cents a yard. Wheat brought fifty cents a bushel in trade. To get cash for it before the Portage Railway was built, it had to be hauled to Milwaukee, a hundred miles away. On the other hand, food was abundant—eggs, chickens, pigs, cattle, wheat, corn, potatoes, garden vegetables of the best, and wonderful melons as luxuries. No other wild country I have ever known extended a kinder welcome to poor immigrants. On the arrival in the spring, a log house could be built, a few acres plowed, the virgin sod planted with corn, potatoes, et cetera, and enough raised to keep a family comfortably the very first year; and wild hay for cows and oxen grew in abundance on the numerous meadows. The American settlers were wisely content with smaller fields and less of everything, kept indoors during excessively hot or cold weather, rested when tired, went off fishing and hunting at the most favorable times and seasons of the day and year, gathered nuts and berries, and in general tranquilly accepted all the good things the fertile wilderness offered.

After eight years of this dreary work of clearing the Fountain Lake farm, fencing it, and getting it in perfect order, building a frame house and the necessary outbuildings for the cattle and horses—after all this had been victoriously accomplished, and we had made out to escape with life—father bought a half-section of wild land about four or five miles to the eastward and began all over again to clear and fence and break up other fields for a new farm, doubling all the stunting, heartbreaking chopping, grubbing, stump-digging, rail-splitting, fence-building, barn-building, house-building, and so forth.

HENRY DAVID THOREAU

Nothing could be a more exuberant, more audacious statement of nascent conservationist themes than Thoreau's eminently quotable essay "Walking," which appeared in the *Atlantic Monthly* in 1862 as Thoreau was dying. A synthesis of two lectures—"Walking" and "The Wild" delivered in 1851 and 1852 and again in 1856 and 1857—Thoreau's amalgam essay was published posthumously in 1873 in the book *Excursions*.

Thoreau's own definition of genius, articulated here as "light which makes the darkness visible, like the lightning's flash, which perchance shatters the temple of knowledge itself," applies emphatically to his work in "Walking," still prescient one hundred and fifty years after its initial crafting. A friend and colleague of Emerson's, and a benefactor of Emerson's generosity as the provider of land for Thoreau's cabin at Walden Pond, Thoreau leapfrogs, in the eyes of writer Edward Abbey and others, his mentor in "Walking," distilling in plain yet poetic terms the esprit de corps for transcendentalism and the burgeoning American conservation movement.

Though not a farmer by trade, Thoreau's occupation as surveyor, a trade to which he turned when his writing failed to win him fame and fortune, brought him into frequent contact with cultivators, encounters he describes somewhat cynically in this essay. Despite Thoreau's somewhat sardonic discussion of the plow as family heirloom and emblem of victory, his true feelings are stated thusly: "Hope and the future for me are not in lawns and cultivated fields, not in towns and cities, but in the impervious and quaking swamps." "In wildness," Thoreau famously writes, "is the preservation of the world." Thoreau himself turned to the wilderness when, after the death of his brother John from lockjaw, he built a small cabin on the north shore of Walden Pond on land owned by Emerson, not far from Concord, Massachusetts, where Thoreau was born in 1817. In the famous passage from the book grown out of his retreat, *Walden*, Thoreau writes "I went to the woods because I wished to live deliberately, to front only the essential facts of life, and see if I could not learn what it had to teach, and not, when I came to die, discover that I had not lived." Thoreau's essay "Succession of Forest Trees," though less accessible than "Walking," is nonetheless viewed by historian Donald Worster and others as an "important contribution to conservation, agriculture, and ecological sciences" as it evidences the technical eye that made Thoreau an able surveyor.

After leaving Walden Pond in 1847, Thoreau wrote and lectured but never traveled as widely or with as much fanfare as his friend Ralph Waldo Emerson, who would long outlive the younger Thoreau. In death, Thoreau was considered something of a failure by the shopkeepers and farmers of his hometown, Concord, though his work is now considered the epitome of the nature essay's ability to express an intimate relationship between man and nature.

Walking[30]

The West of which I speak is but another name for the Wild; and what I have been preparing to say is that in Wildness is the preservation of the world. Every tree sends its fibers forth in search of the wild. The cities import it at any price. Men plow and sail for it. From the forest and wilderness come the tonics and barks which brace mankind. Our ancestors were savages. The story of Romulus and Remus being suckled by a wolf is not a meaningless fable. The founders of every State which has risen to eminence have drawn their nourishment and vigor from a similar wild source. It was because the children of the Empire were not suckled by the wolf that they were conquered and displaced by the children of the northern forests who were.

I believe in the forest, and in the meadow, and in the night in which the corn grows. We require an infusion of hemlock-spruce or arborvitae in our tea. There is a difference between eating and drinking for strength and from mere gluttony. The Hottentots eagerly devour the marrow of the koodoo and other antelopes raw, as a matter of course. Some of our northern Indians eat raw the marrow of the Arctic reindeer, as well as various other parts, including the summits of the antlers, as long as they are soft. And herein, perchance, they have stolen a march on the cooks of Paris. They get what usually goes to feed the fire. This is probably better than stall-fed beef and slaughterhouse pork to make a man of. Give me a wildness whose glance no civilization can endure, as if we lived on the marrow of koodoos devoured raw.

There are some intervals which border the strain of the wood thrush, to which I would migrate—wild lands where no settler has squatted; to which, methinks, I am already acclimated.

The African hunter Cummings tells us that the skin of the eland, as well as that of most other antelopes just killed, emits the most

---

[30] Henry David Thoreau, *Excursions* (Boston: Ticknor and Fields, 1863), 185–197.

delicious perfume of trees and grass. I would have every man so much like a wild antelope, so much a part and parcel of Nature, that his very person should thus sweetly advertise our senses of his presence, and remind us of those parts of Nature which he most haunts. I feel no disposition to be satirical, when the trapper's coat emits the odor of musquash even; it is a sweeter scent to me than that which commonly exhales from the merchant's or the scholar's garments. When I go into their wardrobes and handle their vestments, I am reminded of no grassy plains and flowery meads which they have frequented, but of dusty merchants' exchanges and libraries rather.

A tanned skin is something more than respectable, and perhaps olive is a fitter color than white for a man, a denizen of the woods. "The pale white man!" I do not wonder that the African pitied him. Darwin the naturalist says, "A white man bathing by the side of a Tahitian was like a plant bleached by the gardener's art, compared with a fine, dark green one, growing vigorously in the open fields."

> Ben Jonson exclaims, "How near to good is what is fair!" So I would say, "How near to good is what is wild!"

Life consists with wildness. The most alive is the wildest. Not yet subdued to man, its presence refreshes him. One who pressed forward incessantly and never rested from his labors, who grew fast and made infinite demands on life, would always find himself in a new country or wilderness, and surrounded by the raw material of life. He would be climbing over the prostrate stems of primitive forest trees.

Hope and the future for me are not in lawns and cultivated fields, not in towns and cities, but in the impervious and quaking swamps. When, formerly, I have analyzed my partiality for some farm which I had contemplated purchasing, I have frequently found that I was attracted solely by a few square rods of impermeable and unfathomable bog, a natural sink in one corner of it. That was the jewel which dazzled me. I derive more of my subsistence from the swamps which surround my native town than from the cultivated gardens in the village. There are no richer parterres to my eyes than the dense beds of dwarf andromeda (*Cassandra calyculata*) which cover these tender places on the earth's surface. Botany cannot go farther than tell me the names of the shrubs which grow there—high-blueberry, panicled andromeda, lamb-kill, azalea, and rhodora—all standing in the quaking sphagnum. I often think that I should like to have my house front on this mass of dull red bushes, omitting other flower plots and border, transplanted spruce and trim box, even graveled walks—to have this fertile spot under my windows, not a few imported barrow-fulls of soil only to cover the sand which was thrown out in digging the cellar. Why not put my house, my parlor, behind this plot, instead of behind that meager assemblage of curiosities, that poor apology for a Nature and Art, which I call my front yard? It is an effort to clear up and make a decent appearance when the carpenter and mason have departed, though done as much for the passerby as the dweller within. The most tasteful front yard fence was never an agreeable object of study to me; the most elaborate ornaments; acorn-tops, or what not, soon wearied and disgusted me. Bring your sills up to the very edge of the swamp, then, (though it may not be the best place for a dry cellar,) so that there be no access on that side to citizens. Front yards are not made to walk in, but, at most, through, and you could go in the back way.

Yes, though you may think me perverse, if it were proposed to me to dwell in the neighborhood of the most beautiful garden that ever human art contrived, or else of a dismal swamp, I should certainly decide for the swamp. How vain, then, have been all your labors, citizens, for me!

My spirits infallibly rise in proportion to the outward dreariness. Give me the ocean, the desert, or the wilderness! In the desert, pure air and solitude compensate for want of moisture and fertility. The traveler Burton says of it, "Your morale improves; you become frank and cordial, hospitable and single-minded.... In the desert, spirituous liquors excite only disgust. There is a keen enjoyment in a mere animal existence." They who have been traveling long on the steppes of Tartary say, "On reentering cultivated lands, the agitation, perplexity, and turmoil of civilization oppressed and suffocated us; the air seemed to fail us, and we felt every moment as if about to die of asphyxia." When I would recreate myself, I seek the darkest wood, the thickest and most interminable, and, to the citizen, most dismal swamp. I enter a swamp as a sacred place, a sanctum sanctorum. There is the strength, the marrow of Nature. The wildwood covers the virgin mould, and the same soil is good for men and for trees. A man's health requires as many acres of meadow to his prospect as his farm does loads of muck. There are the strong meats on which he feeds. A town is saved, not more by the righteous men in it than by the woods and swamps that surround it. A township where one primitive forest waves above, while another primitive forest rots below, such a town is fitted to raise not only corn and potatoes, but poets and philosophers for the coming ages. In such a soil grew Homer and Confucius and the rest, and out of such a wilderness comes the Reformer eating locusts and wild honey.

To preserve wild animals implies generally the creation of a forest for them to dwell in or resort to. So it is with man. A hundred years ago they sold bark in our streets peeled from our own woods. In the very aspect of those primitive and rugged trees, there was, methinks, a tanning principle which hardened and consolidated the fibers of men's thoughts. Ah! Already I shudder for these comparatively degenerate days of my native village, when you cannot collect a load of bark of good thickness, and we no longer produce tar and turpentine.

The civilized nations—Greece, Rome, England—have been sustained by the primitive forests which anciently rotted where they stand. They survive as long as the soil is not exhausted. Alas for human culture! Little is to be expected of a nation, when the vegetable mould is exhausted, and it is compelled to make manure of the bones of its fathers. There the poet sustains himself merely by his own superfluous fat, and the philosopher comes down on his marrow-bones.

It is said to be the task of the American "to work the virgin soil," and that "agriculture here already assumes proportions unknown everywhere else." I think that the farmer displaces the Indian even because he redeems the meadow, and so makes himself stronger and in some respects more natural. I was surveying for a man the other day a single straight line one hundred and thirty-two rods long, through a swamp, at whose entrance might have been written the words which Dante read over the entrance to the infernal regions—"Leave all hope, ye that enter"—that is, of ever getting out again; where at one time I saw my employer actually up to his neck and swimming for his life in his property, though it was still winter. He had another similar swamp which I could not survey at all, because it was completely under water, and nevertheless, with regard to a third swamp, which I did survey from a distance, he remarked to me, true to his instincts, that he would not part with it for any consideration, on account of the mud which it contained. And that man intends to put a girdling ditch round the whole in the course of forty months, and so redeem it by the magic of his spade. I refer to him only as the type of a class.

The weapons with which we have gained our most important victories, which should be handed down as heirlooms from father to son, are not the sword and the lance, but the bushwhack, the turf-cutter, the spade, and the bog-hoe, rusted with the blood of many a meadow, and begrimed with the dust of many a hard-fought field. The very winds blew the Indian's cornfield into the meadow, and pointed out the way which he had not the skill to follow. He had no better implement with which to entrench himself in the land than a clamshell. But the farmer is armed with plow and spade.

In literature it is only the wild that attracts us. Dullness is but another name for tameness. It is the uncivilized free and wild thinking in *Hamlet* and the *Iliad*, in all the scriptures and mythologies, not learned in the schools, that delights us. As the wild duck is more swift and beautiful than the tame, so is the wild—the mallard—thought, which 'mid fulling dews, wings its way above the fens. A truly good book is something as natural, and as unexpectedly and unaccountably fair and perfect, as a wildflower discovered on the prairies of the West or in the jungles of the East. Genius is a light which makes the darkness visible, like the lightning's flash, which perchance shatters the temple of knowledge itself, and not a taper lighted at the hearthstone of the race, which pales before the light of common day.

English literature, from the days of the minstrels to the Lake Poets—Chaucer and Spenser and Milton, and even Shakespeare, included—breathes no quite fresh and in this sense wild strain. It is an essentially tame and civilized literature, reflecting Greece and Rome. Her wilderness is a green wood, her wild man a Robin Hood. There is plenty of genial love of Nature, but not so much of Nature herself. Her chronicles inform us when her wild animals, but not when the wild man in her, became extinct.

The science of Humboldt is one thing, poetry is another thing. The poet today, notwithstanding all the discoveries of science, and the accumulated learning of mankind, enjoys no advantage over Homer. Where is the literature which gives expression to Nature? He would be a poet who could impress the winds and streams into his service, to speak for him; who nailed words to their primitive senses, as farmers drive down stakes in the spring, which the frost has heaved; who derived his words as often as he used them, transplanted them to his page with earth adhering to their roots; whose words were so true and fresh and natural that they would appear to expand like the buds at the approach of spring, though they lay half-smothered between two musty leaves in a library—ay, to bloom and bear fruit there, after their kind, annually, for the faithful reader, in sympathy with surrounding Nature.

I do not know of any poetry to quote which adequately expresses this yearning for the Wild. Approached from this side, the best poetry is tame. I do not know where to find in any literature, ancient or modern, any account which contents me of that Nature with which even I am acquainted. You will perceive that I demand something which no Augustan nor Elizabethan age, which no culture, in short, can give. Mythology comes nearer to it than anything. How much more fertile a Nature, at least, has Grecian mythology its root in than English literature! Mythology is the crop which the Old World bore before its soil was exhausted, before the fancy and imagination were affected with blight; and which it still bears, wherever its pristine vigor is unabated. All other literatures endure only as the elms which overshadow our houses; but this is like the great dragon tree of the Western Isles, as old as mankind, and, whether that does or not, will endure as long; for the decay of other literatures makes the soil in which it thrives.

The West is preparing to add its fables to those of the East. The valleys of the Ganges, the Nile, and the Rhine, having yielded their crop, it remains to be seen what the valleys of the Amazon, the Plate, the Orinoco, the St. Lawrence, and the Mississippi will produce. Perchance, when, in the course of ages, American liberty has become a fiction of the past—as it is to some extent a fiction of the present—the poets of the world will be inspired by American mythology.

The wildest dreams of wild men, even, are not the less true, though they may not recommend themselves to the sense which is most common among Englishmen and Americans today. It is not every truth that recommends itself to the common sense. Nature has a place for the wild clematis as well as for the cabbage. Some expressions of truth are reminiscent—others merely sensible, as the phrase is—others prophetic. Some forms of disease, even, may prophesy forms of health. The geologist has discovered that the figures of serpents, griffins, flying dragons, and other fanciful embellishments of heraldry have their prototypes in the forms of fossil species which were extinct before man was created, and hence "indicate a faint and shadowy knowledge of a previous state of organic existence." The Hindus dreamed that the earth rested on an elephant, and the elephant on a tortoise, and the tortoise on a serpent; and though it may be an unimportant coincidence, it will not be out of place here to state that a fossil tortoise has lately been discovered in Asia large enough to support an elephant.

I confess that I am partial to these wild fancies, which transcend the order of time and development. They are the sublimest recreation of the intellect. The partridge loves peas, but not those that go with her into the pot.

In short, all good things are wild and free. There is something in a strain of music, whether produced by an instrument or by the human voice—take the sound of a bugle in a summer night, for instance—which by its wildness, to speak without satire, reminds me of the cries emitted by wild beasts in their native forests. It is so much of their wilderness as I can understand. Give me for my friends and neighbors wild men, not tame ones. The wilderness of the savage is but a faint symbol of the awful ferity with which good men and lovers meet.

EDMUND MORRIS

A classic for small farm managers, hobby farmers, and market gardeners, Edmund Morris' *Ten Acres Enough* was an immensely popular manual used by many back-to-the-landers after its publication in the waning years of the Civil War. Morris, a Philadelphia businessman who abandoned the mercantile world in the early 1800s to buy a farm in the New Jersey countryside, took his business savvy with him to the hinterlands, experimenting with several varieties of fruit trees and crops to support his family. Like Crevecoeur's, Morris' narrative Like Crevecoeur's, Morris' narrative recommends an agrarian experience enriched by close observation of nature.

In "Birds and the Services They Render," Morris sharply distinguishes himself from his neighbors and interlopers, whom he portrays as a "worthless, loafing tribe of gunners." Like Crevecoeur, Morris is particularly fascinated with bird life, pitying the farmer who, amid his work, does not stop to delight in aviary splendors. While contemporary readers may be shocked by Morris' taking of a bird's life to make a point—especially given the era's legislation against the wanton killing of birds—Morris' overall argument is clear "It takes mankind a great while to learn the ways of Providence, and to understand that things are better contrived for him than he can contrive them for himself." In arguing for the reconception of farming to include husbanding native wildlife as well as crops and livestock, Morris stakes out a middle ground between the naturalist dismissive of agrarians and the small farmer bothered by naturalist prescriptions and moralizing. *Ten Acres Enough* responds to the Malthusian theory, which argued that population growth and its demands would consistently outstrip the power of the earth to generate subsistence. Later books of this genre, including two books excerpted in this collection—Tuisco Greiner's *How to Make the Garden Pay* and Bolton Hall's *Three Acres and Liberty*—owe a direct debt to Morris, as does *Five Acres and Independence*, Maurice G. Kain's classic book on small farm management published in 1935.

Birds and the Services They Render[31]

One morning in September, hearing shots fired repeatedly at the further end of my grounds, and proceeding thither to ascertain the cause, I discovered three great, overgrown boobies, with guns in their hands, trampling down my strawberries, and shooting bluebirds and robins. On inquiring where they belonged, they answered in the next township. I suggested to them that I thought their own township was quite large enough to keep its own loafers, without sending them to deprate on me, warned them never to show themselves on my premises again, and then drove them out. This happened to be the only occasion on which I was invaded by any of the worthless, loafing tribe of gunners who roam over some neighborhoods engaged in the manly occupation of killing tomtits and catbirds.

For all such my aversion was as decided as my partiality for the birds was strong. One of the little amusements I indulged in immediately on taking possession of my farm, was to put up at least twenty little rough contrivances about the premises, in which the birds might build. Knowing their value as destroyers of insects, I was determined to protect them, and thus, around the dwelling house, in the garden trees, and upon the sides of the barn, as well as in other places which promised to be popular, I placed boxes, calabashes, and squashes for them to occupy. The wrens and bluebirds took to them, with gratifying readiness, built, and reared their families. But I observed that the wren quickly took possession of every one in which the hole was just large enough to admit himself; and too small to allow the bluebird to enter; while in those large enough to admit a bluebird no wren would build. This was because the bluebird has a standing spite against the wrens, which leads him to enter the nests of the latter, whenever possible, and destroy their eggs. Almost any number of wrens may thus be attracted round the house and garden, where they act as vigilant destroyers of insects.

---

[31] Edmund Morris, *Ten Acres Enough: A Practical Experience, Showing How a Very Small Farm May Be Made to Keep a Very Large Family* (New York: J. Miller, 1866), 131–140.1

These interesting creatures soon hatched out large broods of young to provide food for which they were incessantly on the wing. They became surprisingly tame and familiar, those especially which were nearest the house, and in trees beneath which the family were constantly passing. We watched their movements through the season with increasing interest. No cat was permitted even to approach their nests, no tree on which a family was domiciled was ever jarred or shaken, and the young children, instead of regarding them as game to be frightened off, or hunted, caught, and killed, were educated to admire and love them. Indeed, so carefully did we observe their looks and motions, that many times I felt almost sure that I could identify and recognize the tenants of particular boxes. They ranged over the whole extent of my ten acres, clearing the bushes and vegetables of insects and worms, while the garden in which they sang and chattered from daybreak until sunset was kept entirely clear of the destroyers. I encountered them at the furthest extremity of my domain, peering under the peach leaves, flitting from one tomato vine to another, almost as tame as those at home. They must have known me, and felt safe from harm. I am persuaded that I recognized them. Yet it was at this class of useful birds that the boobies calling themselves sportsmen were aiming their weapons, when I routed them from the premises, and forbid the murderous foray.

Insects are, occasionally, one of the farmer's greatest pests. But high, thorough farming is a potent destroyer. It is claimed by British writers to be a sure one. When the average produce of wheat in England was only twenty bushels per acre, the ravages of the insect tribe were far more general and destructive than they have been since the average has risen to forty bushels per acre. Why may not the cultivation of domestic birds like these, that nestle round the house and garden where insects mostly congregate, be considered an important feature in any system of thorough farming?

Besides the wrens and bluebirds, the robins built under the eaves of the woodshed, and became exceedingly tame. The more social swallow took possession of every convenient nestling place about the barn, while troops of little sparrows came confidingly to the kitchen door to pick up the crumbs of bread which the children scattered on the pavement as soon as they discovered that these innocent little creatures were fond of them. Thus my premises became a sort of open aviary, in which a multitude of birds were cultivated with assiduous care, and where they shall be even more assiduously domesticated, as long as I continue to be lord of the manor. I pity the man who can look on these things, who can listen to the song of wrens, the loud, inspiring carol of the robin on the treetop, as the setting sun gilds its utmost extremities, listening to these vocal evidences of animal comfort and enjoyment, without feeling any augmentation of his own pleasures, and that the lonesome blank which sometimes hangs around a rural residence is thus gratefully filled.

One morning, hearing a great clamor and turmoil in a thicket in the garden where a nest of orioles had been filled with young birds, I cautiously approached to discover the cause. A dozen orioles were hovering about in great excitement, and for some time it was impossible to discover the meaning of the trouble. But remaining perfectly quiet, so as not to increase the disturbance, I at length discovered an oriole, whose wing had become so entangled in one end of a long string which formed part of the nest that she could not escape. The other birds had also discovered her condition, and hence their lamentation over a misfortune they were unable to remedy. But they did all they could, and were assiduously bringing food to a nest full of voracious young ones, as well as feeding the imprisoned parent. I was so struck with the interesting spectacle that my family were called out to witness it; then, having gazed upon it a few moments, I cautiously approached the prisoner, took her in my hands, carefully untied and then cut away the treacherous string, and let the frightened warbler go free. She instantly flew up into her nest, as if to see that all her callow brood were safe, gave us a song of thanks, and immediately the crowd of sympathizing birds, as if conscious that the difficulty no longer existed, flew away to their respective nests.

It takes mankind a great while to learn the ways of Providence, and to understand that things are better contrived for him than he can contrive them for himself. Of late, the people are beginning to learn that they have mistaken the character of most of the little birds, and have not understood the object of the Almighty in creating them. They are the friends of those who plant, and sow, and reap. It has been seen that they live mostly on insects, which are among the worst enemies of the agriculturist and that if they take now and then a grain of wheat, a grape, a cherry, or a strawberry, they levy but a small tax for the immense services rendered. In this altered state of things, legislatures are passing laws for the protection of little birds, and increasing the penalties to be enforced upon the bird-killers.

A farmer in my neighborhood came one day to borrow a gun for the purpose of killing some yellow birds in his field of wheat, which he said were eating up the grain. I declined to loan the gun. In order, however, to gratify his curiosity, I shot one of them, opened its crop, and found in it two hundred weevils, and but four grains of wheat, and in these four grains the weevil had burrowed! This was a most instructive lesson, and worth the life of the poor bird, valuable as it was. This bird resembles the canary, and sings finely. One fact like this affords an eloquent text for sermonizing, for the benefit of the farmers and others who may look upon little birds as inimical to their interests. Every hunter and farmer ought to know that there is hardly a bird that flies that is not a friend of the farmer and gardener.

Some genial spirits have given the most elaborate attention to the question of the value of birds. One gentleman took his position some fifteen feet from the nest of an oriole, in the top of a peach tree, to observe his habits. The nest contained four young ones, well-fledged, which every now and then would stand upon the edge of the nest to try their wings. They were, therefore, at an age which required the largest supply of food. This the parents furnished at intervals of two to six minutes, throughout the day. They lighted on the trees, the vines, the grass, and other shrubbery, clinging at times to the most extreme and delicate points of the leaves, in search of insects. Nothing seemed to come amiss to these sharp-eyed foragers—grasshoppers, caterpillars, worms, and the smaller flies. Sometimes one, and sometimes as many as six, were plainly fed to the young ones at once. They would also carry away the refuse litter from the nest, and drop it many yards off.

A little figuring gives the result of this incessant warfare against the insects. For only eight working hours it will be one thousand worms destroyed by a single pair of birds. But if a hundred pairs be domesticated on the premises, the destruction will amount to one hundred thousand daily, or three million a month!

This may seem to be a mere paper calculation, but the annals of ornithology are crowded with confirmatory facts. The robin is accused of appropriating the fruit which he has protected during the growing season from a cloud of enemies. But his principal food is spiders, beetles, caterpillars, worms, and larvae. Nearly two hundred larvae have been taken from the gizzard of a single bird. He feeds voraciously on those of the destructive worm. In July he takes a few strawberries, cherries, and pulpy fruits generally, more as a dessert than anything else, because it is invariably found to be largely intermixed with insects. Robins killed in the country, at a distance from gardens and fruit trees, are found to contain less stone-fruit than those near villages, showing that this bird is not an extensive forager. If our choicest fruits are near at hand, he takes a small toll of them, but a small one only. In reality, a very considerable part of every crop of grain and fruit is planted, not for the mouths of our children, but for the fly, the curculio, and the cankerworm, or some other of these pests of husbandry. Science has done something, and will no doubt do more, to alleviate the plague. It has already taught us not to wage equal war on the wheat-fly and the parasite which preys upon it, and it will, perhaps, eventually persuade those who need the lesson that a few peas and cherries are well-bestowed by way of dessert on the cheerful little warblers, who turn our gardens into concert rooms, and do so much to aid us in the warfare against the grubs and caterpillars, which form their principal meal.

But if the subject of the value of insect-destroying birds has been so much overlooked in this country, it is not so in Europe. It has been brought formally before the French Senate, and is now before the French government. Learned commissioners have reported upon it, and it is by no means improbable that special legislation will presently follow. The inquiry has been conducted with an elaborate accuracy characteristic of French legislation. Insects and birds have been carefully classified according to their several species, their habits of feeding have been closely observed, and the results ascertained and computed. It has been concluded that by no agency, save that of little birds, can the ravages of insects be kept down. There are some birds which live exclusively upon insects and grubs, and the quantity which they destroy is enormous. There are others which live partly on grubs, and partly on grain, doing some damage, but providing an abundant compensation. A third class—the birds of prey—are excepted from the category of benefactors, and are pronounced, too precipitately we think, to be noxious, inasmuch as they live mostly upon the smaller birds. One class is a match for the other. A certain insect was found to lay two thousand eggs, but a single tomtit was found to eat two hundred thousand eggs a year. A swallow devours about 543 insects a day, eggs and all. A sparrow's nest in the city of Paris was found to contain seven hundred pair of the upper wings of cockchafers, though, of course, in such a place food of other kinds was procurable in abundance. It will easily be seen, therefore, what an excess of insect life is produced when a counterpoise like this is withdrawn, and the statistics before us show clearly to what an extent the balance of nature has been disturbed. A third, and wholly artificial class of destroyers has been introduced. Every *chasseur*, during the season, kills, it is said, from one hundred to two hundred birds daily. A single child has been known to come home at night with one hundred birds' eggs, and it has been calculated and reported that the number of birds' eggs destroyed annually in France is between eighty million and one hundred million. The result is that little birds in that country are actually dying out; some species have already disappeared, and others are rapidly diminishing. But there is another consequence. The French crops have suffered terribly from the superabundance of insect vermin. Not only the various kinds of grain, but the vines, the olives, and even forest trees, tell the same tale of mischief, till at length the alarm has become serious. Birds are now likely to be protected; indeed their rise in public estimation has been signally rapid. Some philosopher has declared, and the report quotes the saying as a profound one, that "the birds can live without man, but man cannot live without the birds."

The same results are being experienced in this country, and our whole agricultural press, as well as the experience of every fruit grower and gardener, testifies to the fact that our fruit is disappearing as the birds upon our premises are permitted to perish. Every humane and prudent man will therefore do his utmost to preserve them.

FREDERICK LAW OLMSTED

The curators of the Library of Congress exhibit *Evolution of the Conservation Movement* call Frederick Law Olmsted's "Yosemite and the Big Tree Grove: A Preliminary Report, 1865" one of the first "systematic expositions in the history of the Western world of the importance of contact with wilderness for the human being, the effect of beautiful scenery on human perception, and the moral responsibility of democratic governments to preserve regions of extraordinary natural beauty for the benefit of the whole people." According to the same note, Olmsted's report met with such skepticism from the Yosemite Commissioners that they unceremoniously tabled the document, though Olmsted would ultimately send a transcription of his report in the form of a letter to the *New York Evening Post* in 1868, the year used to determine the report's chronological placement within this volume.

The "father of American landscape architecture," Olmsted is best known for his design of New York City's Central Park, the grounds at Biltmore, and the "Emerald Necklace" Boston city park system. Earlier in his life, however, Olmstead considered himself a farmer, pouring his substantial energies into cultivating and purchasing, with his father's help, a series of farms, including one in Staten Island. In 1850, Olmsted, who had once moved to New York City at the tender age of eighteen to pursue work as a "scientific farmer," traveled to England to study English agricultural and horticultural practices. The resulting book, *Walks and Talks of an American Farmer in England*, Olmsted prefaced with the following disclaimer: "I expect that all who try to read the book will be willing to come into a warm, good-natured, broad country kitchen fireside with me, and permit me to speak my mind freely, and in such language as I can readily command, on all sorts of subjects that come in my way, forming views from the facts that I gave them and taking my opinions for only just what they shall seem to be worth." The open-endedness of Olmsted's self-granted, carte blanche would be adopted by other rambling nature and agrarian writers of the era, including contemporaries John Burroughs, Walt Whitman, and Henry David Thoreau, and, much later, James Agee and Walker Evans in their famous agrarian collaboration *Let Us Now Praise Famous Men*.

Given the revolutionary and egalitarian bent of the Yosemite report excerpted below, its alleged suppression fails to surprise. Olmsted advocates strenuously for recreational relief for the "agriculture classes" while regarding the growing professional class with something approaching pity, as when he writes of their "diseased condition" caused by "excessive devotion of the mind to a limited range of interests." Called "one of the most profound and original statements to emerge from the American conservation movement," "Yosemite and the Big Tree Grove" perfectly blends an agrarian's consciousness with a naturalist's heart.

Preliminary Report Upon Yosemite and Big Tree Grove[32]

But in this country at least it is not those who have the most important responsibilities in state affairs or in commerce who suffer most from lack of recreation; women suffer more than men, and the agricultural class is more largely represented in our insane asylums than the professional, and for this, and other reasons, it is these classes to which the opportunity for such recreation is the greatest blessing.[33]

---

[32] Frederick Law Olmsted, "Draft of Preliminary Report upon the Yosemite and Big Tree Grove" and "Letter on the Great American Park of the Yosemite." Frederick Law Olmsted Papers, Manuscript Division, Library of Congress, Washington, DC.

[33] This sentence marks the end of the second paragraph of page 5 of a twentieth-century transcription of the Olmsted report made from Olmsted's secretary's original, handwritten transcription.

Forerunners: Readings 1770–1880 71

If we analyze the operation of scenes of beauty upon the mind, and consider the intimate relation of the mind upon the nervous system and the whole physical economy, the action and reaction which constantly occurs between bodily and mental conditions, the reinvigoration which results from such scenes is readily comprehended. Few persons can see such scenery as that of the Yosemite and not be impressed by it in some slight degree. All not alike, all not perhaps consciously, and amongst all who are consciously impressed by it, few can give the least expression to that of which they are conscious. But there can be no doubt that all have this susceptibility, though with some it is much more dull and confused than with others.

The power of scenery to affect men is, in a large way, proportionate to the degree of their civilization and to the degree in which their taste has been cultivated. Among a thousand savaged savages, there will be a much smaller number who will show the least sign of being so affected than among a thousand persons taken from a civilized community. This is only one of the many channels in which a similar distinction between civilized and savage men is to be generally observed. The whole body of the susceptibilities of civilized men and with their susceptibilities their powers, are on the whole enlarged. But as with the bodily powers, if one group of muscles is developed by exercise exclusively, and all others neglected, the result is general feebleness, so it is with the mental faculties. And men who exercise those faculties or susceptibilities of the mind which are called in play by beautiful scenery so little that they seem to be inert with them, are either in a diseased condition from excessive devotion of the mind to a limited range of interests, or their whole minds are in a savage state; that is, a state of low development. The latter class need to he drawn out generally; the former need relief from their habitual matters of interest and to be drawn out in those parts of their mental nature which have been habitually left idle and inert.

But there is a special reason why the reinvigoration of those parts which are stirred into conscious activity by natural scenery is more effective upon the general development and health than that of any other, which is this: the severe and excessive exercise of the mind which leads to the greatest fatigue and is the most wearing upon the whole constitution is almost entirely caused by application to the removal of something to be apprehended in the future, or to interests beyond those of the moment, or of the individual; to the laying up of wealth, to the preparation of something, to accomplishing something in the mind of another, and especially to small and petty details which are uninteresting in themselves and which engage the attention at all only because of the bearing they have on some general end of more importance which is seen ahead.

In the interest which natural scenery inspires there is the strongest contrast to this. It is for itself and at the moment it is enjoyed. The attention is aroused and the mind occupied without purpose, without a continuation of the common process of relating the present action, thought or perception to some future end. There is little else that has this quality so purely. There are few enjoyments with which regard for something outside and beyond the enjoyment of the moment can ordinarily be so little mixed. The pleasures of the table are irresistibly associated with the care of hunger and the repair of the bodily waste. In all social pleasures and all pleasures which are usually enjoyed in association with the social pleasure, the care for the opinion of others, or the good of others largely mingles. In the pleasures of literature, the laying up of ideas and self-improvement are purposes which cannot he kept out of view. This, however, is in very slight degree, if at all, the case with the enjoyment of the emotions caused by natural scenery. It therefore results that the enjoyment of scenery employs the mind without fatigue and yet exercises it, tranquilizes it, and yet enlivens it, and thus, through the influence of the mind over the body, gives the effect of refreshing rest and reinvigoration to the whole system.

Men who are rich enough and who are sufficiently free from anxiety with regard to their wealth can and do provide places of this needed recreation for themselves. They have done so from the earliest periods known in the history of the world, for the great men of the Babylonians, the Persians, and the Hebrews had their rural retreats, as large and as luxurious as those of the aristocracy of Europe at present. There are in the islands of Great Britain and Ireland more than one thousand private parks and notable grounds devoted to luxury and recreation. The value of these grounds amounts to many millions of dollars and the cost of their annual maintenance is greater than that of the national schools; their only advantage to the commonwealth is obtained through the recreation they afford to their owners (except as these extend hospitality to others) and these owners with their families number less than one in six thousand of the whole population.

The enjoyment of the choicest natural scenes in the country and the means of recreation connected with them is thus a monopoly, in a very peculiar manner, of a very few very rich people. The great mass of society, including those to whom it would be of the greatest benefit, is excluded from it. In the nature of the case private parks can never be used by the mass of the people in any country nor by any considerable number even of the rich, except by the favor of a few, and in dependence on them.

Thus without means are taken by government to withhold them from the grasp of individuals, all places favorable in scenery to the recreation of the mind and body will be closed against the great body of the people. For the same reason that the water of rivers should be guarded against private appropriation and the use of it for the purpose of navigation and otherwise protected against obstructions, portions of natural scenery may therefore properly be guarded and cared for by government. To simply reserve them from monopoly by individuals, however, it will be obvious, is not all that is necessary. It is necessary that they should be laid open to the use of the body of the people.

JOHN BURROUGHS

Considered a protégé of Thoreau by some, and a poor imitator by others, history considers John Burroughs second only to Thoreau as a master nature essayist. Wildly popular in his day, and—perhaps because of his popular appeal—oft maligned by critics, Burroughs' patronage, like Thoreau's, was achieved by carefully acquired friendships and truly exceptional powers of nature observation. Burroughs was born on a family farm in New York and married a prosperous New York farmer's daughter, steeping him in agrarianism on all sides. His most life-altering event proved to be a meeting with Walt Whitman in 1864 while Burroughs was in Washington DC looking for conventional work to support his new marriage—work which he found in the form of clerking for the Department of Treasury. After three years under the spell of Whitman, with whom he had become close friends, Burroughs published *Notes on Walt Whitman*, the first of what would be two book-length appreciations of his friend and mentor. Four years later, Burroughs published *Wake-Robin*, a nature study that would launch a long-lived, prolific career in which he would move away from conventional employment to live the life of the retiring writer. Like Thoreau, Burroughs built a waterside nature retreat.

Burroughs ultimately published twenty-three volumes of essays, writings centered on nature and, like Whitman's, ranging to philosophy, literary criticism, and travel, among other topics. Critical praise for Burroughs' popular work could sometimes be curmudgeonly, citing the author's sometimes inaccurate renderings of natural and biographical history, his overzealous Walt Whitman boosterism, and his uncharitable criticism of fellow writers.

In *John Burroughs: An American Naturalist*, Burroughs biographer Edward J. Renehan describes Burroughs' niche as that of a "a literary naturalist with a duty to record his own unique perceptions of the natural world." The essay below, describing farm life in Burroughs' beloved Catskills, resounds with characteristic agrarian nostalgia for a pre-mechanized age, and serves as a valuable documentation of threatened agrarian folkways in the late nineteenth century: "The quill-wheel, and the spinning-wheel, and the loom," Burroughs laments, "are heard no more among us." Like John Muir in his memoir *The Story of My Boyhood and Youth*, Burroughs waxes nostalgic about his own farm roots in this remarkable passage, which intimates the potential for estrangement felt by many farm-born conservationists of the era, especially Burroughs and Muir. "The farmer should be the true naturalist," Burroughs writes in the ultimate paragraph of this excerpt. And yet Burroughs and Muir, despite their yeoman roots, habitually traced the divergence between the naturalist and the agriculturalist. The essay below is presented here in chronological order according to its original publication in *Scrivener's* in 1878 rather than by the date of its publication in book form.

Phases of Farm Life[34]

Machinery, I say, has taken away some of the picturesque features of farm life. How much soever we may admire machinery and the faculty of mechanical invention, there is no machine like a man, and the work done directly by his hands; the things made or fashioned by them, have a virtue and a quality that cannot be imparted by machinery. The line of mowers in the meadows, with the straight swaths behind them, is more picturesque than the "Clipper" or "Buckeye" mower, with its team and driver. So are the flails of the threshers, chasing each other through the air, more pleasing to the eye and the ear than the machine, with its uproar, its choking clouds of dust, and its general hurly-burly.

Sometimes the threshing was done in the open air, upon a broad rock, or a smooth, dry plat of greensward, and it is occasionally done there yet, especially the threshing of the buckwheat crop, by a farmer who has not a good barn floor, or who cannot afford to hire the machine. The flail makes a louder thud in the fields than you would imagine, and in the splendid October weather it is a pleasing spectacle to behold the gathering of the ruddy crop, and three or four lithe figures beating out the grain with their flails in some sheltered nook, or some grassy lane lined with cedars. When there are three flails beating together, it makes lively music, and when there are four, they follow each other so fast that it is a continuous roll of sound, and it requires a very steady stroke not to hit or get hit by the others. There is just room and time to get your blow in, and that is all. When one flail is upon the straw, another has just left it, another is halfway down, and the fourth is high and straight in the air. It is like a swiftly revolving wheel that delivers four blows at each revolution. Threshing, like mowing, goes much easier in company than when alone; yet many a farmer or

---

[34] John Burroughs, *In the Catskills* (Boston: Houghton Mifflin, 1910), 56–62.

laborer spends nearly all the late fall and winter days shut in the barn, pounding doggedly upon the endless sheaves of oats and rye.

When the farmers made "bees," as they did a generation or two ago much more than they do now, a picturesque element was added. There was the stone bee, the husking bee, the "raising," the "moving," et cetera. When the carpenters had got the timbers of the house or the barn ready, and the foundation was prepared, then the neighbors for miles about were invited to come to the "raisin'." The afternoon was the time chosen. The forenoon was occupied by the carpenter and the farm hands in putting the sills and "sleepers" in place ("sleepers," what a good name for the those rude hewn timbers that lie under the floor in the darkness and silence!). When the hands arrived, the great beams and posts and joists and braces were carried to their place on the platform, and the first "bent," as it was called, was put together and pinned by oak pins that the boys brought. Then pike poles were distributed, the men, fifteen or twenty of them, arranged in a line abreast of the bent; the boss carpenter steadied and guided the corner post and gave the word of command, "Take holt, boys!" "Now, set her up!" "Up with her!" "Up she goes!" When it gets shoulder high, it becomes heavy, and there is a pause. The pikes are brought into requisition; every man gets a good hold and braces himself, and waits for the words. "All together now!" shouts the captain; "Heave her up!" "He-o-he!" (*heave-all, heave*), at the top of his voice, every man doing his best. Slowly the great timbers go up; louder grows the word of command, till the bent is up. Then it is plumbed and stay-lathed, and another is put together and raised in the same way, till they are all up. Then comes the putting on the great plates, timbers that run lengthwise of the building and match the sills below. Then, if there is time, the putting up of the rafters.

In every neighborhood there was always some man who was especially useful at "raisin's." He was bold and strong and quick. He helped guide and superintend the work. He was the first one up on the bent, catching a pin or a brace and putting it in place. He walked the lofty and perilous plate with the great beetle in hand, put the pins in the holes, and, swinging the heavy instrument through the air, drove the pins home. He was as much at home up there as a squirrel.

Now that balloon frames are mainly used for houses, and lighter sawed timbers for barns, the old-fashioned raising is rarely witnessed.

Then the moving was an event, too. A farmer had a barn to move, or wanted to build a new house on the site of the old one, and the latter must be drawn to one side. Now this work is done with pulleys and rollers by a few men and a horse; then the building was drawn by sheer bovine strength. Every man that had a yoke of cattle in the country round about was invited to assist. The barn or house was pried up and great runners, cut in the woods, placed under it, and under the runners were placed skids. To these runners it was securely chained and pinned; then the cattle—stags, steers, and oxen, in two long lines, one at each runner—were hitched fast, and while men and boys aided with great levers, the word to go was given. Slowly the two lines of bulky cattle straightened and settled into their bows; the big chains that wrapped the runners tightened, a dozen or more "gads" were flourished, a dozen or more lusty throats urged their teams at the top of their voices, when there was a creak or a groan as the building stirred. Then the drivers redoubled their efforts; there was a perfect Babel of discordant sounds; the oxen bent to the work, their eyes bulged, their nostrils distended; the lookers-on cheered, and away went the old house or barn as nimbly as a boy on a hand-sled. Not always, however; sometimes the chains would break, or one runner strike a rock, or bury itself in the earth. There were generally enough mishaps or delays to make it interesting.

In the section of the state of which I write, flax used to be grown, and cloth for shirts and trousers, and towels and sheets, woven from it. It was no laughing matter for the farm boy to break in his shirt or trousers those days. The hair shirts in which the old monks used to mortify the flesh could not have been much before then in this mortifying particular. But after the bits of shives and sticks were subdued, and the knots humbled by use and the washboard, they were good garments. If you lost your hold in a tree and your shirt caught on a knot or limb, it would save you.

But when has any one seen a crackle, or a swingling-knife, or a hatchel, or a distaff, and where can one get some tow for strings or for gun-wadding, or some swingling-tow for a bonfire? The quill-wheel, and the spinning-wheel, and the loom are heard no more among us. The last I knew of a certain hatchel, it was nailed up behind the old sheep that did the churning, and when he was disposed to shirk or hang back and stop the machine, it was always ready to spur him up in no uncertain manner. The old loom became a hen-roost in an outbuilding, and the crackle upon which the flax was broken—where, oh, where is it?

When the produce of the farm was taken a long distance to market, that was an event, too; the carrying away of the butter in the fall, for instance, to the river, a journey that occupied both ways four days. Then the family marketing was done in a few groceries. Some cloth, new caps, and boots for the boys, and a dress, or a shawl, or a cloak for the girls were brought back, besides news and adventure, and strange tidings of the distant world. The farmer was days in getting ready to start; food was prepared and put in a box to stand him on the journey, so as to lessen the hotel expenses, and oats were put up for the horses. The butter was loaded up overnight, and in the cold November morning, long before it was light, he was up and off. I seem to hear the wagon yet, its slow rattle over the frozen ground diminishing in the distance. On the fourth day toward night all grew expectant of his return, but it was usually dark before his wagon was heard coming down the hill, or his voice from before the door summoning a light. When the boys got big enough, one after the other accompanied him each year, until all had made the famous journey and seen the great river and the steamboats, and the thousand and one marvels of the faraway town. When it came my turn to go, I was in a great state of excitement for a week beforehand, for fear my clothes would not be ready, or else that it would be too cold, or else that the world would come to an end before the time fixed for starting. The day previous I roamed the woods in quest of game to supply my bill of fare on the way, and was lucky enough to shoot a partridge and an owl, though the latter I did not take. Perched high on a "springboard" I made the journey, and saw more sights and wonders than I have ever seen on a journey since, or ever expect to again.

But now all this is changed. The railroad has found its way through or near every settlement, and marvels and wonders are cheap. Still, the essential charm of the farm remains and always will remain: the care of crops, and of cattle, and of orchards, bees, and fowls; the clearing and improving of the ground; the building of barns and houses; the direct contact with the soil and with the elements; the watching of the clouds and of the weather; the privacies with nature, with bird, beast, and plant; and the close acquaintance with the heart and virtue of the world. The farmer should be the true naturalist; the book in which it is all written is open before him night and day, and how sweet and wholesome all his knowledge is!

Part I:

The Gilded Age, Readings 1880–1899

Introducing the Gilded Age

As a colloquialism, *golden age* encompasses, for many lay readers, the roughly fifty-year period of agricultural productivity and prominence occurring between Reconstruction and the Treaty of Versailles. Historians, however, are more likely to separate the era's aggregate farm productivity into smaller units, the Gilded Age, covering roughly the years 1870 to 1899, and the Golden Age, approximately 1900 to 1920.

The diverse readings in this first section come exclusively from the Gilded Age—the term coined in 1873 by co-authors Mark Twain and Charles Dudley in their satiric novel of the same name. *The Gilded Age*, a thinly veiled lampoon of the alleged greed and graft of the Grant administration, features a motley crew of politicians, speculators, and socialites on the loose and on the make. In fact, the "gilded" in Gilded Age better represents the opportunistic spirit created by the end of the Civil War than widespread wealth or decorum. Indeed, the sarcastic sting in Twain's *Gilded Age* was less intended for the well-planted, well-rooted farmer than it was for the robber baron, the monopolist, and the industrialist. Even so, the farmer was not immune from the capitalistic excess that characterized the era, as he doubled, often at the expense of the environment, the value of his products over the roughly thirty-year period from 1870 to 1900—the Age of Industrialism. Still, the farmer's gains paled in comparison to those recorded in the industrial sector, where the value of manufactured products more than quadrupled.[35]

Twain did have a particular kind of "farmer" in his crosshairs in *The Gilded Age*—the land speculator and "land-skinner"—and for this class the falsities inherent in the appellation *gilded* were well deserved, if not long overdue. The exploitative happenings of the Gilded Age that so riled Twain had been going on in the Ohio Valley and elsewhere long before the Civil War, as historian John Lauritz Larson reminds in his suggestively titled gloss of the period "Pigs in Space; Or What Shapes America's Regional Cultures." Focusing on the Ohio Valley and on Indiana in particular, Larson portrays the pioneer class as "not so much seeking the land that had been 'promised' to them as searching for an eligible place where they could steal (or if, necessary purchase) some land on which they could impose their will and live out their private story of getting and spending."[36] It is at the abuses of "getting and spending," prevalent in Twain's Missouri and neighboring states, at which the author aims his sardonic sobriquet "Gilded Age." Though both the Gold Rush and the Native American westward migration had slowed by time *The Gilded Age* was published, an estimated 300,000 people had already traveled the Oregon-California

---

[35] David Danbom, *Born in the Country: A History of Rural America* (Baltimore: Johns Hopkins, 1995), 133.

[36] John Lauritz Larson, "Pigs in Space; Or What Shapes America's Regional Cultures" in *The American Midwest: Essays on Regional History*, ed. Andrew R.L. Cayton and Susan E. Gray (Bloomington: Indiana University Press, 2001), 71.

road through the great American interior between 1841 and 1859, and many more had traversed the Santa Fe Trail.[37] The result was a "significant ecological disaster" created by overgrazing, overcutting, and overtramping (ibid) and, by 1893, the closing of the American frontier according to historian *Frederick Jackson Turner* in his essay "The Significance of the Frontier in American History." In fact, the crush of settlers crossing the Great Plains exacerbated dustbowl conditions in the 1890s that equaled and, in some cases rivaled, the Dust Bowl of the Great Depression.

To be a sure, any "gilded age"—suggesting, on a literal level, the gold leaf rather than gold bullion—bespeaks, both in name and in spirit, the agrarian and conservationist commitment to the underlying, unadulterated environmental truth. Indeed, a world view unfamiliar to many yeoman farmers and wilderness advocates—that nature existed solely for economic advantage—came to dominate the Gilded Age, as profits soared to all-time highs in both agricultural and manufacturing sectors. Environmental historian Ted Steinberg frames the ironic coalescence of greater productivity and greater waste thusly: "It is no coincidence that the most destructive period in the nation's wildlife history—replete with ruthless and systematic annihilation of some entire animal species—coincided with decades when conservation gripped the nation's political imagination. Efficiency and extermination went hand in hand."[38] On the positive side, the view that nature existed for the benefit of humanity made of the United States the world's fourth largest industrial nation.[39] By the numbers, a doubling of the population from approximately 26 million to 76 million and an influx of immigrants meant expanding markets for America's farmers. The same economic growth proved disastrous, though, for American forests, whose supplies of wood and game were seriously depleted

---

[37] Ted Steinberg, *Down to Earth: Nature's Role in American History* (Oxford: Oxford University Press, 2002), 122.
[38] Steinberg, *Down to Earth*, 144.
[39] Benjamin Kline, *First Along the River: A Brief History of the U.S. Environmental Movement*, (Lanham: Acada Books, 2000), 38.

by 1899, a condition whose grave consequences are well detailed in *Edwin J. Houston's* essay "Destruction of the Forest." Game proved equally vulnerable in times of economic scarcity, especially the wild passenger pigeon, a species slaughtered to near extinction by the close of the Gilded Age (ibid 44). As a young *Theodore Roosevelt* reports in "Ranching the Badlands," the buffalo was likewise endangered and, along with it, the culture of the open range. Several years after Roosevelt left his Dakota ranch, William Temple Hornaday published his report to the secretary of the Smithsonian, *The Extermination of the American Bison*, documenting the species' near extinction in the West and demanding government intervention. Prior to the advent of the market gunners Hornaday despised, the Plains had supported an estimated 27 million bison.[40]

And yet despite the considerable and, in some cases permanent, losses of flora and fauna of the Gilded Age, the era saw the creation of what has been called "the world's first instance of large-scale wilderness preservation in the public interest" in 1872, when President Grant designated by law over 2 million acres of northwestern Wyoming as Yellowstone National Park.[41] And thirteen years later, 715,000 acres of Adirondack forest were likewise set aside with the proviso that the Adirondacks remain forever wild. The influences behind the conservation of Yellowstone, in particular, are various, and yet agriculture proved key among them. When debate on the bill to carve out Yellowstone began in 1871, proponents focused on Yellowstone's unsuitability for agriculture as a primary reason for its protection. "The strategy," writes Roderick Frazier Nash, "was not to justify the park positively as a wilderness, but to demonstrate its uselessness to civilization" (ibid 112). Arguments for the park founded on economic gain or loss were many, but a minority of voices focused on the spiritual and physical benefits of the forest, including Missouri

---

[40] Steinberg, *Down to Earth*, 125.
[41] Roderick Frazier Nash, *Wilderness and the American Mind*, 4th Edition. (New Haven: Yale University Press), 108.

Congressman George G. Vest, who claimed that Yellowstone would be a "great breathing-place for the national lungs" (ibid 114) and Representative McAdoo of New Jersey who later went on record as preferring "the beautiful and the sublime … to the heartless mammon and the greed of capital" (ibid 115). And though a number of acts were passed in Congress creating national parks—including a single week in which Sequoia National Park and Yosemite National Parks were established—the agrarian model of cash cropping and harvesting still held sway in Washington. A case in point: the creation of the first forest reservation 1891 lagged behind the establishment of Yellowstone itself by twenty years, strongly suggesting the national parks' early raison d'etre as money-making tourist destinations rather than pristine nature preserves. The debate over preservation versus economic capitalization of the forests would intensify in 1898 when Gifford Pinchot was appointed chief of the Division of Forestry, which was then housed within the Department of Agriculture. Pinchot and John Muir, once friends united in their love of the forests, would become rivals, with Pinchot advocating forest management using an agricultural paradigm—utilitarianism—and John Muir, an advocate of preservationism, arguing for the defense of wilderness for its own sake.

The desire to escape American cities and recreate the conditions of native man—in a nutshell, primitivism—prompts several of the selections in this section, beginning with *John Burroughs'* essay "Footpaths," which hints at a growing land-use conflict between urbanites, suburbanites, and farmers. Burroughs writes, "The plow must respect it [the footpath], and the fence or hedge make way for it." Noting their infatuation with roads and highways instead of simple footpaths, Burroughs criticizes his countrymen for their base preoccupation with escape, with "getting out." The recreational use of nature, embodied by Frederick Law Olmsted's design of New York City's Central Park and other public areas in the East, was becoming a widely accepted notion, embraced by Burroughs himself despite his disdain for the recreational walking and biking craze then amplified in the pages of new publications such as *The*

Wheelman. In "Footpaths" Burroughs summarizes the advantages of the nature constitutional thusly: "When the exercise of your limbs affords you pleasure, and the play of your senses upon the various objects and shows of nature quickens and stimulates your spirit, your relation to the world and to yourself is what it should be—simple and direct and wholesome." In sympathy with Burroughs' privileging of the natural over the unnatural, *Helen Gilbert Ecob* and *Charlotte Perkins Gilman* document the rise of nervous disorders, particularly in women, a phenomenon Ecob attributes to the specialization and stress of the machine age and, consequently, to the absence of restorative physical exercise in a natural setting. *Walt Whitman*, so emblematic of the vigorousness Helen Gilbert Ecob champions, is represented in this section by several prose passages from *Specimen Days*, including Whitman's tribute to "the freedom and vigor and sane enthusiasm of ... perfect western air and autumn sunshine" experienced on his first visit to Kansas.

In their writings, Ecob, Whitman, and others responded to a variety of social pressures falling under the heading of Social Darwinism, a term coined by Englishman Herbert Spencer to suggest the "survival of the fittest" doctrine as applied to humans. This Darwinistic zeitgeist affected the agrarian and conservation writers of the Gilded Age variously. Charlotte Perkins Gilman anticipated, for example, a new egalitarianism in educational and occupational options for women, opportunities sure to break down the old patriarchal order. To a young Teddy Roosevelt out on the range, the idea of Social Darwinism seemed self-evident, on the one hand, and sinister on the other. Life on the range daily proved for Roosevelt the truism that superior bodies displaying superior skill made superior livings. At the same time, it was clear to TR that the fraternity and camaraderie of the collective cattle drive and the communal cowboy lifestyle would soon be extinct in a dog-eat-dog world. The pervasiveness of Social Darwinism, as well as its unforgiving nature, is particularly evident in *Edward Francis Adams'* essay "The Agricultural Colleges," in which Adams, a former businessman, advises his readers, "We do not need to concern ourselves with the lazy, the dishonest, or the unthrifty. Nature will take care of them. She will kill them." Similarly, even Helen Gilbert Ecob's otherwise healthful encouragement of physical fitness occasionally belies a Darwinistic class consciousness, as when she observes, "The laboring class, whether the work be manual or mental, have a lean, hungry look. The denizens of society are fat and flourishing."

Questions of naturalness and unnaturalness preoccupy the Gilded Age, as the readings that follow attest. Roosevelt, for instance, in "Ranching the Badlands" considers the genuine article, the real cowboy, while *Thorstein Veblen* in "Conspicuous Consumption" theorizes the rise of what he dubs the "leisure class" with its predatory spending and artificed recreation. Against this backdrop, narratives of escape, including John Burroughs' book *Pepaction*, and the title essay of the same name, depict adventurous retreats from a maddening crowd of harried professionals. Comparing the rising white-collar class to sensitive "hothouse plants" desperately in need of recouping their losses in a backyard garden, *Tuisco Greiner's* "The Home Garden" suggests a work-from-home scheme as well as a psychological cure. In "Women as Farmers," *Frances Elizabeth Willard* conjures a hopeful image of a feminist ecotopia based on an ideal spirit of feminine cooperation among agrarians. Walt Whitman's joie de vivre and wanderlust serve, likewise, as a call to arms in defense of America's threatened agrarianism. The great poet, himself recovered from illness by virtue of his convalescence on a friend's farm, writes thusly in "Nature and Democracy, Morality":

> American Democracy, in its myriad personalities, in factories, workshops, stores, offices—through the dense streets and houses of cities, and all their manifold sophisticated life—must either be fibred, vitalized, by regular contact with outdoor light and air and growths, farm-scenes, animals, fields, trees, birds, sun-warmth and free skies, or it will certainly dwindle and pale.

And of course John Muir's magisterial accounts of the West were, in the Gilded Age, eagerly consumed by readers in the urbanizing East, who found in works such as "A Windstorm in the Forest" an immediacy lacking in their everyday lives.

Paradoxically, Americans sought both to cultivate and to criticize notions of industrial efficiency in the Gilded Age. In this ambivalent milieu, agricultural and environmental interests existed uneasily. In the Midwest, yeomen had largely transcended subsistence farming thanks to the efficiencies of new railroads, better plows, grain drills, reapers and binders, combines and, towards the very end of the century, steam tractors. Though many small-scale farmers could not afford this new technology, the efficiency of the steam tractor, reckoned at 35 to 45 acres plowed per day, dwarfed the horse and moldboard plow.[42] Malthusians, who claimed that resources would grow scarce with population increase, continued to have the urban public's ear, prompting back-to-the-land experiments in self-sufficiency similar to those described by Tuisco Greiner and Frances Elizabeth Willard. In the factories, the advent of large-scale, machined production improved efficiency even as it disheartened many workers. Some four decades before the mass-produced Model T, Frederick Taylor of Midvale Steel Works in Pennsylvania developed a means of measuring worker productivity called Taylorism. The impact of Taylor's efficiency program was felt well beyond the workplace, as utilitarian conservationists, particularly Gifford Pinchot, indulged analogous instincts to increase order and efficiency in nature. In an era of "Taylor-made" products, reactionaries increasingly embraced the subjectivities and sublimities of "natural men" like John Muir, who, recollecting his thoughts from atop a pine tree in the High Sierras wrote, "We hear much nowadays concerning the universal struggle for existence, but no struggle in the common meaning of the word was manifest here; no recognition of danger by any tree; no deprecation; but rather an invincible gladness as remote from exultation as from fear." Another of Muir's benedictions—"climb the mountains and get good tidings"

---

[42] R. Douglas Hurt, *American Agriculture: A Brief History* (West Lafayette: Purdue University Press, 2002) 202.

expressed, in a nutshell, the transcendental foundations of the Sierra Club, which Muir founded in 1892. The popular appeal of Muir's message in *The Mountains of California* eventually sold some 10,000 copies,[43] attesting to his growing constituency. Muir's enduring objective, first articulated in 1874 was to "live only to entice people to nature's loveliness."[44]

Efficiency as a theme would come to dominate the Progressive Era, with its talented cast of social scientists, religious reformers, and government agents. But the seeds for a fully-flowered Progressivism were planted in the institutionalization and bureaucratization of 1880 to 1899. In the Gilded Age, the land provided what Ralph Waldo Emerson called "Commodity," the raw materials which, in turn, stoked the industrial engine. Thus, when Secretary of Agriculture was elevated to a cabinet level position by Congressional approval in 1889, the change celebrated agriculture's new amenability to central control and scientific reasoning as much or more than it acknowledged farming's classical virtues or unique spiritual value. Still, the Secretary of Agriculture, despite what advocacy and visibility he might lend the cultivator's cause, did not reach the farmer where he lived as much as did the Morrill Land-Grant College Act of 1862. The Act succeeded in creating agricultural colleges across the nation, beginning with Iowa in 1862, where farmer *Isaac Phillips Roberts* built a pioneering agricultural program while serving as superintendent of the college grounds. But many farmers' sons and daughters were wary of what critics charged were second rate schools with inferior faculty, and those students who did enroll seldom returned to the farm,[45] a fact Roberts learned quickly in his sink-or-swim, first-in-the-nation experiment. "So intently have you longed to have some great corporation brand and number you," Roberts empathizes in "The Fertility of the Land: A Chat with a Young Farmer," "when the tasks at home

---

[43] Library of Congress, "The Evolution of the Conservation Movement, 1850–1920," http://memory.loc.gov/ammem/amrvhtml/cnchron3.html.
[44] Nash, *Wilderness and the American Mind*, 129.
[45] Hurt, *American Agriculture*, 193.

were hard." With "free" land available via the Homestead Act and new provisions made for land acquisitions via the Timber Culture Acts of 1873 and 1878, the temptation to pioneer-farm proved irresistible for both skilled and unskilled rural youth. The "irrational exuberance," to use a contemporary term, with which young men and women regarded the West is well represented by *Lucius "Lute" Wilcox's* essay the "Advantages of Irrigation," wherein the author describes the promise of a West reanimated with irrigated water. "In traveling in the far West over long stretches of parched and dusty plains or through mountain gorges," Wilcox gushes, "the writer has often seen fields, orchards, vineyards and gardens all dressed in living green." Wilcox's abiding belief in the redemptive power of widespread irrigation persisted in the face of a record-breaking drought in the Great Plains that began in the late 1880s and produced reports of drought-induced malnutrition and starvation as well as tumbling wheat yields.[46]

The initial unpopularity of the agricultural colleges risked orphaning agriculture, and conservation for that matter, from practice grounded in the scientific principle, and meant that the agricultural consensus and collective action that *Thomas Shaw* seeks in his essay "The Prevalence of Weeds" never materialized. Shaw, affiliated with the Ontario Agriculture College experiment farm, expresses the frustration of many faculty members at ag colleges and experimental research stations newly created by the Hatch Act of 1887 when he writes, "It would probably be found a greater task to correct ... the minds of many farmers than to uproot the weeds themselves from their fields." A fuller treatment of the controversy surrounding the creation of the agricultural colleges and research stations is offered by Edward Francis Adams' "The Agricultural Colleges," wherein Adams confirms the greatest fears of many of the era's farmers—that the graduate of the agricultural institution might well lose his or her taste for hard physical labor. The education offered by these new institutions of higher learning, Adams crows, approximates a professional

---

[46] Steinberg, *Down to Earth*, 134.

education of the sort given to doctors and lawyers, and is better suited to "superintendency" of farms than yeomanry in practice.

The business model of farming, or, more accurately, the divorce of agricultural theory from on-farm practice, profoundly affected conservation in the Gilded Age. Prior to the period beginning 1880, many American yeoman had been "mixed farmers," combining cash or subsistence cropping with animal husbandry, a method that maintained soil fertility holistically via manure and composts produced in the farmer's own pasture and barnyard.[47] Now, with the development of artificial fertilizers such as superphosphates, the management of the soil was increasingly ceded to chemists and college-trained agriculturalists and labeled science rather than art. The "ecology" of the farm, indeed the ecology of the wilderness, was thus broken—the mutual understanding of man and beast having been undermined, in large part, by the specialization of the Industrial Revolution.

The changes in life on the farm from 1880 to 1899—coupled with harsh winters and droughts in the Plains and Midwest, the growing power of the railroads, the fluctuation of prices, surplus production, and high debt loads—eventually pushed farmers toward organizational cooperation and agitation: a phenomenon know as the "Agrarian Revolt."[48] But while some historians present the age of Agrarian Revolt as a product solely of farmer dissatisfaction, such a one-sided portrayal neglects the tremendous productivity gains of a Gilded Age in which agriculture doubled in size in almost every measure. In an era when the number of farms increased from 2.66 million to 5.74 million,[49] the advent of farmers' alliances is as much a barometer of the farmer's sense of his own collective power as an outlet for his unmet needs.

Oliver Hudson Kelly, a Minnesota farmer and USDA clerk, created what would become the template for Gilded Age farm organizations, the National Grange of the Patrons of the Husbandry, after touring

---

[47] Kline, *First Along the River*, 40.
[48] Hurt, *American Agriculture*, 203.
[49] Danbom, *Born in the Country*, 132.

the farms of the war-torn, economically depressed South in 1866. After returning to Washington DC, Kelley determined that collection action would be possible if northern and southern farmers could be made to see their common interests. After its establishment the following year, the Grange, as it came to be abbreviated, flourished, though not primarily as an educational and social organization—as Kelley imagined it would[50]—but as a political, economic union of common interests bent on disempowering exploitive middlemen while empowering farm families, including women, within the organization and without. After peaking in the mid 1870s, the Grange declined and many of its members migrated to an upstart alliance, likewise born in the South, the Texas Farmers' Alliance, which spread quickly through Dixie based on a platform of cooperative stores, warehouses, gins and elevators—all intended to reduce the grip of the furnishing merchants who held liens against farmers' crops in return for "furnishing" food, textiles, and tools (ibid 169). Primarily of interest to southern cotton farmers, sharecroppers, and tenant farmers, the Southern Alliance eventually courted Great Plains wheat farmers similarly hampered by overproduction and low prices. Ultimately, the National Farmers' Alliance, also known as the Northern Farmers' Alliance, founded in Chicago in 1880 by Milton George, appealed more successfully to the Plains audience, (ibid 206), raising the specter of sectionalism here addressed in conciliatory essays by Southern Alliance members *Leonidas K. Polk* and *Benjamin Hutchinson Clover*.

Unsuccessful attempts to merge the Southern and Northern Alliances were made in a reform meeting in St. Louis in 1889, but the southern sect, now called the National Farmers' Alliance and Industrial Union, balked at the North's insistence that African American farmers be admitted into the organization and its demand that the organization cease its secret status.[51] The Farmers' Alliance, though egalitarian in its inclusion of planters, tenants, farm workers, and

---

[50] Hurt, *American Agriculture*, 204.
[51] Hurt, 208.

owners, was limited to rural whites and primarily served the interests of men. The organization's attempts to widen its political base without compromising its ideals are well demonstrated here by several additional readings from the *Farmers' Alliance History and Agricultural Digest* of 1891. In "History of the Colored Farmers' Alliance and Cooperative Union" *R. M. Humphrey* documents the history of the Colored Farmers' Alliance, a parallel organization formed in March 1888 in Houston, Texas. Essays by *Bettie Gay* and *Jennie E. Dunning* show the diversity of women's role and goals within the Alliance. Bettie Gay makes surprisingly strident calls for women's leadership inside and outside the Alliance while Jennie E. Dunning makes more traditional, but no less astute, claims for women's providence as homemakers and caregivers.

In 1892, just one year after the publication of the *Farmer's Alliance History and Agricultural Digest*, Plains Alliancemen and agrarian fundamentalists gathered at a national nominating convention to field a third party to be based on Southern Alliance goals enacted via the partisan political machine of the Northern Alliance.[52] Those gathered, including Bettie Gay, would nominate North Carolinian Leonidas K. Polk for President, who would die before he could officially accept the nomination, leaving Iowan James B. Weaver to bring home 22 electoral votes and carry four states as the People's candidate in the Presidential election of that followed (ibid). Encouraged by their electoral inroads, the Populists vied for the highest office in the land once more in 1896, nominating Williams Jennings Bryan, who ultimately ran on a Democratic ticket that had co-opted many of the key planks in the Populist platform (ibid 212).

After 1896, the Populists would cease to be a viable third party. And while its rural base was more interested in economic rather than environmental issues facing farmers, its emphasis on government protections of the public good and its dismissal of Social Darwinism lent itself to a conservationist mindset. Though more closely aligned

---

[52] Hurt, *American Agriculture*, 209.

with Democratic interests, the Populist agenda would be taken up next by a Republican, Theodore Roosevelt, the "accidental president" borne of McKinley's assassination in 1901. Roosevelt's tenure would be the one to usher in a truly golden, rather than merely gilded, age for agrarians and conservationists.

Farm Pages: The Gilded Age

The following suite of images, selected from hundreds of pages of original farm and conservation writings for their reproduction quality as well as their historical relevance, represent the Gilded Age in pictures. These farm pages, though far from comprehensive, distill the dynamic equipoise of an era suspended between animal power and traditional practice (here represented in diagrams of Indian Corn and horse) and rapid industrialization—the very "progress" John Burroughs laments in "Phases of Farm Life." In particular, the agricultural triumphalism of the Chicago World's Fair of 1893 pervades these images, which celebrate on-farm ingenuity and entrepreneurship (the hay press and silo cross-section, for example), and an almost Darwinistic evolution of farm equipment and practice (plow and drainage).

Published images such as these were exceptional in and of themselves during the Gilded Age of farm and conservation writing, as illustrations added both complexity and cost to book production. For many subsistence farmers and homesteaders, such books as these would have been out of reach, both financially and geographically.

In order of appearance, the image sources for these plates include Edward Enfield's *Indian Corn: Its Values, Culture, and Uses* (D. Appleton, 1886); George Martin's *Farm Appliances: A Practical Manual* (Orange Judd, 1887); Armand Goubaux's *The Exterior of the Horse* (J. B. Lippincott, 1892); Manly Miles' *Land Draining ...* (Orange Judd, 1892); James Sanders' *Practical Hints About Barn Building ...* (J. H. Sanders Pub., 1893); and Robert L. Ardey's *American Agricultural Implements ...* (Robert L. Ardey, 1894).

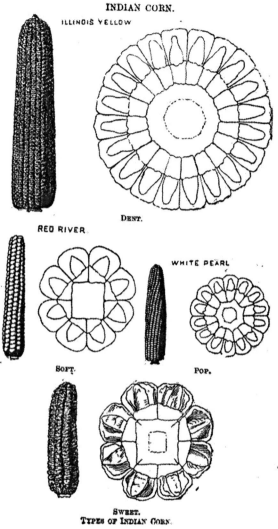

TYPES OF INDIAN CORN.

FARM APPLIANCES.

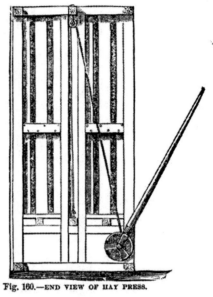

Fig. 160.—END VIEW OF HAY PRESS.

Fig. 161.—MOVABLE BOTTOM.

THE HEAD.

6th. **Superior Extremity.**—It comprises the structures intermediary to the head and neck : above, the *poll* or *nape*; below, the *throat* or *pharyngo-laryngeal* region ; on each side, the *parotid* region.

Such are the different regions of the head, which we will examine presently.

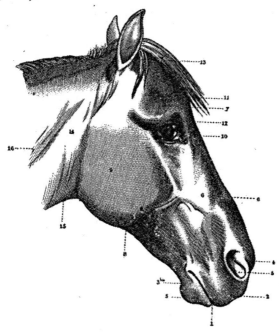

Fig. 19.

1. Mouth.
2. Superior lip.
3. Inferior lip.
3 *bis*. Chin.
4. Extremity of the nose.
5. Nostril.
6-6. Face.
7. Forehead.
8. Inferior maxilla.
9. Cheeks.
10. Eye.
11. Supra-orbit.
12. Temples.
13. Ear.
14. Parotid region.
15. Throat.
16. Neck.

DISCOVERY AND INVENTION.

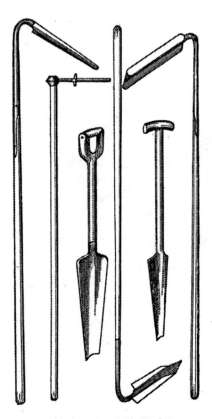

FIG. 23. OBSOLETE DRAINING TOOLS.

SILOS AND ENSILAGE.

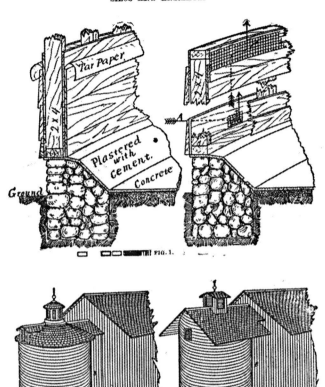

AMERICAN AGRICULTURAL IMPLEMENTS.

PRAIRIE BREAKING PLOW OF 50 YEARS AGO.

ANOTHER STYLE PRAIRIE BREAKING PLOW.

THE FIRST STEEL PLOW, 1833.

JOHN LANE'S PATENT, 1868, "SOFT CENTER" STEEL.

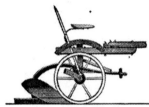

GILPIN MOORE, JUNE 29, 1875.

GILPIN MOORE'S PATENT, JUNE 29, 1875.

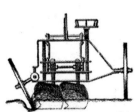

W. L. CASADAY'S PATENT, MAY 2, 1876.

W. L. CASADAY'S PATENT, SEPT. 6, 1881.

JOHN BURROUGHS

In a review of *Pepaction*, the book from which the following essay, "Footpaths," is drawn, a reviewer writing in the July 7, 1881 *Nation* declares, "The truth is—and it may as well be told frankly, as the one drawback found, by most readers in the very pleasant books of Mr. Burroughs—that he allows himself an amount of dogmatizing, in presence of nature, which is not only unattractive but quite unsafe." The same intrepid reviewer goes on criticize Burroughs' utter lack of "a punctilious habit of accuracy" and implies that he is second-rate Thoreau. In "Footpaths," though, Burroughs stays well clear of overt sentimentalism towards nature, writing fluidly and insightfully on the American obsession with highway travel by comparison with the British love of the walking path. "The devil is in the horse to make men proud and fast and ill-mannered," Burroughs writes, "only when you go afoot do you grow in the grace of gentleness and humility." Revealing an affinity for Thoreau's essay, "On Walking," Burroughs capitalizes on a trend towards pedestrianism then sweeping across America to chastise road-bound Americans and exhort them to the "wholesomeness" of the nature trail.

Footpaths[53]

An intelligent English woman, spending a few years in this country with her family, says that one of her serious disappointments is that she finds it utterly impossible to enjoy nature here as she can at home—so much nature as we have and yet no way of getting at it, no paths, or byways, or stiles, or footbridges, no provision for the pedestrian outside of the public road. One would think the people had no feet and legs in this country, or else did not know how to use them. Last summer she spent the season near a small rural village in the valley of the Connecticut, but it seemed as if she had not been in the country: she could not come at the landscape; she could not reach a wood or a hill or a pretty

---
[53] John Burroughs, *Pepaction*, Riverside Press Edition, (Boston: Houghton Mifflin & Company, 1881, 1895, 1899), 175–183.

nook anywhere without being a trespasser, or getting entangled in swamps or in fields of grass and grain, or having her course blocked by a high and difficult fence; no private ways, no grassy lanes; nobody walking in the fields or woods, nobody walking anywhere for pleasure, but everybody in carriages or wagons.

She was staying a mile from the village, and every day used to walk down to the post office for her mail, but instead of a short and pleasant cut across the fields, as there would have been in England, she was obliged to take the highway and face the dust and the mud and the staring people in their carriages.

She complained, also, of the absence of bird voices—so silent the fields and groves and orchards were compared with what she had been used to at home. The most noticeable midsummer sound everywhere was the shrill, brassy crescendo of the locust.

All this is unquestionably true. There is far less bird music here than in England, except possibly in May and June, though, if the first impressions of the Duke of Argyle are to be trusted, there is much less even then. The duke says, "Although I was in the woods and fields of Canada and of the States in the richest moments of the spring, I heard little of that burst of song which in England comes from the blackcap, and the garden warbler, and the whitethroat, and the reed warbler, and the common wren, and (locally) from the nightingale." Our birds are more withdrawn than the English, and their notes more plaintive and intermittent. Yet there are a few days here early in May, when the house wren, the oriole, the orchard starling, the kingbird, the bobolink, and the wood thrush first arrive, that are so full of music, especially in the morning, that one is loath to believe there is anything fuller or finer even in England. As walkers, and lovers of rural scenes and pastimes, we do not approach our British cousins. It is a seven days' wonder to see anybody walking in this country except on a wager or in a public hall or skating rink, as an exhibition and trial of endurance.

Country men do not walk except from necessity, and country women walk far less than their city sisters. When city people come to the country they do not walk, because that would be conceding too much to the country; beside, they would soil their shoes, and would lose the awe and respect which their imposing turn-outs inspire. Then they find the country dull; it is like water or milk after champagne; they miss the accustomed stimulus, both mind and body relax, and walking is too great an effort.

There are several obvious reasons why the English should be better or more habitual walkers than we are. Taken the year round, their climate is much more favorable to exercise in the open air. Their roads are better, harder, and smoother, and there is a place for the man and a place for the horse. Their country houses and churches and villages are not strung upon the highway as ours are, but are nestled here and there with reference to other things than convenience in "getting out." Hence the grassy lanes and paths through the fields. Distances are not so great in that country; the population occupies less space. Again, the land has been longer occupied and is more thoroughly subdued; it is easier to get about the fields; life has flowed in the same channels for centuries. The English landscape is like a park, and is so thoroughly rural and mellow and bosky that the temptation to walk amid its scenes is ever present to one. In comparison, nature here is rude, raw, and forbidding, has not that maternal and beneficent look, is less mindful of man, and runs to briers and weeds or to naked sterility.

Then as a people the English are a private, domestic, homely folk: they dislike publicity, dislike the highway, dislike noise, and love to feel the grass under their feet. They have a genius for lanes and footpaths; one might almost say they invented them. The charm of them is in their books; their rural poetry is modeled upon them. How much of Wordsworth's poetry is the poetry of pedestrianism! A footpath is sacred in England; the king himself cannot close one; the courts recognize them as something quite as important and inviolable as the highway.

A footpath is of slow growth, and it is a wild, shy thing that is easily scared away. The plow must respect it, and the fence or hedge make way for it. It requires a settled state of things, unchanging habits among the people, and long tenure of the land; the rill of life that finds its way there must have a perennial source, and flow there tomorrow and the next day and the next century.

When I was a youth and went to school with my brothers, we had a footpath a mile long. On going from home after leaving the highway, there was a descent through a meadow, then through a large maple and beech wood, then through a long stretch of rather barren pastureland which brought us to the creek in the valley, which we crossed on a slab or a couple of rails from the near fence; then more meadow land with a neglected orchard, and then the little gray schoolhouse itself toeing the highway. In winter our course was a hard, beaten path in the snow visible from afar, and in summer a well-defined trail. In the woods it wore the roots of the trees. It steered for the gaps or low places in the fences, and avoided the bogs and swamps in the meadow. I can recall yet the very look, the very physiognomy of a large birch tree that stood beside it in the midst of the woods; it sometimes tripped me up with a large root it sent out like a foot. Neither do I forget the little spring run near by, where we frequently paused to drink, and to gather "crinkle-root" (*Dentaria*) in the early summer; nor the dilapidated log fence that was the highway of the squirrels; nor the ledges to one side, whence in early spring the skunk and coon sallied forth and crossed our path; nor the gray, scabby rocks in the pasture; nor the solitary tree; nor the old weather-worn stump; no, nor the creek in which I plunged one winter morning in attempting to leap its swollen current. But the path served only one generation of schoolchildren; it faded out more than thirty years ago, and the feet that made it are widely scattered, while some of them have found the path that leads through the Valley of the Shadow. Almost the last words of one of these schoolboys, then a man grown, seemed as if he might have had this very path in mind, and thought himself again returning to his father's house: "I must hurry," he said; "I have a long way to go up a hill and through a dark wood, and it will soon be night."

We are a famous people to go "'cross lots," but we do not make a path, or, if we do, it does not last; the scene changes, the currents set in other directions, or cease entirely, and the path vanishes. In the South one would find plenty of bridlepaths, for there everybody goes horseback, and there are few passable roads, and the hunters and lumbermen of the North have their trails through the forest following a line of blazed trees; but in all my acquaintance with the country—the rural and agricultural sections—I do not know a pleasant, inviting path leading from house to house, or from settlement to settlement, by which the pedestrian could shorten or enliven a journey, or add the charm of the seclusion of the fields to his walk.

What a contrast England presents in this respect, according to Mr. Jennings' pleasant book *Field Paths and Green Lanes*. The pedestrian may go about quite independent of the highway. Here is a glimpse from his pages:

> A path across the field, seen from the station, leads into a road close by the lodge gate of Mr. Cubett's house. A little beyond this gate is another and smaller one, from which a narrow path ascends straight to the top of the hill and comes out just opposite the post-office on Ranmore Common. The Common at another point may be reached by a shorter cut. After entering a path close by the lodge, open the first gate you come to on the right hand. Cross the road, go through the gate opposite, and either follow the road right out upon Ranmore Common, past the beautiful deep dell or ravine, or take a path which you will see on your left, a few yards from the gate. This winds through a very pretty wood, with glimpses of the valley here and there on the way, and eventually brings you out upon the carriage-drive to the house. Turn to the right and you will soon find yourself upon the Common. A road or path opens out in front of the upper lodge gate. Follow that and it will take you to a small piece of water from whence a green path strikes off to the right, and this will lead you all across the Common in a northerly direction.

Thus we may see how the country is threaded with paths. A later writer, the author of *The Gamekeeper at Home* and other books, says:

> Those only know a country who are acquainted with its footpaths. By the roads, indeed, the outside may be seen; but the footpaths go through the heart of the

land. There are routes by which mile after mile may be traveled without leaving the sward. So you may pass from village to village; now crossing green meadows, now cornfields, over brooks, past woods, through farmyard and rick 'barken.'

The conditions of life in this country have not been favorable to the development of byways. We do not take to lanes and to the seclusion of the fields. We love to be upon the road, and to plant our houses there, and to appear there mounted upon a horse or seated in a wagon. It is to be distinctly stated, however, that our public highways, with their breadth and amplitude, their wide grassy margins, their picturesque stone or rail fences, their outlooks, and their general free and easy character, are far more inviting to the pedestrian than the narrow lanes and trenches that English highways for the most part are. The road in England is always well-kept, the roadbed is often like a rock, but the traveler's view is shut in by high hedges, and very frequently he seems to be passing along a deep, nicely graded ditch. The open, broad landscape character of our highways is quite unknown in that country.

The absence of the paths and lanes is not so great a matter, but the decay of the simplicity of manners, and of the habits of pedestrianism which this absence implies, is what I lament. The devil is in the horse to make men proud and fast and ill-mannered; only when you go afoot do you grow in the grace of gentleness and humility. But no good can come out of this walking mania that is now sweeping over the country, simply because it is a mania and not a natural and wholesome impulse. It is a prostitution of the noble pastime.

It is not the walking merely; it is keeping yourself in tune for a walk, in the spiritual and bodily condition in which you can find entertainment and exhilaration in so simple and natural a pastime. You are eligible to any good fortune when you are in the condition to enjoy a walk. When the air and the water taste sweet to you, how much else will taste sweet! When the exercise of your limbs affords you pleasure, and the play of your senses upon the various objects and shows of nature quickens and stimulates your spirit, your relation to the world and to yourself is what it should be—simple and direct and wholesome. The mood in which you set out on a spring or autumn ramble or a sturdy winter walk, and your greedy feet have to be restrained from devouring the distances too fast, is the mood in which your best thoughts and impulses come to you, or in which you might embark upon any noble and heroic enterprise. Life is sweet in such moods, the universe is complete, and there is no failure or imperfection anywhere.

WALT WHITMAN

The following five sections, collected from Whitman's *Prose Works* and presented together here in thematic grouping, demonstrate the great poet's agrarian and conservation fidelities. *Specimen Days*, from which four of the following five passages are excerpted, was issued as a prose counterpart to Whitman's classic book of poetry *Leaves of Grass* in the 1881–1882 edition—the date used to determine the work's chronological placement in this volume. Whitman described the result as "the most wayward, spontaneous, fragmentary book ever printed." Although much of this unconventional autobiography is devoted to family genealogy (Whitman's father farmed near Long Island before moving the family to Brooklyn), the Civil War, and the literary life, a significant portion is made up of nature-inspired reflections born of a period of convalescence Whitman spent at the Safford Farm at Timber Creek, where he convalesced from a stroke by purportedly "wrestling" with saplings and taking "mud baths." *Specimen Days* also details his 1879 trip to attend the quarter-centennial celebration of the Kansas settlement, an event and a locale that evokes ecstatic, laudatory prose equal to the energy and vivacity of *Leaves of Grass*.

Though Whitman suffered ambivalent feelings about Transcendentalists Thoreau, and, particularly, Emerson, who greatly admired Whitman's work, he begins, in these writings to embrace their conservationist, agrarian, and egalitarian notions. In fact, his dismissal of Old Europe explicitly echoes Thoreau, while Whitman's vision of an America sustained by innumerable small farms and farmers—"on which God seems to smile"—is an almost direct paraphrase of the Jeffersonian representation of cultivators as the "chosen people of God." An autodidact in the best agrarian sense, Whitman was not a naturalist to the extent of contemporaries John Muir or John Burroughs, but his quick, poetic intuition and his keen, journalistic powers of observation lead him to a grand appreciation of the unique ecology of the Great Plains as well as an abiding concern for the consequences of its cultivation, as expressed in "Prairie Analogies. The Tree Question." Drawn always to the farms and woodlands—the kind of Long Island countryside his grandparents inhabited and which he always regarded as an escape from Brooklyn—Whitman combines in *Specimen Days* a rare combination of agrarian subjectivity, journalistic notation, and prophetic prediction.

Selections from Specimen Days *and* Notes Left Over[54]

Entering a Long Farm-Lane. As every man has his hobby-liking, mine is for a real farm lane fenced by old chestnut rails gray-green with dabs of moss and lichen, copious weeds and briers growing in spots athwart the heaps of stray-picked stones at the fence bases—irregular paths worn between, and horse and cow tracks—all characteristic accompaniments marking and scenting the neighborhood in their seasons—apple-tree blossoms in forward April—pigs, poultry, a field of August buckwheat, and in another the long flapping tassels of maize—and so to the pond, the expansion of the creek, the secluded-beautiful, with young and old trees, and such recesses and vistas![55]

The Prairies and an Undelivered Speech. At a large popular meeting at Topeka—the Kansas State Silver Wedding, fifteen or twenty thousand people—I had been erroneously billed to deliver a poem. As I seemed to be made much of, and wanted to be good-natured, I hastily

---

[54] Walt Whitman, *Prose Works* (Philadelphia: David McKay, 1892), 83, 141, 149, 151, 200, 337.

[55] This and the following five sections are considered, for accuracy of chronological organization, dated 1881–1882, the original date of publication of *Specimen Days*. "Our Real Culmination" originates in the *Prose Works* edition and is best dated 1892.

penciled out the following little speech. Unfortunately, (or fortunately) I had such a good time and rest, and talk and dinner, with the U. boys, that I let the hours slip away and didn't drive over to the meeting and speak my piece. But here it is just the same:

> My friends, your bills announce me as giving a poem; but I have no poem—have composed none for this occasion. And I can honestly say I am now glad of it. Under these skies resplendent in September beauty—amid the peculiar landscape you are used to, but which is new to me—these interminable and stately prairies—in the freedom and vigor and sane enthusiasm of this perfect western air and autumn sunshine—it seems to me a poem would be almost an impertinence. But if you care to have a word from me, I should speak it about these very prairies; they impress me most, of all the objective shows I see or have seen on this, my first real visit to the West. As I have rolled rapidly hither for more than a thousand miles, through fair Ohio, through bread-raising Indiana and Illinois—through ample Missouri, that contains and raises everything; as I have partially explored your charming city during the last two days, and, standing on Oread hill, by the university, have launched my view across broad expanses of living green, in every direction—I have again been most impressed, I say, and shall remain for the rest of my life most impressed, with that feature of the topography of your western central world—that vast Something, stretching out on its own unbounded scale, unconfined, which there is in these prairies, combining the real and ideal, and beautiful as dreams.
>
> I wonder indeed if the people of this continental inland West know how much of first-class art they have in these prairies—how original and all your own—how much of the influences of a character for your future humanity, broad, patriotic, heroic and new—how entirely they tally on land the grandeur and superb monotony of the skies of heaven, and the ocean with its waters—how freeing, soothing, nourishing they are to the soul?

Then is it not subtly they who have given us our leading modern Americans, Lincoln and Grant?—vast-spread, average men—their foregrounds of character altogether practical and real, yet (to those who have eyes to see) with finest backgrounds of the ideal, towering high as any. And do we not see, in them, foreshadowings of the future races that shall fill these prairies?

Not but what the Yankee and Atlantic States, and every other part—Texas, and the states flanking the southeast and the Gulf of Mexico—the Pacific shore empire—the Territories and Lakes, and the Canada line (the day is not yet, but it will come, including Canada entire) are equally and integrally and indissolubly this Nation, the sine qua non of the human, political and commercial New World. But this favored central area of (in round numbers) two thousand miles square seems fated to be the home both of what I would call America's distinctive ideas and distinctive realities.

The Prairies and Great Plains in Poetry (After traveling Illinois, Missouri, Kansas and Colorado). Grand as the thought that doubtless the child is already born who will see a hundred millions of people, the most prosperous and advanced of the world, inhabiting these prairies, the great Plains, and the valley of the Mississippi, I could not help thinking it would be grander still to see all those inimitable American areas fused in the alembic of a perfect poem, or other esthetic work, entirely western, fresh and limitless—altogether our own, without a trace or taste of Europe's soil, reminiscence, technical letter or spirit. My days and nights, as I travel here—what an exhilaration!—not the air alone, and the sense of vastness, but every local sight and feature. Everywhere something characteristic—the cactuses, pinks, buffalo grass, wild sage—the receding perspective, and the far circle-line of the horizon all times of day, especially forenoon—the clear, pure, cool, rarefied nutriment for the lungs, previously quite unknown—the black patches and streaks left by surface conflagrations—the deep-plowed furrow of the "fireguard"—the slanting snow-racks built all along to shield the railroad from winter drifts—the prairie dogs and the herds of antelope—the curious "dry rivers"—occasionally a "dugout" or corral—Fort Riley and Fort Wallace—those towns of the northern Plains, (like ships on the sea,) Eagle-Tail, Coyote, Cheyenne, Agate, Monotony, Kit Carson—with ever the anthill and the buffalo-wallow—ever the herds of cattle and the cowboys ("cowpunchers") to me a strangely interesting class, bright-eyed as hawks, with their swarthy complexions and their broad-brimmed hats—apparently always on horseback, with loose arms slightly raised and swinging as they ride.

Prairie Analogies—The Tree Question. The word *prairie* is French, and means literally meadow. The cosmical analogies of our North American plains are the Steppes of Asia, the Pampas and Llanos of South America, and perhaps the Saharas of Africa. Some think the plains have been originally lake beds; others attribute the absence of forests to the fires that almost annually sweep over them—the cause, in vulgar estimation, of Indian summer. The tree question will soon become a grave one. Although the Atlantic slope, the Rocky Mountain region, and the southern portion of the Mississippi Valley, are well wooded, there are here stretches of hundreds and thousands of miles where either not a tree grows, or often useless destruction has prevailed; and the matter of the cultivation and spread of forests may well be pressed upon thinkers who look to the coming generations of the prairie states.

Nature and Democracy—Morality. Democracy most of all affiliates with the open air, is sunny and hardy and sane only with Nature—just as much as art is. Something is required to temper both—to check them, restrain them from excess, morbidity. I have wanted, before departure, to bear special testimony to a very old lesson and requisite. American democracy, in its myriad personalities, in factories, workshops, stores, offices—through the dense streets and houses of cities, and all their manifold sophisticated life—must either be fibered, vitalized, by regular contact with outdoor light and air and growths, farm scenes, animals, fields, trees, birds, sun warmth and free skies, or it will certainly dwindle and pale. We cannot have grand races of mechanics, work people, and commonalty, (the only specific purpose of America,) on any less terms. I conceive of no flourishing and heroic elements of democracy in the United States, or of democracy maintaining itself at all, without the Nature element forming a main part—to be its health element and beauty element—to really underlie the whole politics, sanity, religion and art of the New World.

Finally, the morality: "Virtue," said Marcus Aurelius, "what is it, only a living and enthusiastic sympathy with Nature?" Perhaps indeed the efforts of the true poets, founders, religions, literatures, all ages, have been, and ever will be, our time and times to come, essentially the same—to bring people back from their persistent strayings and sickly abstractions, to the costless average, divine, original concrete.

Our Real Culmination. The final culmination of this vast and varied republic will be the production and perennial establishment of millions of comfortable city homesteads and moderate-sized farms, healthy and independent, single separate ownership, fee simple, life in them complete but cheap, within reach of all. Exceptional wealth, splendor, countless manufactures, excess of exports, immense capital and capitalists, the five-dollar-a-day hotels well filled, artificial improvements, even books, colleges, and the suffrage—all, in many respects, in themselves, (hard as it is to say so, and sharp as a surgeon's lance) form, more or less, a sort of anti-democratic disease and monstrosity, except as they contribute by curious indirections to that culmination—seem to me mainly of value, or worth consideration, only with reference to it.

There is a subtle something in the common earth, crops, cattle, air, trees, et cetera, and in having to do at firsthand with them, that forms the only purifying and perennial element for individuals and for society. I must confess I want to see the agricultural occupation of America at first hand permanently broadened. Its gains are the only ones on which God seems to smile. What others—what business, profit, wealth, without a taint? What fortune else—what dollar—does not stand for, and come from, more or less imposition, lying, unnaturalness?

Theodore Roosevelt

Though perhaps less glamorous than charging with the Rough Riders or stalking big-game in the far West, Theodore Roosevelt's stint as a rancher in the Dakota badlands added credibility to his later agricultural and environmental policies. As a title, *Hunting Trips of a Ranchman: Sketches of Sport on the Northern Cattle Plains* is something of a misnomer, suggesting undue emphasis on hunting. In fact, in his chapter "Ranching the Badlands," Roosevelt, just a half dozen years removed from Harvard, returns to the precocious scientific and ethnographic curiosity that had caused him to publish, as a teen, a study of summer birds in the Adirondacks.

Part memoir, part travelogue, part nature essay, and part sociological study, "Ranching the Badlands" offers the historian a peculiarly comprehensive documentary of life on the Plains, encompassing wildlife survey, cowboy culture, geographical study, and environmental exigency. Roosevelt's observations, while thoroughly subjective and occasionally sentimental, square well with the historical record of the open range on the Great Plains in the mid-1880s. In particular, his admiration for the roping skills of the Texas cowboys in the Dakotas substantiates the southern cattlemens' expansion into the northern range. Aside from Roosevelt's characteristic belief in the equality of men and his love of hard, physical labor and the fraternity it created—both instincts that found a true home on the range—TR's assertion that latter-day cowboys would be "too late" to enjoy any real quality of life in this "ephemeral" business proved prescient. His prediction that the lifestyle he so thoroughly enjoyed would not last into the next century indicates the growing sobriety of ranchmen then under pressure from settlers, farmers, and large, foreign-owned ranches. Roosevelt proved absolutely right in his dire forecast, as barbed wire effectively ended open range culture just over a decade after the publication of *Hunting Trips of a Ranchman* in 1885. In light of the "boomer" mentality that increased the population of Kansas, Nebraska, and the Dakota Territory by almost 2.5 million from 1870 to 1890, Roosevelt's dispute of Native American land claims represents a dubious historical rationalization caused by TR's sometimes too-literal egalitarianism.

Ranching the Badlands[56]

My own ranches, the Elkhorn and the Chimney Butte, lie along the eastern border of the cattle country, where the Little Missouri flows through the heart of the Badlands. This, like most other Plains rivers, has a broad, shallow bed, through which in times of freshets runs a muddy torrent, that neither man nor beast can pass; at other seasons of the year it is very shallow, spreading out into pools, between which the trickling water may be but a few inches deep. Even then, however, it is not always easy to cross, for the bottom is filled with quicksands and mudholes. The river flows in long sigmoid curves through an alluvial valley of no great width. The amount of this alluvial land enclosed by a single bend is called a bottom, which may be either covered with cottonwood trees or else be simply a great grass meadow. From the edges of the valley the land rises abruptly in steep high buttes whose crests are sharp and jagged. This broken country extends back from the river for many miles, and has been called always, by Indians, French voyagers, and American trappers alike, the "Badlands," partly from its dreary and forbidding aspect and partly from the difficulty experienced in traveling through it. Every few miles it is crossed by creeks which open into the Little Missouri, of which they are simply repetitions in miniature, except that during most of the year they are almost dry, some of them having in their beds here and there a never-failing spring or muddy alkaline water hole. From these creeks run coulees, or narrow, winding valleys, through which water flows when the snow melts; their bottoms contain patches of brush, and they lead back into the heart of the Badlands. Some of the buttes spread out into level plateaus, many miles in extent; others form chains, or rise as steep isolated masses. Some are of volcanic

---

[56] Theodore Roosevelt, *Hunting Trips of a Ranchman: Sketches of Sport on the Northern Cattle Plains* (New York: G. P. Putnam's Sons, 1885), 22–40.

origin, being composed of masses of scoria; the others, of sandstone or clay, are worn by water into the most fantastic shapes. In coloring they are as bizarre as in form. Among the level, parallel strata which make up the land are some of coal. When a coal vein gets on fire it makes what is called a burning mine, and the clay above it is turned into brick; so that where water wears away the side of a hill sharp streaks of black and red are seen across it, mingled with the grays, purples, and browns. Some of the buttes are overgrown with gnarled, stunted cedars or small pines, and they are all cleft through and riven in every direction by deep narrow ravines or by canyons with perpendicular sides.

In spite of their look of savage desolation, the Badlands make a good cattle country, for there is plenty of nourishing grass and excellent shelter from the winter storms. The cattle keep close to them in the cold months, while in the summer time they wander out on the broad prairies stretching back of them or come down to the river bottoms.

My home ranch house stands on the river brink. From the low, long veranda, shaded by leafy cottonwoods, one looks across sand bars and shallows to a strip of meadowland, behind which rises a line of sheer cliffs and grassy plateaus. This veranda is a pleasant place in the summer evenings when a cool breeze stirs along the river and blows in the faces of the tired men, who loll back in their rocking chairs (what true American does not enjoy a rocking chair?), book in hand—though they do not often read the books, but rock gently to and fro, gazing sleepily out at the weird-looking buttes opposite, until their sharp outlines grow indistinct and purple in the afterglow of the sunset. The story-high house of hewn logs is clean and neat, with many rooms, so that one can be alone if one wishes to. The nights in summer are cool and pleasant, and there are plenty of bearskins and buffalo robes, trophies of our own skill, with which to bid defiance to the bitter cold of winter. In summertime we are not much within doors, for we rise before dawn and work hard enough to be willing to go to bed soon after nightfall. The long winter evenings are spent sitting round the hearthstone, while the pine logs roar and crackle, and the men play checkers or chess in the firelight. The rifles stand in the corners of the room or rest across the elk antlers which jut out from over the fireplace. From the deer horns ranged along the walls and thrust into the beams and rafters hang heavy overcoats of wolfskin or coonskin, and otter fur or beaver fur caps and gauntlets. Rough board shelves hold a number of books, without which some of the evenings would be long indeed. No ranchman who loves sport can afford to be without Van Dyke's *Still Hunter*, Dodge's *Plains of the Great West*, or Caton's *Deer and Antelope of America*; and Coues' *Birds of the Northwest* will be valued if he cares at all for natural history. A western plainsman is reminded every day, by the names of the prominent landmarks among which he rides, that the country was known to men who spoke French long before any of his own kinsfolk came to it, and hence he reads with a double interest Parkman's histories of the early Canadians. As for Irving, Hawthorne, Cooper, Lowell, and the other standbys, I suppose no man, East or West, would willingly be long without them; while for lighter reading there are dreamy Ike Marvel, Burroughs' breezy pages, and the quaint, pathetic character sketches of the southern writers—Cable, Cradock, Macon, Joel Chandler Harris, and sweet Sherwood Bonner. And when one is in the Badlands he feels as if they somehow look just exactly as Poe's tales and poems sound.

By the way, my books have some rather unexpected foes, in the shape of the pack rats. These are larger than our house rats, with soft gray fur, big eyes, and bushy tails, like a squirrel's; they are rather pretty beasts and very tame, often coming into the shacks and log cabins of the settlers. Woodmen and plainsmen, in their limited vocabulary, make great use of the verb "pack," which means to carry, more properly to carry on one's back; and these rats were christened pack rats, on account of their curious and inveterate habit of dragging off to their holes every object they can possibly move. From the hole of one, underneath the wall of a hut, I saw taken a small revolver, a hunting knife, two books, a fork, a small bag, and a tin cup. The little shack mice are much more common than the rats, and among them there is a wee pocket mouse, with pouches on the outside of its little cheeks.

In the spring, when the thickets are green, the hermit thrushes sing sweetly in them; when it is moonlight, the voluble, cheery notes of the thrashers or brown thrushes can be heard all night long. One of our sweetest, loudest songsters is the meadowlark; this I could hardly get used to at first, for it looks exactly like the eastern meadowlark, which utters nothing but a harsh, disagreeable chatter. But the Plains air seems to give it a voice, and it will perch on the top of a bush or tree and sing for hours in rich, bubbling tones. Out on the prairie there are several kinds of Plains sparrows which sing very brightly, one of them hovering in the air all the time, like a bobolink. Sometimes in the early morning, when crossing the open, grassy plateaus, I have heard the prince of them all, the Missouri skylark. The skylark sings on the wing, soaring over head and mounting in spiral curves until it can hardly be seen, while its bright, tender strains never cease for a moment. I have sat on my horse and listened to one singing for a quarter of an hour at a time without stopping. There is another bird also which sings on the wing, though I have not seen the habit put down in the books. One bleak March day, when snow covered the ground and the shaggy ponies crowded about the empty corral, a flock of snow buntings came familiarly round the cowshed, clambering over the ridgepole and roof. Every few moments one of them would mount into the air, hovering about with quivering wings and warbling a loud, merry song with some very sweet notes. They were a most welcome little group of guests, and we were sorry when, after loitering around a day or two, they disappeared toward their breeding haunts.

In the still, fall nights, if we lie awake we can listen to the clanging cries of the waterfowl, as their flocks speed southward; and in cold weather the coyotes occasionally come near enough for us to hear their uncanny wailing. The larger wolves, too, now and then join in, with a kind of deep, dismal howling, but this melancholy sound is more often heard when out camping than from the ranch house.

The charm of ranch life comes in its freedom, and the vigorous, open-air existence it forces a man to lead. Except when hunting in bad ground, the whole time away from the house is spent in the saddle, and there are so many ponies that a fresh one can always be had. These ponies are of every size and disposition, and rejoice in names as different as their looks. *Hackamore, Wire Fence, Steel-trap, War Cloud, Pinto, Buckskin, Circus,* and *Standing Jimmie* are among those that, as I write, are running frantically round the corral in the vain effort to avoid the rope, wielded by the dexterous and sinewy hand of a broad-hatted cowboy.

A ranchman is kept busy most of the time, but his hardest work comes during the spring and fall roundups, when the calves are branded or the beeves gathered for market. Our roundup district includes the Beaver and Little Beaver creeks (both of which always contain running water, and head up toward each other), and as much of the river, nearly two hundred miles in extent, as lies between their mouths. All the ranches along the line of these two creeks and the river space between join in sending from one to three or four men to the roundup, each man taking eight ponies; and for every six or seven men there will be a four-horse wagon to carry the blankets and mess kit. The whole, including perhaps forty or fifty cowboys, is under the head of one first-class foreman, styled the captain of the roundup. Beginning at one end of the line, the roundup works along clear to the other. Starting at the head of one creek, the wagons and the herd of spare ponies go down it ten or twelve miles, while the cowboys, divided into small parties, scour the neighboring country, covering a great extent of territory, and in the evening come into the appointed place with all the cattle they have seen. This big herd, together with the pony herd, is guarded and watched all night, and driven during the day. At each home-ranch (where there is always a large corral fitted for the purpose) all the cattle of that brand are cut out from the rest of the herd, which is to continue its journey, and the cows and calves are driven into the corral, where the latter are roped, thrown, and branded. In throwing the rope from horseback, the loop, held in the right hand, is swung round and round the head by a motion of the wrist; when on foot, the hand is usually held by the side, the loop dragging on the ground. It is a pretty sight to see a man who knows how, use the rope; again and again an expert will catch fifty animals by the leg without making a misthrow. But unless practice is begun very young it is hard to become really proficient.

Cutting out cattle, next to managing a stampeded herd at night, is that part of the cowboy's work needing the boldest and most skillful horsemanship. A young heifer or steer is very loath to leave the herd, always tries to break back into it, can run like a deer, and can dodge like a rabbit; but a thorough cattle pony enjoys the work as much as its rider, and follows a beast like a four-footed fate through every double and turn. The ponies for the cutting-out or afternoon work are small and quick; those used for the circle-riding in the morning have need rather to be strong and rangy.

The work on a roundup is very hard, but although the busiest it is also the pleasantest part of a cowboy's existence. His food is good, though coarse, and his sleep is sound indeed; while the work is very exciting, and is done in company, under the stress of an intense rivalry between all the men, both as to their own skill, and as to the speed and training of their horses. Clumsiness, and still more the slightest approach to timidity, expose a man to the roughest and most merciless raillery; and the unfit are weeded out by a very rapid process of natural selection. When the work is over for the day, the men gather round the fire for an hour or two to sing songs, talk, smoke, and tell stories; and he who has a good voice, or, better still, can play a fiddle or banjo, is sure to receive his meed of most sincere homage.

Though the ranchman is busiest during the roundup, yet he is far from idle at other times. He rides round among the cattle to see if any are sick, visits any outlying camp of his men, hunts up any band of ponies which may stray—and they are always straying—superintends the haying, and, in fact, does not often find that he has too much leisure time on his hands. Even in winter he has work which must be done. His ranch supplies milk, butter, eggs, and potatoes, and his rifle keeps him, at least intermittently, in fresh meat; but coffee, sugar, flour, and whatever else he may want, has to be hauled in, and this is generally done when the ice will bear. Then firewood must be chopped, or, if there is a good coal vein, as on my ranch, the coal must be dug out and hauled in. Altogether, though the ranchman will have time enough to take shooting trips, he will be very far from having time to make shooting a business, as a stranger who comes for nothing else can afford to do.

There are now no Indians left in my immediate neighborhood, though a small party of harmless Grosventres occasionally passes through; yet it is but six years since the Sioux surprised and killed five men in a log station just south of me, where the Fort Keogh trail crosses the river; and, two years ago, when I went down on the prairies toward the Black Hills, there was still danger from Indians. That summer the buffalo hunters had killed a couple of Crows, and while we were on the prairie a long-range skirmish occurred near us between some Cheyenne and a number of cowboys. In fact, we ourselves were one day scared by what we thought to be a party of Sioux; but on riding toward them they proved to be half-breed Cree, who were more afraid of us than we were of them.

During the past century a good deal of sentimental nonsense has been talked about our taking the Indians' land. Now, I do not mean to say for a moment that gross wrong has not been done the Indians, both by government and individuals, again and again. The government makes promises impossible to perform, and then fails to do even what it might toward their fulfillment; and where brutal and reckless frontiersmen are brought into contact with a set of treacherous, revengeful, and fiendishly cruel savages a long series of outrages by both sides is sure to follow. But as regards taking the land, at least from the western Indians, the simple truth is that the latter never had any real ownership in it at all. Where the game was plenty, there they hunted; they followed it when it moved away to new hunting grounds, unless they were prevented by stronger rivals; and to most of the land on which we found them they had no stronger claim than that of having a few years previously butchered the original occupants. When my cattle came to the Little Missouri the region was only inhabited by a score or so of white hunters; their title to it was quite as good as that of most Indian tribes to the lands they claim; yet nobody dreamed of saying that these hunters owned the country. Each could eventually have kept his own claim of 160 acres, and no more. The Indians should be treated in just the same way that we treat the white settlers. Give each his little claim; if, as would generally happen, he declined this, why then let him share the fate of the thousands of white hunters and trappers who have lived on the game that the settlement of the country has exterminated, and let him, like these whites, who will not work, perish from the face of the earth which he cumbers.

The doctrine seems merciless, and so it is; but it is just and rational for all that. It does not do to be merciful to a few at the cost of justice to the many. The cattlemen at least keep herds and build houses on the land; yet I would not for a moment debar settlers from the right of entry to the cattle country, though their coming in means in the end the destruction of us and our industry.

For we ourselves, and the life that we lead, will shortly pass away from the Plains as completely as the red and white hunters who have vanished from before our herds. The free, open-air life of the ranchman, the pleasantest and healthiest life in America, is from its very nature ephemeral. The broad and boundless prairies have already been bounded and will soon be made narrow. It is scarcely a figure of speech to say that the tide of white settlement during the last few years has risen over the west like a flood, and the cattlemen are but the spray from the crest of the wave, thrown far in advance, but soon to be overtaken. As the settlers throng into the lands and seize the good ground, especially that near the streams, the great fenceless ranches, where the cattle and their mounted herdsmen wandered unchecked over hundreds of thousands of acres, will be broken up and divided into corn land, or else into small grazing farms where a few hundred head of stock are closely watched and taken care of. Of course, the most powerful ranches, owned by wealthy corporations or individuals, and already firmly rooted in the soil, will long resist this crowding; in places, where the ground is not suited to agriculture, or where, through the old Spanish land-grants, title has been acquired to a great tract of territory, cattle ranching will continue for a long time, though in a greatly modified form; elsewhere I doubt if it outlasts the present century. Immense sums of money have been made at it in the past, and it is still fairly profitable; but the good grounds—aside from those reserved for the Indians—are now almost all taken up, and it is too late for new men to start at it on their own account, unless in exceptional cases, or where an Indian reservation is thrown open. Those that are now in will continue to make money; but most of those who hereafter take it up will lose.

TUISCO GREINER

Called "the very best and most practical work ever written for the benefit of the American vegetable garden" by Philadelphia businessman and publisher William Henry Maule, Tuisco Greiner's *How to Make the Garden Pay* entered an 1890s book and magazine market already glutted with manuals for the backyard gardener. Greiner, however, aimed for a philosophic as well as practical treatise, an advice book written, according to its author's dictates, in everyday language for the everyday gardener. In so doing, Greiner aligns himself with the yeoman and against professors of horticulture, whom he condemns for their flowery language and obtuse instruction.

Tuisco composed *How to Make the Garden Pay* to, as he intimates in his preliminary remarks, "convince people in rural districts, and in the suburbs of the cities, that gardening in reality is a very strong combination of pleasure, health, and profit, and to point out the ways and means how to relieve the task of all semblance of drudgery." In the passage that follows, the author's arguments in favor of the therapeutic values of the garden as an antidote to greasy, fattening foods seem remarkably contemporary. Greiner's work is perhaps most notable for its tone, a combination of indignation at the farmer's dull lifestyle, intolerance for incompetent market gardeners, and zealotry regarding the garden's many virtues. In "Home Gardening" the attentive reader also notes the author's references to the low crop prices of the early 1890s, an opportunity that, in Greiner's, competitive, Darwinistic mind, would only help separate the wheat from the chaff among market gardeners.

The Home Garden[57]

The physician, the lawyer, the preacher, the bookkeeper, the bank clerk—in short all people whose life occupation confines them to study or office for a large part of the day, and who for this reason are in danger of waxing tender and sensitive like hothouse plants—will find the gratification of the greatest need of their lives in a little garden of their own, namely, contact with nature, unadulterated air, relaxation and recreation, pleasure, health and ruggedness, not to speak of the more substantial and more immediate results: freshly plucked berries (not the stale fruit of the market stands in the first or more advanced stages of decay, in other words, half-rotten), crisp lettuce and radishes (not the wilted stuff of the dealer), peas and beans, with the morning dew still on them, and melons in all their perfection, freshness and lusciousness. With people of this class the question of profit may have little weight; but the home garden affords a combination of pleasure and health which nobody, and be he a millionaire, can well afford to overlook or ignore. The greatest luxuries of the garden cannot be bought with mere money. For the hardworking mechanic, on the other hand, who passes so many hours daily in the dust-laden, gas-impregnated atmosphere of the shop, the point of profit enters more largely into this question, with that of recreation in open air and pleasurable contact with nature still prominent. The garden need only be small, for much manual exercise is not often desirable, although as it comes in a different way from that of the shop, resting the muscles already tired, and giving exercise to those not called in operation by the regular shop work (thus serving to produce the natural balance of the life forces and muscles in the same way as garden work served to establish the equilibrium between the mental and physical functions of the office man), the work of the garden may only come as a pleasant change to the mechanic and not at all appear tiresome. His good spouse, less occupied with household duties than the farmer's wife,

---

[57] Tuisco Greiner, *How to Make the Garden Pay* (Philadelphia: W. H. Maule, 1890), 11–19.

will also find a needed change from indoor life and kitchen routine in the fragrant atmosphere of the home garden, and the manual labor for both should not be feared, for an abundant supply of superior vegetables can be produced on a small piece of ground, if proper tools and methods are used.

With the farmer the question of raising vegetables is chiefly one of profit, although other points are not unimportant. Many farmers who till plenty of good land concentrate all their efforts upon the production of wheat, corn, oats, wool, cattle or other so-called "money crops," and pay little or no attention to the home garden. So we have the astonishing and deplorable fact that a majority of American farmers have no garden worthy to be called a "family garden," unless so named because it is entirely given into the care of the already overworked farmer's wife and other members of the family, especially of the half-grown boys, if they in true appreciation of the good things to be had in compensation, consent to spend an extra working hour now and then in hoeing and pulling weeds. Outraged nature, unappeased hunger for vegetable food often makes them submit without grumbling to the lesser outrage of imposing an extra amount of work on their young shoulders.

Fried Pork, fried potatoes, poor bread from poorly ground flour, lardy pies, and rich cakes—these, with hardly a variation, are the chief articles of food for thousands of farmer families. Can you draw health from a pork barrel? No more than you can gather grapes from a thorn bush. Many a farmer, having sown a half acre or so of black-eye, marrowfat, or Canadian field peas, from which his family may have an abundant supply of green peas for a whole week, and given them the privilege to help themselves to all the roasting ears they may desire from the cornfield (half a mile away) for another whole week, is self-satisfied with his generosity, and boasts that his full duty is done. According to statistics taken in Illinois in 1888, only 17 percent of the farmers had the luxury of a strawberry patch. Think of this. Only one boy in every six knew what it was to pluck the luscious fruit from the vine, and eat to his heart's content! Without the stimulating, cooling, and cheering effect of fruit and vegetable diet, what wonder that the blood of so many becomes sluggish and laden with impurities; what wonder the stomach revolts at the excess of grease, and becomes nauseated from want of change; what wonder the race degenerates, dyspepsia, scrofula, and similar afflictions are becoming alarmingly frequent and general, while the concocters and venders of patent quack medicines are making fortunes! What wonder the sons leave the farm, and rush to the city, and the daughters have no desire to sell themselves into new bondage and deprivations by marrying farmers! Boy nature (and girl nature either) will not long submit to the daily farm routine of "All work and no play. All pork and no pay."

Even the dullest kind of a Jack will remonstrate against and resent this treatment. I have been a boy once, and I have learned the irresistible attraction that luscious strawberries, raspberries, gooseberries, currants, plums, pears, nuts, et cetera have for young people and old ones too, for that matter. Nature only claims her rights, and will not be outraged with impunity. I have learned the charms hidden, in crisp lettuce, radishes, green peas, and the like, in spring when the human internal machinery is clogged with a winter's excess of animal food.

There is nothing in this wide world, that with just and fair treatment otherwise will keep the farmer's boys and girls content with rural life, and make them appreciate the great natural advantages of their situation as does a good home garden and a bountiful supply of good fruits, and nothing that will bring the bloom and happy smile on the good wife's face as the assistance she will receive from the same source in solving the problem how to provide the three daily meals to the satisfaction of all.

I have already alluded to the moral side of the question. The half-starved, lean-faced street gamin standing in front of the baker's show window, and longingly contemplating the loaves, pies, cakes, and other dainties displayed in tempting array before his eyes, is not an uncommon sight, and it has often filled my inmost soul with pity. Imagine the youngster with an intense longing for fruit and vegetables peeking through the picket fence which divides his brute father's possessions from the garden of his neighbor whose fortunate children he can watch as they are gathering strawberries, or pulling crisp radishes in joy and glee. There is the luscious and coveted fruit almost within his reach, and temptingly displayed. Will you wonder if the boy, the first chance he gets to do so unobserved, removes a picket, and crawls into what to him is paradise beyond, and helps himself to what really is his due? If the father refuses to grow these things in his garden, and has "no money to spare for such luxuries," the boy will have no scruples to take surreptitiously what is so temptingly put before him. Average human nature is not built that way, to be strong enough against such odds. You cannot extract purity from glittering temptation, or morality from undue restriction, no more than health from the pork barrel. The man who willfully and needlessly deprives his family of the privileges of a good vegetable garden fails in one of his foremost duties. He cannot possibly be a good husband, nor a good father, and he certainly is not a good Christian!

Neither does he deserve to be called a good manager; for the question of profit also enters in this combination. Self-interest is a strong motive power. Here I wish I were able to convince every farmer in this glorious country of the great truth that an acre of vegetable or fruit garden, properly taken care of, will be the most profitable acre on the farm, a fact as undeniable as it is important, and one that will bear the most rigid investigation.

The amount of "green stuff" that can be produced on a single acre, well-tilled, in a single summer, is simply incredible wagon loads upon wagon loads; and there need not be a single meal from early spring until winter that is not made more cheerful, more palatable, more wholesome, and altogether more enjoyable by the presence of some good dishes from the garden, not to say anything about the canned tomatoes, peas, berries and the crisp stalks of celery, et cetera, during the winter months. I and my family live almost exclusively on the product of garden and poultry yard during the entire summer, and we enjoy pretty good health generally. No meat bills to pay, no nausea caused by greasy food, no dyspepsia! Think of sixty meals with big plates of strawberries, and sixty more with raspberries and blackberries! Think of the wholesome dishes of asparagus, of the young onions, radishes, the various salads, the green peas and beans, the pickles and cucumbers, the tomatoes, squashes, melons, et cetera! And all this practically without expense, at least, without cash outlay. There is plenty of good manure in the barnyard; horses stand in the stable more or less unused during the gardening season, and the needed labor can also be had in an emergency. At the same time few farmers will have difficulty to sell or trade off the surplus to advantage. The village blacksmith may take part if not all of his pay in good vegetables. The wagon-maker, the carpenter, the storekeeper, the physician, the banker—all of them need vegetables, and often are glad to take what good things you have to offer in exchange for money, goods, or services. If the working forces on the farm are insufficient, it will often be advisable to reduce the area of wheat or oats, and grow an acre of garden stuff instead; for the same work devoted to the garden will pay you 500 percent profit above that realized from grain culture.

Market Gardening and Truck Farming. To produce is one thing, to sell another. Money and money alone is the object of the market gardener; and the considerations of pleasure, health and morality are necessarily subordinate to that of profit.[58] Business, not pleasure—that is gardening for the man who tries to support himself and family by growing vegetables for market. To be successful it often requires a rare combination of skill and experience, with a thorough understanding of the wants of his available market, and considerable tact, if not shrewdness, in the sale of articles produced. It is no business for the careless, the lazy, or the stupid.

Neither is it a royal road to fortune, and I feel it my duty to dispel the cherished delusions of people who wish to engage in market gardening as an easy and sure way of making a comfortable living. Before me is a letter received some time ago from a "preacher of the gospel," thirty-five years of age, who having been compelled to resign his position on account of throat affliction, has hit upon the idea of

---

[58] This sentence marks the beginning of Chapter II in the 1890 edition, "Market Gardening and Truck Farming" which is here represented as a subheading.

growing garden stuff for market. "Is it possible," he asks, "to make a living on three acres of ground, 115 miles from Philadelphia? Soil good, and in town, near railroad station. I am happiest when I am hard at work, and oh! I love to work in the soil! This alone gives me renewed vigor, and a degree of health. Yet I am not willing to become a market boy, and I cannot peddle out what I raise off the soil." Here, evidently, we have met with a wrong conception of market gardening; but it is a somewhat common one. I know of localities where three acres of good ground well-managed would afford quite a respectable living to a small family, with a market right at the door, and grocers in the near town willing to take almost any good garden produce brought them at fair prices. Advantage might often be taken of a local demand for certain productions, as berries, onions, celery, et cetera, and such articles grown on a larger scale, for sale to retailers, thus avoiding the "peddling" feature. But kid-glove and silk-hat gardening will under no consideration fit into successful market gardening or truck farming; "barter and trade" is one of the essentials of the business anywhere, and the grower must be in readiness, if an emergency arises, to take hold and become merchant or peddler. This feature is an indispensable part of the business in most cases.

Gardening for money requires unceasing attention, close and thorough management, considerable hard labor, and often more or less exposure to the vicissitudes and inclemencies of the seasons. Nevertheless it is true that the majority of the profession make altogether too much work of it, especially by neglecting to make use of the newer improved implements of tillage. The hand hoe is yet left to play a by far too prominent part in garden culture, and the advantages of the wheel hoe are not yet recognized and made use of as they deserve.

There was a time when even the rudest methods combined with hard work insured to the market gardener near large cities a good income. But competition has grown with the demand, and with cheapened and increased production, prices have gradually declined until now they are far below what only a few years ago growers would have considered mere cost of production. It is not so many years since the main crop of strawberries sold at twenty-five cents per quart; and when the price first dropped down to twenty cents, the cry went forth that "strawberry growing does not pay." Then thousands of growers abandoned the business in disgust. At present, strawberries are grown at six and eight cents per quart in many localities, and people are satisfied with the profits. So with vegetables. We have learned to produce much cheaper than formerly, and we can afford to produce and sell at figures which did not cover first cost ten or twenty years ago, and yet realize a fair profit. Hence people who continue to grow garden crops in the old laborious and unsatisfactory ways, and with old-style implements, who produce inferior vegetables and fruits at old-time cost, cannot successfully meet the competition of their progressive brethren. This is simply a question of the "survival of the fittest," and the fittest is the man who by taking advantage of the latest labor-saving methods and devices manages to raise the best produce at the smallest cost, thus preserving or even widening the narrow margin of profit which at the present time characterizes all legitimate branches of business. The spade must give way to the plow; the rake, and often cultivator also, to the harrow; hand and fingers in sowing seeds to the drill; the hand hoe to the wheel hoe, et cetera. These changes are imperative and unavoidable, if the business is to be made profitable. The grower who has learned to produce most cheaply and can offer the earliest or best articles in his line is the one who succeeds, and efforts to excel must be made continuously to prevent a getting left in this race. This requires the exercise of thought, studying short of brains as well as of muscle. Excellence will have its reward; but he who neglects a single point, who allows himself to be excelled by others, is not likely to receive a prize.

Special vegetable crops are often grown on a large scale in localities especially adapted to their cultivation, or having special market facilities for such crops. So we have the celery fields of Kalamazoo, Michigan, the onion patches of Wethersfield, Connecticut, and Danvers, Massachusetts, and other places, the cauliflower gardens of Long Island, the tomato fields of New Jersey, the melon patches of Virginia, et cetera. To produce is often much easier than to sell the product at a profit, and it is not safe to engage in a business of this kind on an extensive scale, or invest much money in it, unless a local demand is assured for the produced articles. Wagon- and carloads of good vegetables are yearly thrown away for want of chance to sell them in time at an acceptable price. Where the enterprise is carried on in colonies, however, there is always a local market; for the center of production is also the center of demand.

LEONIDAS L. POLK AND BENJAMIN HUTCHINSON CLOVER

The following two readings from *The Farmers' Alliance History and Agricultural Digest* appear here under the umbrella heading "Sectionalism," and are presented together to represent the dialogue between two of the leading figures of the Alliance.

Published within a year of Leonidas Polk's death and what would likely have been his nomination for President in 1892 by the newly-formed People's Party, Polk's "Sectionalism and the Alliance" ably displays the National Farmers' Alliance brand of passionate, post Civil War reconciliation. Polk's writing here reveals the consolidation of his organization's power and the stated benevolence of its ends. And while Polk, a southern colonel, cannot resist condemning Yankee greed, he also does not lament the passing of slavery per se or the defeat of ardent sectionalism.

Polk's "Sectionalism and the Alliance" shows a man at the height of his political and rhetorical powers, spurred on by an increasingly disgruntled and impoverished southern tenant farmer. Polk, the North Carolina based editor of the *Progressive Farmer*, proves particularly sensitive to the plight of the cotton farmer whose frustration fueled the rise of Farmers' Alliance in the first place. Polk articulates herein the southern-based, anti-industrialist critique later echoed by the Vanderbilt Agrarians, as he characterizes the North's capitalism and carpetbaggery as a new kind of slavery. "Whatever may be said of chattel slavery, with all its acknowledged evils," Polk writes, "history nowhere records that it ever made a millionaire."

At the time Benjamin Hutchinson Clover penned "Sectionalism," the second reading in this grouping, he had recently been elected to represent Kansas in the Fifty-Second Congress as a member of an upstart agrarian party, the Populists. Vice President of the Farmers' Alliance and Industrial Union headed by Leonidas Polk, Clover's sentiments closely follow Alliance party line, though his second-in-command position perhaps affords him greater freedom. A product of the Kansas common schools, Clover served some fifteen years on the Kansas board of school commissioners before winning his seat in Congress.

Though he served under Polk, a North Carolinian, Clover the Kansan exhibits in the following essay the wished-for national unity of the Farmers' Alliance. Offering a classic critique of the American two-party system and the monopolistic industries that empower it, Clover writes, "Sectional hate and its other self, party prejudice, have been the means by which monopoly has been enabled to bind the people; and a blind subserviency to party and designing party leaders has been the means by which it has accomplished what in other countries it obtained by violence, bloodshed, conquest, and other forms of oppression." Clover's sentiments here—firmly against the usurer, the party politician, and the greedy industrialist—privilege the hard-won wisdom of the tiller over the demagoguery of the idle rich. In part due to his belief in the citizen-legislator, Clover would serve out a single two-year term in Congress as a Populist, returning to agricultural pursuits immediately thereafter.

Taken together, Polk's and Clover's essays privilege the kind of everyday heroism born of the disenfranchisement suffered by many struggling farmers of the era. While the association produced an unprecedented agrarian momentum and political clout in the early 1890s, better economic times and ongoing sectional differences would mean the Alliances' virtual disappearance as a political player by 1900.

On Sectionalism: Selections from The Farmers' Alliance History

Sectionalism and the Alliance by Leonidas K. Polk. The year 1865 witnessed the culmination of the mightiest contest of modern times.[59] The brave and heroic men of the two armies, worn and wearied with

---

[59] N. A. Dunning, ed., *The Farmers' Alliance History and Agricultural Digest* (Washington DC: Alliance Publishing Company, 1891), 249–253.

war, returned to their homes, and beating "their spears into pruning hooks, and their swords into plowshares," addressed themselves, with a faith and a devotion that were sublime, to the solution of problems which would have appalled the hearts of any but those who had been educated in the terrible ordeal through which they had passed. The happy greetings of welcome of the loved ones at the threshold were more thrilling and inspiring than were the wild shouts of triumph in victorious battle.

As a rule, the soldiers of the North and the South were willing and anxious to accept and abide by the result, in good faith. They knew they had fought like men, and they were willing to accept the result like men. Slavery was gone, and all true patriots fondly hoped that the prejudices, animosities, and divisions which were born of its existence would go with it.

But the selfish, sectional agitator again appeared upon the scene, and, with unholy purpose, spared not even the sacred dust of the heroic dead that he might inflame and keep alive the bitter recollections and animosities of the past. Social and financial chaos was abroad in the land, and he gloated in the opportunity thus afforded to prosecute his wicked designs. Ordinarily, he was the man, North and South, who had failed to see, in four years of war, any opportunity to prove his devotion to his section. Ordinarily, he was the man, North and South, who was "invisible in war, and had become invincible in peace." The liberation and enfranchisement of four millions of human beings, the confusion incident to a long protracted and terrible struggle, presented conditions peculiarly favorable to the propagation and perpetuation of sectionalism. Even our industrial development and expansion evolved conditions which were made to contribute to this unnatural and unfortunate estrangement between the sections. The rich, powerful, and densely populated East must needs have an outlet for its aggressive enterprise, its rapidly accumulating wealth, and its growing population. The dense forests and fertile plains of the magnificent and inviting West were transformed into rich and powerful states. Lines of immigration and enterprise, of wealth and of general development, were pushed forward with marvelous rapidity and success to the shores of the Pacific. Along these lines were transplanted from the East the prejudices and animosities engendered for a half century. The South, traversed by no transcontinental line of communication, sullen and humiliated in her great and crushing losses, and by defeat in war, most naturally nursed the sectional animosities and prejudices of the past.

Their fields were devastated, their homes desolate, their household goods destroyed; without money, without food, without implements with which to work; their credit gone, their labor utterly destroyed, their industrial systems wiped out, the accumulated wealth of generations swept away as by a breath; in the shadow of drear desolation and the blackened ruins of once happy homes, they were left friendless and unaided, to depend on those qualities of true manhood which are always evolved by terrible emergencies. Theirs was the noble and heroic task to remove the ghastly wreck which marked the feast of war gods, who had reveled in their high carnival of blood, of carnage, and of death.

What an inviting condition was thus presented for wicked sectional agitators, and how assiduously they utilized it, let the shameful sectionalism of the past quarter of a century, with its baneful fruits, tell. Whatever may be said of chattel slavery, with all its acknowledged evils, history nowhere records that it ever made a millionaire. Whatever may have been its effect upon our society and civilization, it produced no tramps. But we have developed another system of slavery, the slavery of honest labor, a slavery of sweat, and brawn, and brain, a slavery more terrible and degrading in its effects than the African ever knew, and the legitimate outgrowth of which has cursed our country with an army of three millions of tramps, and has placed the greater part of the wealth of this great nation in the hands of one two-thousandth part of its population. It has made the eight millions of American farmers—once the proud possessors of the most princely heritage that God ever gave to man—virtually a nation of tenants, whose every possession, and whose every day of toil and labor, is forced to pay tribute to exacting, domineering, legalized monopoly. In all the discriminating partisan legislation which has disgraced the annals of the nation for the last quarter of a century, and in all the machinations and intrigues which have conspired to destroy that essential equipoise between the great industries of the country, and which has robbed the many to enrich the few, and thus placed our republic and its institutions in imminent peril, no factor has been more potent than the wicked spirit of sectionalism.

We have thus been brought to confront forces, social, industrial, moral, and political, which are dangerous alike to the liberty of the citizen and to the life of the republic, and we stand today in the crucial era of our free institutions, of our republican form of government, and of our Christian civilization. Mighty problems confront us, and they must be met in a spirit of fairness, of manliness, of justice, and of equity.

The evils under which the great laboring millions of America are suffering are national in their character, and can never be corrected by sectional effort or sectional remedies. In all the broad field of our noble endeavor as an order, there is no purpose grander in its design, more patriotic in its conception, or more beneficent in its possible results to the whole country and to posterity, than the one in which we declare to the world that henceforth there shall be no sectional lines across Alliance territory. Failing in all else we may undertake as an organization, if we shall accomplish only a restoration of fraternity and unity, and obliterate the unnatural estrangement which has unfortunately so long divided the people of this country, the Alliance will have won for itself immortal glory and honor. In the spirit of a broad and liberal patriotism, it recognizes but one flag and one country. Confronted by a common danger, afflicted with a common evil, impelled by a common hope, the people of Kansas and Virginia, of Pennsylvania and Texas, of Michigan and South Carolina, make common cause in a common interest. The order recognizes the fact that the war ended in 1865, that chattel slavery is gone, and that the prejudices and divisions, born of its existence, should go with it.

Happily for the country and posterity, the great mass of the people have become aroused to this truth, and they have severed sectional lines in twain. The ex-slave holder of the South, who believed that he held the slaves not only by constitutional but by divine right, and who bravely imperiled his life to defend the institution, today stands hand-in-hand with him who was born and reared an abolitionist, and who regarded slavery as an unmitigated evil and curse, and disregarding sectional folly and madness, they have solemnly pledged their alliance in a common cause the cause of a common country. We cannot fail to see the opportunity of the hour, and, recognizing that opportunity, we must not forget that it carries with it corresponding responsibilities. The opportunity is for the great, conservative, law-abiding, patriotic masses to assert and establish a perpetual union between the people. The sequent obligation is that these great masses must discourage, discountenance, and discard from their councils the wicked, demagogical agitators, who for the last twenty-five years have sought to foster discord and dissension, that they themselves might thrive. We are told in sacred history, that, in the olden time, one Jeroboam was crowned a king in Israel. He conceived the absurd idea that to strengthen his people he should divide them, that to fraternize them he should destroy their unity, and he forbade and abolished their annual national meeting at the city of Jerusalem. He erected a golden calf at a place in the north, and one at a place in the south, and directed that the people of the two sections should hold their annual meetings at these places, respectively. We are told that even in that remote age Jeroboam adopted some of the methods of modern politics, in that "he made high priests of the lowest people." The avenging hand of outraged justice was laid upon him. Does history repeat itself? Sectionalism, for purposes of greed and gain, decreed that the people of these United States should be divided, and to perpetuate that division it directed that idols should be erected for the people of the North, and for the people of the South. And has it not "made high priests of the lowest people"? And shall it not be rebuked and destroyed? Divided, we stand as a Samson shorn of his locks; united, we stand, a power that is invincible. Cato fired and thrilled the Roman Senate with the fierce cry, "Carthage must be destroyed." Must we, as citizens of this great republic, emulate such a vengeful spirit? Hannibal, while yet a tender youth, was placed by his father on his knees, and made to swear eternal vengeance against the Romans. Must we, as Christian parents, entail upon our children the bitter legacy of hate? Hundreds of thousands of noble, aspiring, and patriotic young men, all over the land, are manfully undertaking the responsible duties of American citizenship. Born since the war—thank God—their infant vision was first greeted by the light of heaven, unobscured by the smoke of battle, and their infant ear first caught the sweet sound of hallowed peace, unmingled with the hoarse thundering of hostile cannon. Shall they be taught to cherish, foster, and perpetuate that prejudice and animosity, whose fruits are evil, and only evil?

"Let the dead past bury its dead," and let us, with new hope, new aspirations, new zeal, new energy, and new life, turn our faces toward the rising sun of an auspicious and inviting future, and reconsecrate ourselves to the holy purpose of transmitting to our posterity a government "of the people, by the people, and for the people," and which shall be unto all generations the citadel of refuge for civil and religious liberty.

Sectionalism by Benjamin Hutchinson Clover. "In peace there is nothing so much becomes a man as modest stillness and humility." Following this thought of the famous Roman orator Rienzi, I would fain maintain a "modest stillness"; but I see in our country a condition that never could have existed but for the false and pernicious teachings of those who stir up strife and keep alive the fires of sectional hate.[60]

Do you ask for what purpose is this ceaseless arraignment of the North against the South, and the South against the North, kept up? One who has been chief in the strife, and loudest in his demands for "a solid North against a solid South," says that they have been "alienated by those who sought to prey upon them."

This is surely a frank admission. He further says that "invidious discriminations have robbed them of their substance, and unjust tariffs have repressed their industries." The objects of sectional agitators can

---

[60] N. A. Dunning, ed., *The Farmers' Alliance History and Agricultural Digest* (Washington DC: Alliance Publishing Company, 1891), 253–256.

not be more fully and tersely stated. Some of them, possibly by reason of their ignorance, were honest in their belief, but with the great majority self-aggrandizement, and the service of an oppressive and unscrupulous combination of public robbers, was the sole end in view.

So successful have been their efforts that the money power of the world has laid tribute upon honest industry, and the laborer, once king, finds himself a pauper, a wanderer, a homeless, nameless stranger in the land of his fathers. Samson, while listening to the siren song, of the party Delilah, was shorn of his locks, of his strength, of his manhood, and virtually of his freedom. But some may say, *Has sectionalism done all this*? Gentle reader, let me ask, could any other thing have kept the people so blinded to their interest, that, having the ballot in their hands, they would have allowed the soul-and-body-destroying, monopolistic influences to wrap their slimy folds around each and every industry, and send the honest toiler shivering to a hovel, and elegant idleness to a palace? Sectional hate and its other self, party prejudice, have been the means by which monopoly has been enabled to bind the people, and a blind subserviency to party and designing party leaders has been the means by which it has accomplished what in other countries it obtained by violence, bloodshed, conquest, and other forms of oppression.

The favorite method of those who "toil not neither do they spin," is to array those whom they wish to rob against each other. This once accomplished, the rest is easy. Nor is the robbery of industry and a virtual enslavement of the laboring people all the harm that has come from this the most blighting curse that ever came upon the people of free America. It has arrayed brother against brother, and made enemies of those who, by every tie that binds men's hearts together, should have been friends.

Neither time nor space will allow a detail of methods resorted to by those who "alienate the people only to prey upon them." It is through false politics, and politicians more false and designing, that they seek to accomplish their ends, and they have so far succeeded. All have heard the cry of the campaign howler. I shall not attempt to describe him. He is the bane of civilization, the enemy of liberty and humanity. His mission is to stir up old animosities, engender new strifes, fight over dead issues, and write platforms to be read before election and disregarded and forgotten afterwards. He is an "oily" fellow. He has been selected for his fitness for the work he is to perform. With him "it is lawful to deceive, to hire Hessians, to purchase mercenaries, to kill, to mutilate, to destroy"—anything for success. It makes him exceedingly "weary" should any one suggest that the politics of our country be placed upon a higher plane. He worships no god but his own ambition, and that ambition is to be the "cutest" trickster and slyest deceiver of his party; for well he knows that those who prey upon the people and wealth producers will see his "transcendent ability" and pay well for his treason to the interests he is supposed to represent, and heap "honors" upon him.

There is no sympathy in his heart for the miseries of the millions who, by reason of his infamous schemes, are robbed of home, happiness, and all hope of the future. There is no tear in his eye as the hapless family—the heartbroken father, the sad-faced and weeping mother, and the sorrowing children—find themselves driven from their home to become helpless wanderers up and down the earth. He has never heard the sigh come up from the bosom of his wife as she listens to the reading of the foreclosure summons. Little cares he though tears may fall like rain, though hearts may break, though hope may go out forever from the hearts and homes of his victims. In his mad rush for office and spoils he has forgotten that there is a just God, who has said: *Vengeance is mine, I will repay.*

It is indeed a gloomy picture that the past thirty years present, in this so-called free land of America. Designing demagogues, sustained by the money and monopolistic power of the world, have so far succeeded in deceiving the people, and arraying them against one another, and despoiling them of their homes, the fruits of their labor, and their hope of the future. Liberty, with the great mass, has become an empty farce, and American independence an "iridescent dream."

This for the past; but what of the future? The early fathers told us that "eternal vigilance is the price of liberty." Have we been vigilant? Do not political sins bring political death, as surely as moral sins bring moral death, or a violation of the laws of health brings on physical disease and death? The fathers taught us that in unity is strength. Have we as a people obeyed their injunction? Sectionalism, with its agitators, has stood guard over the bursting vaults of the public plunderer, and if any one raised his voice in protest against the infamous robbery, the "bloody shirt" was brought out on one side, and the "Yankee hireling" howlers split the air on the other, and the robbery of the people went on.

But I wanted to take a peep into the future; I wanted to write of the time when sectionalism shall be buried deep. Its grave is being dug now. The "great common people" of the South and of the North are realizing their condition and its cause, and they are meeting together, becoming acquainted, and wondering why they ever should have been enemies.

The stock in trade of the sectional agitator is going below par. He will soon be a thing of the past. He is now in his dying throes, and while some of them are bowing to the inevitable, others are nerving themselves for a last supreme effort. But their time has come. The people are awaking from their lethargy, and they find themselves made beggars while they slept. They are fast learning the truth. The "alienator" is out of a job. The "white rose of peace" is being planted over the grave of sectionalism. It is being watered by the repentant tears of the victims of this hideous monster sectional strife. The old leaders, who have been responsible for the sectional hate of the past, are being sent into retirement. New blood and new ideas are coming. The people are looking to the future instead of brooding over the past. They know that they have been robbed by infamous legislation, and that righteous legislation will give them back their homes and happiness again. They are refusing to be mere hewers of wood and drawers of water for a favored class of moneychangers. When the happy time comes that sectionalism is dead and buried out of sight, and is remembered only as a hideous nightmare; when the toiling masses of both North and South shall join hands and remember only that they are brothers, children of a common father, citizens of a common country, with one flag, one destiny, and that they are "Americans all"; and when patriots and not partisans shall rule in legislation, then shall the brotherhood of man be acknowledged, and fraternity, peace, and goodwill will come among the people.

When I think of the past, and contemplate the present, and anticipate what may be in store for the common people in the future, if they will be friends and act wisely and contend for, instead of against, each other, I am constrained to quote again from the grand Roman, who, when he found his beloved country ruined and desolate, and his fellow citizens ground down by the heel of oppression, cried out: "Rouse ye, Romans! Rouse ye, slaves! Our country yet remains!"

Then he told them of that "elder day," when to be a "Roman was greater than to be a king." Shall not we look back with a patriotic longing to that elder day, when to be an American was greater than to be a king?

Though poor, though crouching at the feet of as arrogant and unscrupulous oppressors as ever robbed a widow or starved an orphan, let us remember that our country yet remains.

Brothers of the sunny South, after thirty years, is it not time that the past should be buried? Grant is dead. Lee is no more. Stonewall Jackson and William McPherson gave up their lives on the field of battle, and fill soldiers' graves. Almost the last one of the great commanders, and a majority of their followers, have gone where war is not known; and why should not we, in our memories, let them lie side by side, and over their graves clasp hands and say to each of them, "Soldier rest, thy warfare's o'er; Sleep the sleep that knows not breaking; Dream of battlefields no more: Days of danger, nights of waking"? Will not the proudest monument we can build to their memory be a just and righteous government, that will protect the weak, do justice to all, and be of for, and by the people? Shall we not build a temple of liberty wherein the poorest and humblest shall have a seat, as well as the rich and arrogant, and where he can feel that he is heir to all the glories which the wisdom of the fathers and the unselfish patriotism of our country can give us? "Let us have peace."

BETTIE GAY AND JENNIE E. DUNNING

Just as Leonidas Polk and Benjamin Clover intimate the concerns among the leadership of the Farmers' Alliance, Texas' Bettie Gay and Washington DC's Jennie E. Dunning present contrasting views of a woman's role within the Alliance. As with Polk and Clover, they are presented here as a dialogic pair.

As with so many women in the Alliance, neither Gay nor Dunning appears to have held official title in the organization, though Gay did serve as a delegate from Texas to the all-important Farmers' Alliance and Industrial Union meeting in St. Louis in 1892. Still, the sentiments expressed in their essays, and in the letters to the editor Gay wrote to the official newspaper of the Farmers' Alliance, the *Southern Mercury*, suggest a level of empowerment and self-expression allowed if not encouraged by the organization. While women comprised a mere 25 percent of the membership of the Alliance, their influence, as Bettie Gay suggests, proved disproportionate. Unlike the like-minded Grange organization, the Farmers' Alliance did not advocate stridently for women's suffrage, though women members were valued, as Gay writes below, as "companion and helpmeet to man." Much like the Grange, the Alliance viewed the farmer's problems as principally economic, hence Gay's impatience with women too timid to become involved in the economic reform movement. "Poverty and want are the chief causes of crime," Gay writes, "and the reason why so many people are found occupying unnatural conditions is because of the violation of the principles of justice and right." In the essay that follows, Gay wisely plies her audience with agrarian themes and metaphors, arguing convincingly for the "cultivation" of women's minds and for women's peacemaking role at the organizational table. Gay's characterization of women as "chief sufferers" imbued with unusual powers of empathy, temperance, and politeness, would have resonated with Alliancemen who knew of Gay's own suffering. Widowed, Gay had been left a 1700-acre farm to manage and a son, James Jehu Bates Gay, to raise. Gay would later become a prominent Populist in his mother's image. Bettie Gay was an Alabama Baptist by birth and a socialist by choice, partially explaining though not excusing her use of the phrase "savage races" as well as her contempt for agrarian women concerned merely with the health and welfare of their nuclear family.

Jennie E. Dunning's "The Home and Flower Garden" makes for a compelling comparison-contrast with Bettie Gay's impassioned calls for women's rights and responsibilities. Dunning, likewise a member of the Alliance, hailed from Washington DC, and her contribution to the Alliance annual bespeaks a level of urbaneness.

Dunning's historical approach to domestic affairs links the passage that follows to the increasingly popular home economics curricula enabled by the Morrill Land Grant College and Hatch Acts. For Dunning, a historical lens magnifies the role of women as economic and spiritual partners as well as estate managers and homekeepers. Women, by the author's reckoning, are responsible for loveliness in personal appearance and in household decoration and are vested thereby with particular ideological powers—to bring order and to encourage fidelity and continuity. Deceptively progressive, Dunning advocates for the literacy of both farm girls and boys and suggests that beauty comes from the inside rather than the outside, from inner decorum rather than outer ornamentation—from a women's creative force rather than her economic wherewithal. Conservationist readers will note Dunning's emphasis on thriftiness and recycling, while students of literature will note, with some amusement, the example made of a resourceful homemaker who, like Scarlet O'Hara, turns dresses into curtains and back again.

On Roles for Women: Selections from The Farmers' Alliance History

The Influence of Women in the Alliance by Bettie Gay. In the past, woman has been secondary as a factor in society.[61] She has been placed in this position because the people have been educated to believe that she is mentally inferior to the sterner sex. Only of late has the discussion of her social and political rights been brought prominently before the country. The male portion of our population, through a false gallantry, have assumed that they are the protectors of the "weaker sex": women have been led to believe that they had no political or social rights to be respected, and a very large majority of them have bowed in quiet submission.

---

[61] N. A. Dunning, ed., *The Farmers' Alliance History and Agricultural Digest* (Washington DC: Alliance Publishing Company, 1891), 308–312.

History proves that the more crude and savage society is, the lower women are placed in the social scale. The men of savage races compel their women to do all the work; in fact, to be their slaves. When this social question is investigated from a scientific standpoint, the wonder is that man has ever been able to emerge from his original condition, while the situation of the mothers of the race has been such as to naturally impede intellectual progress. Only the plain manifestation of the laws of nature and the human mind has enabled man to raise himself above the crude forms of barbarism, and establish what is now termed civilized society.

Education concerning the effects of social conditions is demonstrating that most of the moral evils which afflict society are produced by the unnatural conditions which are imposed upon women. Nature has endowed her with brains; why should she not think? If she thinks, why not allow her to act? If she is allowed to act, what privilege should men enjoy of which she should be deprived? These are pertinent questions which society should begin to consider.

Go into the rural districts and look at the position occupied by the wives and daughters of the farmers. They have, until of late, occupied a social position which tended only to discourage intellectual effort. In most of the churches women have been allowed no voice, and the very moment some brainy woman in a community would rise above her surroundings and take an interest in public questions, the men, as well as the women, would begin to discourage her efforts. She would be told by her father, brother, or husband, that such questions are not the concern of women. But the Alliance has come to redeem woman from her enslaved condition, and place her in her proper sphere. She is admitted into the organization as the equal of her brother, and the ostracism which has impeded her intellectual progress in the past is not met with, and men have begun to recognize the fact that, when the women are educated, the battle for human rights will have been fought and won.

Her position in the Alliance is the same as it is in the family, the companion and helpmeet of man. In it she is given the opportunity to develop her faculties. She is made to feel that she is the equal of man, and that she can make herself useful in every department of human affairs; that her mission in the world is more than merely to be called wife or mother (both of which are honorable), but her work is one of sympathy and affection, and her help is as much needed in the great work of reform.

Only in late years have women been considered a necessary factor in reform movements. This has been brought about by advanced thinkers, who have studied sociology and the science of intellectual and moral development. Society seems never to have thought of the fact that there is no progress without opportunity, and that depriving women of their social and political rights has taken from them the inducement to become educated upon great questions. The Alliance contemplates the opening of every avenue of intelligence, which will induce women to become educated, and capable of taking care of themselves in the struggle for existence, and the establishment of a social system which will guarantee to every human being the results of his labor. The condition of the wives and daughters of the farmers is but little better than that of the women who work in factories. In probably a majority of instances, in the South and Southwest, the women assist in cultivating and gathering the crops. Such a condition of industrial serfdom the Alliance, with other reform organizations, expects to overthrow.

In the effort for reform, none can be more interested than women, as they are the chief sufferers whenever poverty or misfortune overtakes the family. They are the ones to look after the welfare of the children of the family. They, more readily than the fathers, see what is necessary to make the family happy and comfortable. But, having been educated to believe that bad conditions are caused by divine providence, or are the result of mismanagement, many of them have borne the social evils in silence, and trusted for happiness after they shall have crossed "the silent river."

Through the educational influence of the Alliance, the prejudice against woman's progress is being removed, and within the last five years much has been accomplished in that direction. Women are now recognized as a prominent factor in all social and political movements. In the meetings of the Alliance she comes in contact with educated reformers, whose sympathies she always has. Her presence has a tendency to control the strong tempers of many of the members, and places a premium upon politeness and gentility. She goads the stupid and ignorant to a study of the principles of reform, and adds an element to the organization, without which it would be a failure. Being placed upon an equality with men, and her usefulness being recognized by the organization in all of its work, she is proud of her womanhood and is better prepared to face the stern realities of life. She is better prepared to raise and educate her offspring by teaching the responsibility of citizenship and their duty to society.

The meetings give recreation to the mind, and the physical being is for a time relieved from incessant toil. The entire being is invigorated, and the mind is prepared for the reception of such truths as fit her to be companion, mother, and citizen. As stated above, woman has not been considered a factor in great movements until of late years, but she comes prominently to the front in the Alliance, and demands that she be allowed to render service in the great battle for human rights, better conditions, happier homes, and a higher civilization generally. In fact, she has come to the conclusion that she has some grievances for which remedies should be found, and that she owes it as a duty to herself and society to help work out the social and political salvation of the people. I believe that there are remedies for most of the evils which afflict society; that poverty and want are the chief causes of crime; and the reason why so many people are found occupying unnatural conditions is because of the violation of the principles of justice and right, by the government allowing the few to monopolize the land, money, and transportation, which deprives a large portion of the people of their natural right to apply their labor to the gifts of nature. Under such conditions, the people become dependent, hopeless slaves, a condition which drives the last spark of manhood and womanhood from their bosoms, and they become outcasts and criminals and fill our jails and penitentiaries and other places of shame.

It is the duty of the Alliance to consider these questions, and none others are so much interested in the regeneration of society as women. When the battles of life are to be fought, she is always a valiant soldier, and many of them bear upon their faces the scars of the battle with poverty and want. The faces and forms of many of the farmers' wives bear marks of premature age. Their sensibilities are deadened with the cares and toils of life. They have enjoyed but few of the benefits of modern civilization, and but few of the luxuries of life which they have helped to create. They have plodded along, while conscienceless greed has fattened upon their labor, and deprived them of the conditions which are necessary to make them happy and good, their lives a blessing, their homes a heaven.

But this is a new era in human progress, when woman demands an equal opportunity in every department of life. She is no longer to be considered a tool, a mere plaything, but a human being, with a soul to save and a body to protect. Her mind must be cultivated, that she may be made more useful in the reform movement and the development of the race. It is an acknowledged principle in science that cultivated and intelligent mothers produce brainy children, and the only means by which the minds of the human race can be developed is to strengthen, by cultivation, the intellectual capacities of the mothers, by which means a mentally great race may be produced. When I look into the hard and stolid faces of many of the mothers of the present, and know that they have been deprived of the opportunities which would have improved them, I am not surprised that we are surrounded by people who are the advocates of a system but little better than cannibalism.

Through a system of education, in the Alliance and kindred organizations, we are slowly but surely eradicating the false doctrines of the Dark Ages, and the traditions of the pagans, handed down to us through false teaching. To remove these evils is the grandest work of the age, and the woman who holds herself aloof from reform organizations, either through false pride or a lack of moral courage, is an object of pity, and falls far short of the duty she owes to herself society, and posterity.

If I understand the object of the Alliance, it is organized not only to better the financial condition of the people, but to elevate them socially, and in every other way, and make them happier and better, and to make this world a fit habitation for man, by giving to the people equal opportunities. Every woman who has at heart the welfare of the race should attach herself to some reform organization, and lend her help toward the removal of the causes which have filled the world with crime and sorrow, and made outcasts of so many of her sex. It is a work in which all may engage, with the assurance that they are entering upon a labor of love, in the interest of the downtrodden and disinherited, a work by which all mankind will be blessed and which will bless those who are to come after for all time.

The education of the masses is the hope of the world, and a healthy public sentiment must be created in the interest of labor. Poverty must be abolished, and the natural rights of the people must be respected. It is unnecessary for me to pay any tribute to, or heap any abuse upon, woman. She is precisely what her opportunities have made her, whether she is found in a palace or a hovel. She is flesh and blood, and whatever virtues or vices she may possess, can only be attributed to environment and opportunity. What we need, above all things else, is a better womanhood, a womanhood with the courage of conviction, armed with intelligence and the greatest virtues of her sex, acknowledging no master and accepting no compromise. When her enemies shall have laid down their arms, and her proper position in society is recognized, she will be prepared to take upon herself the responsibilities of life, and civilization will be advanced to that point where intellect instead of brute force will rule the world. When this work is accomplished, avarice, greed, and passion will cease to control the minds of the people, and we can proclaim, "Peace on earth, good will toward men."

The Home and Flower Garden by Jennie E. Dunning. A parlor or living room should bear a decided resemblance to its mistress.[62] Endow it with a marked personality; let her choicest flowers, her favorite poem and song, be found upon the table. The very bric-a-brac and furniture should speak of her refinement and tastes, instead of the depth of the family pocketbook. Furniture may be substantial and inexpensive, and

---

[62] N. A. Dunning, ed., *The Farmers' Alliance History and Agricultural Digest* (Washington DC: Alliance Publishing Company, 1891), 637–646.

at the same time possess those home-like attractions which are so captivating to average humanity.

It is said that furniture is the story of the race, from sumptuous Egypt down through the Dark Ages; and it may be interesting to note the result of some research in this direction. The Greeks perfected Egyptian suggestions and ideas, but seem to have produced nothing new. They lived largely in public temples, theaters, groves, and porticos. Holding their women in slight esteem, and having little home life, they expended their wealth and energies on public sculpture, paintings, architecture, caring little for home arts and comforts. With their artistic ideas, their articles of domestic furniture were perfect, but few in number. The Romans paid more attention to household arts, and the position of woman was somewhat advanced, but they borrowed their ideas of household articles from the Greeks, as they, before them, borrowed from the Egyptians. The first chair was a thing of state, and doubtless was developed from the Egyptian throne. At the downfall of Rome, whatever household art had accomplished went with it. The barbarians destroyed nearly everything, and the industrial arts no longer existed in the western empire. All there was of convenience, comfort, or splendor came from the East. Silks, spices, gems, ivory, and smaller articles of furniture, reached the West and middle Europe, at first through Egypt, and afterward through a commerce established between the pilgrims who visited Jerusalem, and the Arabs who visited Mecca. Some writers have intimated that this trade, and the profits from it, had much to do with the zeal with which pilgrims sometimes undertook this long and perilous journey. Down through the Dark Ages, every lord of a castle was a sovereign, liable at any time to be obliged to yield to stronger forces. If he went abroad, he was uncertain of his ability to return, and his home was a kind of fortress. At this period, his furniture consisted of little else than chests, which, in the castle, served as seat, bed, table, and treasury, and if he was overpowered by an enemy, his valuables were hastily packed in the chest and easily moved.

The curule stool (camp or folding chair) was handy for camp life, was used between the Roman and modern sway, and probably never went quite out of use. In the Dark Ages, the home was doubtless adapted to the conditions of life in the strong, rough houses, which were intended as safeguards against attack. The family lived in one great hall. It was sleeping room, dining room, living room. If a guest came, his bed was screened off for him. There was but one chair, a mere box, with a six-inch railing around three sides. It belonged to the master, and was a seat of honor. If a superior visited the castle, it was relinquished to him; if an inferior, the master retained his seat, and the guest seated himself upon a bench, which was only a plank supported by side pieces. The table which succeeded the bench appears to have been a number of boards bound together and laid upon folding trestles; and this is perhaps the reason why the word board is used as synonymous with table. The horseshoe form of table had been preserved from the conquered southern race, and was spread upon occasions of great ceremony. As times grew more quiet, the lord of an estate could afford to increase the evidences of his wealth, so the chest grew into the cabinet, the bench into the chair, and was enriched with carvings and expensive coverings and cushions, until, in the fifteenth century, we find the beautiful and useful combined in a pleasing and artistic manner.

Woman's influence was powerful in effecting these changes. The priests prevailed upon the men of the northern countries to practice monogamy, and celebrated the marriage with the most sacred ceremony. They honored woman, and made her honorable in the husband's estimation, and through her obtained an influence over her husband which they would not otherwise have had. Under the feudal system the husband was compelled to make his wife a partner, because, while away from the castle, subduing the enemy, he must necessarily leave all his interests in her hands, and make her thoroughly acquainted with his business. History gives many accounts of the bravery of woman in defending her husband's possessions in those perilous times. Her lord, realizing her ability to manage affairs, allowed her to remain in command while she whispered in his ear the rumor of some fortunate dame who possessed a square of carpet, which had come all the way from Persia, or of the flowered leather and carved chair; and she gently intimated that his own wife was as deserving and noble a helpmate. Lovely woman and patient perseverance were as successful in past ages as in the present generation, and as early as the thirteenth century, we begin to find a change in the home life. Gentle pleasures, wealth, and elegance are often met with, and woman makes herself and her home lovely, with lovely surroundings.

The word *parlor* is obtained from the Norman word *parloir*. In primitive times the Normans entertained their friends in bedrooms, but, as time and civilization advanced, the reception room or *parloir*, which means talking-room, was added to the house, and it now seems a necessity to every housewife's happiness. Here it is that we find the choicest treasures, the dainty bric-a-brac, the pretty tidies, scarves, et cetera; and if one has time for embroidery, or is even in a small way an artist, many pretty devices will be continually suggested to the mind. Many persons entertain the mistaken idea that beauty can only be obtained by a profuse outlay of money. On the contrary, beauty is largely independent of expense. When one is possessed of a moderate amount of what is called taste or aptitude, very satisfactory results can be obtained with a small outlay of money. Select furniture best designed and best made that can be afforded, all of it being intended for use. These wants being provided for, then admit the ornaments of life. A piano or organ adds greatly to the attractiveness of a parlor, and much to the enjoyment of the household. A few pictures, engravings, and books are a necessity. They need not be many or expensive, but use the greatest care in making a selection, and choose only those that contain true worth.

There is some danger of depending too much upon furniture and bric-a-brac for ornament, and not enough upon things permanent and interesting. Seek individual expression of one's own way of living, thinking, acting, more than doing as other people are doing, and having what other people are having. Harmony should prevail in colors; also in the entire furnishing. The decorations of the walls, or papering, the furniture, the entire room, should blend together in a way that is pleasing, and will give a feeling of rest and happiness to the home circle, and whoever is fortunate enough to be entertained within its walls.

When there is an abundant supply of money, a dining room and a library should be distinct and separate features of the home, and this same abundance will furnish these rooms in the approved manner. While these are desirable, do not entertain the idea that they are necessary for comfort or happiness. Expend money carefully, with a view of obtaining service and durability from your investment, *and avoid debt.* A lady who left a father's elegant home for one of her own, whose first furniture bill was only fifty dollars, and whose parlor, sitting, and living room in one was covered with a rag carpet, while from the windows hung curtains made from an old white dress, has been heard to say that, while the home soon outgrew its modest surroundings, and better and more expensive furniture entered it, there never entered with it more of comfort or real happiness than came with the unpretending rag carpet.

When economy is an object, and it becomes necessary to combine the dining room, library and living room in one, it should be the largest, sunniest room the house contains. The furniture should be solid, substantial, and serviceable. A window of growing plants is a great attraction. A broad shelf, from which is suspended a pretty lambrequin, answers admirably for a sideboard, and here can be placed the choice and dainty pieces of silver or china. A bookcase, which can be constructed of a narrow dry goods box, with curtains hung from a pole, to hide its roughness, will do very well. At all events, begin the foundation of the home and library simultaneously. The living room is an important agent in the education of life, and it is no trifling matter that worthless books occupy the tables and shelves, and that poor pictures and engravings are hung on the walls. As well say that it makes no difference what friends you select. Interest the children in newspaper clippings and scrapbooks. Much useful information can be preserved in this inexpensive way. If not very inconvenient, allow them space for a cabinet, which may consist of a few drawers or shelves, where they can, in time, collect many curiosities, which will be an unfailing source of entertainment. Whatever retrenchment is necessary, make it an infallible rule to add a few good books to the home each year, for it is a wise economy. An early cultivation and love of good reading have saved many a boy from the enticing snares of the saloon, many a girl from a light, frivolous life. Through the medium of books, vast chasms of ages may be spanned, and an acquaintance becomes possible with the mighty intellects who lived, moved, and had their being when the world was young. Socrates, Plato, Demosthenes, Solon and his wisdom, Pericles and the brilliancy of his age, may all be ours, even though that wondrous thing, the spirit, has vanished like the morning dew.

R. M. Humphrey

Though the following "History of the Colored Farmers' National Alliance and Cooperative Union" was recorded by white superintendent and former Alabama infantry Captain General R. M. Humphrey, the accounting offered here is a bold one. Humphrey's history recollects, with some editorializing, a period from 1886, when the Colored Farmers' Alliance was officially organized, to 1890, when the Colored Alliance, boasting 1.2 million members, met with the National Farmers' Alliance in Ocala, Florida. In fact, Humphrey paid dearly for his modest efforts to represent the cause of African American cultivators, losing his bid for the second Congressional seat in Texas to W. H. Martin, who accused Humphrey of "slipping around at night with lantern in his hand trying to organize the Negroes into alliances."

Like the Farmers' Alliance he represented, General Humphrey's accounting of the Colored Alliance is both explicitly racist and admirably egalitarian, as evidenced in the following, conflicted description, which simultaneously acknowledges the oppression of slavery and codifies the racial inferiority of African Americans: "The new order had no money, no credit, few friends, and was expected to reform and regenerate a race which, from long endurance of oppression and chattel slavery, had become exceedingly besotted and ignorant." Historical bigotry aside, the story of the Colored Farmers' Alliance, does, in part, evidence an Alliance committed to brotherly aid and progressive racial attitudes, as evidenced by Humphrey's hopeful but naive claim that the Alliance would be "known in future ages as the burial of race conflict, and finally of race prejudice." A document of significant historical import, Humphrey's narrative documents the successful establishment of agricultural exchanges, where cash capital was made readily available to African American farmers, and details the establishment of black newspapers and agricultural academies that would predate widespread national movements along the same lines.

History of the Colored Farmers' National Alliance and Cooperative Union[63]

The Colored Farmers' Alliance had its origin in Texas. The first subordinate Colored Alliance was organized in Houston County, in that state, on the eleventh day of December, 1886. Immediately following this, a number of others were organized in Houston and adjoining counties. The necessity for general organization soon became apparent. Accordingly, these several Alliances chose delegates to a central convention, which assembled in the Good Hope Baptist Church, at Weldon, on the twenty-ninth day of the same month. After some discussion and earnest prayer, it was unanimously agreed that union and organization had become necessary to the earthly salvation of the colored race. The convention then proceeded to adopt the following declaration of principles: "1) To create a body corporate and politic, to be known as 'The Alliance of Colored Farmers of Texas.' 2) The objects of this corporation shall be: a) To promote agriculture and horticulture; b) To educate the agricultural classes in the science of economic government, in a strictly nonpartisan spirit, and to bring about a more perfect union of said classes; c) To develop a better state mentally, morally, socially, and financially; d) To create a better understanding for sustaining our civil officers in maintaining law and order; e) To constantly strive to secure entire harmony and goodwill to all mankind, and brotherly love among ourselves; f) To suppress

---

[63] N. A. Dunning, ed., *The Farmers' Alliance History and Agricultural Digest* (Washington DC: Alliance Publishing Company, 1891), 288–291.

personal, local, sectional, and national prejudices, and all unhealthful rivalry and selfish ambition; g) To aid its members to become more skillful and efficient workers, promote their general intelligence, elevate their character, protect their individual rights; the raising of funds for the benefit of sick or disabled members, or their distressed families; the forming a closer union among all colored people who may be eligible to membership in this association."

This declaration was promptly signed by the following colored men, being all the delegates present: H. J. Spencer, William Armistead, R. M. Saddler, Anthony Turner, T. Jones, N. C. Crawley, J. W. Peters, Israel McGilbra, G. W. Coffey, Green Lee, J. J. Shuffer, Willis Nichols, Jacob Fairfax, Abe Fisher, S. M. Montgomery, John Marshall.

J. J. Shuffer was elected President, and H. J. Spencer, Secretary. Suitable committees were appointed to draft a constitution and bylaws, a ritual, and a form of charter. After receiving the reports of these committees, it was agreed that the Colored Farmers' Alliance should be a secret association. R. M. Humphrey of Lovelady was elected general superintendent, and to him was committed the work of organization. The new order had no money, no credit, few friends, and was expected to reform and regenerate a race which, from long endurance of oppression and chattel slavery, had become exceedingly besotted and ignorant.

On the 28th of February, 1887, a charter was obtained under the laws of Texas, and the organization assumed definite shape as The Alliance of Colored Farmers. The work now spread with great rapidity over the State of Texas, and was soon introduced into several of the neighboring states. The colored people everywhere welcomed the organizers with great delight, and received the Alliance as a sort of second emancipation.

On the 14th of March, 1888, a meeting of the states convened at Lovelady, Texas, and after some discussion, agreed to charter as a trades union, in accordance with the laws of the United States. The new association adopted the Texas state work, with only such changes as were necessary to give it national character. The new charter was duly filed in the office of the recorder of deeds for the District of Columbia, in compliance with the laws of Congress, and will be found recorded in Book IV., at page 354, "Acts of Incorporation, United States of America." Under this new arrangement, the Alliance continued to thrive.

About this time, leading minds among the colored people in the South began to realize the importance of a better system of cooperation. They were desirous, too, of utilizing and, as far as possible, extending the benefits of their organization. The national trustees addressed the following communication to the general superintendent:

> *Lovelady, Texas, July 20, 1888*
>
> *To the General Superintendent of the Colored Farmers' National Alliance:*
>
> *Sir: Upon receipt of this order you will at your earliest convenience proceed to establish such trading post, or posts, or exchanges, for the use and benefit of our order in the several States, as in your judgment will be most conducive to the interest of the people. We leave you to adopt such plans as in your opinion will he most effective.*
>
> *With much respect yours,*
>
> *J. J. Shuffer, President*
>
> *H. J. Spencer, Secretary, Colored Farmers' National Alliance and Cooperative Union*

In compliance with this order, exchanges were established in Houston, Texas; New Orleans, Louisiana; Mobile, Alabama; Charleston, South Carolina; and Norfolk, Virginia. These institutions, with varying success, are still in existence, and have accomplished great things for the elevation of the colored race. Occupying as these posts do, the great centers of the country's commerce, we are not without hope that they will be, in the future as in the past, well-supported by the people. Our method in their establishment is this: an assessment of $2.00 is levied upon each male member of the order, within prescribed boundaries, for the benefit of the exchange within his territory. These small amounts paid by each member become a cash capital for the basis of our business operations. The money may be used to buy a stock of bacon, or to pay off a mortgage, and being at once replaced, is ready the next week for some similar investment. Being thus often turned over, it will, in a year, save many times its value as against the speculator, who always reckons the term of a credit at twelve months, and the rate of interest at 50 to 100 percent, though the actual time of such credit may be only from August till September.

Again, this kind of cash basis is not exhausted nor exhaustible; fifty or a hundred years hence it may be still present to do the same work it is now doing; or should the Colored Alliance cease or become extinct, the funds on hand might be turned to the endowment of schools or colleges for colored youths, and so render a perpetual service during all time.

With the beginning of 1889 the Alliance established a weekly newspaper, called *The National Alliance*. They designed it for the practical education of their members. It has been reasonably well-supported, and is still published weekly, at Houston, Texas, each of its editions reaching many thousand colored families.

At this writing, Texas, Louisiana, Mississippi, Alabama, Florida, Georgia, South Carolina, North Carolina, Virginia, and Tennessee have State Colored Alliances, working under State charters. Several other States expect to be chartered at an early day, while organizations of greater or less extent exist in more than twenty States. The total membership is nearly 1,200,000, of whom 300,000 are females, and 150,000 males under twenty-one years of age, leaving 750,000 adult males.

It is freely admitted by all that the colored people have made great strides forward in intelligence, morals, and financial standing during these years of organization. Thousands of their public free schools have been wonderfully improved in character of teaching, and the duration of their sessions much extended by the combining of the people, and the payment by each member of the Alliance of a small sum in the form of tuition. Very many Alliance academies and high schools have been opened in various sections of the country. In not a few communities the people, impelled by the higher cultivation of their social instincts, have built new places of worship, while the intellectual and moral grade of their pastors and teachers has been immeasurably advanced.

The relation of the colored people in the South to their white neighbors had been long a question of the last importance to both races. There were not wanting those who believed in race conflict, race war, and even race extermination. These beliefs and opinions were shared by some of the best people on both sides, as, perhaps, painfully inevitable results which must follow from existing conditions, but there were others who were in apparent haste to put their views into practical operation, and who, if judged by their own testimony, were ready to baptize their prejudices in the blood of their fellow beings, and dishonor themselves by the destruction of their country. The Alliances, both colored and white, were organized from the first largely with a view to the suppression of all prejudices, whether national, local, sectional, or race, and to create conditions of peace and goodwill among all the inhabitants of our great nation. On this account the "race question" was from the beginning a matter of profoundest interest to the order. At the first practicable moment steps were taken looking to the peaceful solution of that much vexed and intricate problem.

On December 3, 1889, the representatives of the Colored Farmers' National Alliance convened in the city of St. Louis. During this session they were visited by committees of fraternal regard from the Farmers and Laborers' Union, the Farmers Mutual Benefit Association, and the National Farmers' Alliance. These visits were acknowledged with the utmost goodwill, so that the messengers from the several brotherhoods were looked upon rather as ministers of light and salvation. Like committees were appointed from our body to visit and bear our goodwill and fraternal greetings to these several organizations.

Again, in Ocala, Florida, at which place their National Council was held in December, 1890, they were visited by committees from the Farmers and Laborers' Union, and by officers of the Knights of

Labor, and by members of other labor associations. They appointed committees to each of these bodies as bearers of their goodwill and fraternal regard. They further proposed the holding of a joint meeting by these committees to form an association or confederation of the several orders represented, for purposes of mutual protection, cooperation, and assistance. The committees, in their joint session, found themselves able to agree, and the matter of their agreement being reported back to their several orders, was heartily endorsed by all concerned. It recognizes common citizenship, assures commercial equality and legal justice, and pledges each of the several organizations for the common protection of all. This agreement will be known in future ages as the burial of race conflict, and finally of race prejudice. Its announcement has fired many hearts with renewed hope, has given a new impetus to progress among the people, and will exert tremendous influences in the healing of sectional and national misconceptions and prejudices throughout the entire country.

HELEN GILBERT ECOB

The wife of a prominent clergy member in Denver, Helen Gilbert Ecob played an important role in the suffragist movement, an influence acknowledged in *The Blue Book* published in 1917 by the National American Suffrage Movement Association. Thus, though the title of Ecob's *The Well-dressed Woman* implies a superficial focus on appearance, the vision she presents in the chapter below, "Physical Development," is both proto-feminist and prophetic. Predicting a world where "physical culture will soon be summarily insisted on," she foresees, in 1892, pervasive physical education in the schools, at home, and at the workplace, for women as for men. Her unequivocal and sobering statement about conditions in the nineteenth century strangely echoes current claims: "But the ideal life, which is also the natural life, is not possible in the nineteenth century. From the cradle to the grave we are surrounded by deteriorating influences."

In fact "Physical Development" addresses only briefly women's dress, as it focuses instead on a woman's overall health and well-being, serving as an expression of Ecob's larger aim, expressed in the book's preface, to help "the struggling minds and hearts of my countrywomen toward a true emancipation of body, as well as of intellect and soul." Some fifteen years ahead of a Country Life Commission Report that reached similar conclusions, Ecob characterizes "women's work," so-called, as inadequate for her physical, emotional, and spiritual life. Challenging the agrarian myths of healthful living relative to the cities, Ecob points out that farm women remain "mewed up within four walls" and, more dramatically, that "the rosy-cheeked country girl is a poetic fiction."

Veblen-like in its critique of a leisure class destined, by its very excess, to invite poor health, and anticipatory of twenty-first century simplicity movements, Ecob's meditation on physical fitness stands up well more than a century later.

Physical Development[64]

> *Le physique gouverne toujours le moral.*
>
> *—Voltaire*
>
> *It is impossible to repress luxury by legislation, but its influence may be counteracted by athletic games.*
>
> *—Solon*
>
> *Abashed the devil stood, And felt how awful goodness is, And saw Virtue in her shape how lovely.*
>
> *—Milton*

Were it practicable to live an ideal life no study of physical culture would be necessary. Our daily activities would furnish the proper development. This is Ruskin's picture: "An ancestry of the purest race, trained from infancy constantly but not excessively in

---

[64] Helen Gilbert Ecob, *The Well-dressed Woman: A Study in the Practical Application to Dress of the Laws of Health, Art, and Morals* (New York: Fowler & Wells, 1892), 160–175.

all exercises of dignity—not in twists and straining dexterities, but in natural exercises of running, casting, or riding; practiced in endurance, not of extraordinary hardship, for that hardens and degrades the body, but of natural hardships, vicissitudes of winter and summer and cold and heat, yet in a climate where none are too severe; surrounded also by a certain degree of right luxury, so as to soften and refine the forms of strength." But the ideal life, which is also the natural life, is not possible in the nineteenth century. From the cradle to the grave we are surrounded by deteriorating influences. The well-born child is our only embodiment of physical beauty; muscles relaxed, body perfectly poised, every movement is full of unconscious grace. At an early age, inherited muscular and nervous peculiarities begin to show themselves. Through ignorance, indifference, the subtle power of imitation, and defects of character he loses the erect carriage and falls into slouchy ways of sitting, standing, and walking. Physical faults are seldom corrected. The occasional injunction, "Throw back your shoulders," does not secure erect carriage, and the persistent care exercised to secure right-handedness results in one-sided development. Teachers of physical culture mention cases in which the left shoulder had fallen two inches through lack of exercise. Not only the left hand, but the whole left side of the body, is crippled by the unfortunate supremacy given to the right hand. Dr. Richardson tells us that right-handedness and right-sidedness are also registered in the brains; "the left side of the brain, which supervises the right side of the body, being generally the larger."

In childhood, when physical development should be of paramount importance, the precedence is usually given to brain culture. Children are shut up in stuffy schoolrooms, cramped into badly constructed seats, allowed to stoop over books in a way which depresses the chest, while the grading system, the examinations, and other stimuli used to excite ambition are a great draught on the nervous system. Very few children enter adult life with well-knit, harmoniously developed bodies, and the formative period is as vital for physical as for moral well-being.

The occupations of adult life necessitate, on the one hand, excessive and monotonous manual labor, and, on the other, insufficient exercise and mental strain. A familiar example of the first class is the farmer, and nearly one third of our population are farmers, whose work develops the muscles of the back and arms, while the chest is contracted and narrow. We seldom see a farmer who stands erect or walks with elastic step. The farmer's son develops the same characteristics before he reaches his teens. There is now an increasing tendency to city life.

Multitudes crowd the factory, the counting-room, the artisan, mercantile and professional life, where they are shut away from sunshine and fresh air. In these avocations there is a tendency to chest contraction, and no opportunity for muscle-making exercise. The use of machines, the specialization of labor, the increase of wealth with its attendant luxury, the competition and complication of business life, the intellectual and emotional character of our amusements, are pointed out as causes of physical deterioration and increase of nervous disease. It is urged that systematic physical training be made compulsory in our public schools. The end sought is not heavy gymnastic feats, but harmonious development, strong nerves, and clear brains.

Well-built bodies do not come by chance. The physical superiority of the ancient Greeks was not an accident. For years the entire nation had given itself to the training of the body as a religious service. Their women also were exercised in games of running, throwing, and casting. When the deteriorating influence of increasing wealth made itself felt, the public authorities endeavored to counteract the threatening evil by physical education. Should we today undertake the education of the body with the same earnest spirit, the same results would be accomplished. It is the purpose of physical culture to correct the deviations from the law of our being which result from abnormal conditions. In its full sense, the subject of physical culture includes the study of food, clothing, ventilation, bathing, exercise, and rest. In a technical sense it concerns the development, relaxation, and guidance of the muscles, and the development and strengthening of the nerve centers which control them; the needs of each individual are studied, defective attitudes are corrected, and appropriate exercise is given.

Systematic exercise, under intelligent supervision, is as necessary for physical as for mental growth. We recognize the necessity of exercise in the physical development of boys, but practically ignore it in the development of girls. Herbert Spencer contrasts the playground of a boys' and girls' school:

> In the case of a boys' school nearly the whole of a large garden is turned into an open, graveled space, affording ample scope for games, and supplied with poles and horizontal bars for gymnastic exercises. Every day before breakfast, again at midday, again in the afternoon, and once more after school is over, the neighborhood is awakened by a chorus of shouts and laughter as the boys rush out to play; and for as long as they remain both eyes and ears give proof that they are absorbed by enjoyable activity which makes the pulse bound and insures the healthful activity of every organ. How unlike is the picture offered by the establishment for young ladies! Until the fact was pointed out, we actually did not know that we had a girls' school as close to us as the school for boys. The garden, equally large with the other, affords no sign whatever of any provision for juvenile recreation, but is entirely laid out with prim grass plots, gravel walks, shrubs, and flowers, after the usual suburban style. During five months we have not once had our attention drawn to the premises by a shout or a laugh. Occasionally girls may be observed sauntering along the paths with their lesson-books in their hands, or else walking arm in arm. Once, indeed, we saw one chase another around the garden; but, with this exception, nothing like vigorous exertion has been visible.

In our cities we cannot, it is true, have a free, open-air life, but dumbbells and calisthenics, we must remember, are poor substitutes for natural outdoor pastimes.

The contrast in activities, which result in vigorous or enervated life, marks every period a woman's existence. Her work gives a tendency to chest contraction, is sedentary, and is uniformly conducted within doors. Few American women know what active exercise is, for "the laboring class" is usually of foreign birth. The great portion of American women are housewives and do their own work. Only about one family in twelve employs domestic service. Then we have art army of shopgirls, factory hands, seamstresses, teachers, whose nerve-exhausting work has no antidote of exercise. These, with a few in the well-appointed homes of the upper classes, who are physically starved by luxurious idleness, make up the bulk of female population.

Whether housework affords a sufficient amount of exercise depends somewhat on the character of the work. The washtub and a Brussels carpet certainly call forth muscle-developing action, but dishwashing, dusting, "picking up," mending, and the little round which keeps the housewife eternally occupied, require no blood-stirring exertion. She may be wearied to the point of exhaustion with this monotonous routine, yet have no vigorous muscular action. Housework is better than no work. Tolstoy is right: manual employment is necessary even for brain workers. It is better to cobble second-rate shoes than to do no work. Housework is an important part of the domestic economy, yet it is despised by all classes of women. Those who are obliged to do it usually chaff under the drudgery. Those who can, delegate it to others. It is to be hoped that the present interest in physical culture will give dignity to this department of labor. The gymnastics of bedmaking, which even a society girl can indulge in, exercise the muscles of the arms, side, and back. Sweeping the parlor carpet or shaking the rugs will start the circulation as well as a chest weight, and has the advantage of being useful.

Childhood is the period ordained for health-giving pursuits. Freedom from care, exuberance of spirits, elastic muscles, are nature's indications of the golden age for physical culture. Realizing the importance of the formative period of life, Plato, in his ideal republic, prescribes for the pastimes of youth. With girls active, open-air sports are especially necessary as a counterbalance for the shut-in life of maturity, yet these recreations society tacitly frowns upon. Girls are trained from babyhood into sedentary habits; the doll is considered her legitimate plaything. It is tacitly understood that girls are born with a passion for dolls, and boys with a passion for animals. The boy who finds amusement in the caressing and dressing of dolls is called a girl-baby. He is looked upon with anxiety as of a maudlin nature, and as not possessing the animal spirit which calls for active amusement. This maudlin nature is just as lamentable in a girl, but custom blinds our eyes to the fact. The care of a doll is said to develop the mother instinct and to teach the care of the wardrobe. Why the mother instinct should require premature cultivation in girls and not the father instinct in boys society does not say. Amusements which are the prerogative of boys are also the prerogative of girls. To run, jump, climb, hammer, go fishing, play ball, slide down hill, skate, if they were allowed freedom of choice, would be as instinctive in girls as in boys. Instead of calling the girl who loves these sports a tomboy, we ought to rejoice that she has the physical strength to enjoy them. If she lacks this strength, the same effort to improve her vitality should be made as is made for boys.

Much of the indoor life of women is willful. Even in the country, with nature's beautiful invitation to ozone and sunlight constantly in view, women are usually mewed up within four walls. The rosy-cheeked country girl is a poetic fiction. Women show the same willful neglect of exercise; in fact the woman who is called upon for active muscular exertion feels herself abused, for heavy lifting she always appeals to masculine aid. Men would doubtless lose muscular power if they refused to use the means by which strength is developed. A certain woman of sedentary habits and predisposed to pulmonary troubles tried the experiment of chopping wood a short time every day, and the woodpile saved her from the grave. Five-finger exercises at the piano do not compensate for lack of exercise in the thousand muscles of the body.

The proper amount of fresh air, exercise, and rest is possible only to a privileged class; the majority cannot regulate their lives to meet the requirements of hygiene. An army of women fighting for daily bread with only the weapon of their hands, an army of women prematurely aged by drudgery has no time for aesthetic physical culture. It must be admitted that the problem of the working classes lies in the readjustment of social questions. Much can be done meantime to alleviate unfavorable conditions. If women are compelled to live within doors, the most careful attention should be given to the ventilation of their dwellings. Carpets and draperies which retain dust and impure emanations should be banished. Sunshine and fresh air should be freely admitted. A third part of our life is spent in sleep. By opening the windows of the sleeping room, one can practically sleep outdoors the year round. Opening the windows of the bedchamber does not mean one window grudgingly thrown up a quarter of an inch; it means windows open to the full height. We should rise superior to the old woman's theory that night air is bad air. It is the only air we have at night, and it is usually purer than day air. Our double windows and our battened doors cheat us of a heaven-given birthright. No woman is so destitute that she cannot have fresh air and sunshine. The real difficulty is that all of us, rich and poor alike, prefer hot, stuffy dwelling rooms.

An important means of alleviating woman's burden of drudgery is to distinguish work which is really important from that which is superfluous. A great deal of woman's work is unnecessary; it is the frills and furbelows of cooking and housekeeping and sewing. Much undue labor is caused by foolish ambition and the sordid desire for accumulating property; we are not content having food and raiment. Much is due to ignorance of the laws of muscular economy; the seamstress laboring to keep the wolf from the door dares not allow herself a moment for recreation, when to draw a long breath, to stretch a tired muscle, would impart fresh power to continue.

The chief obstacle in the way of introducing correct dress is the physical condition of women. The stout wear tight clothing because of a mistaken idea that pressure diminishes apparent bulk. The thin wear the corset to conceal and fill out a meager body. The present interest in dress contemplates not only improvement in clothing, but the formation of erect bodies under the control of flexible muscles. A corpulent body is not to be stayed up by lacing, but must be taught to sustain itself and must have room for easy movement. Defects of the body are not to be concealed by dress, but overcome by healthful exercise. A narrow chest is not to be padded, but developed.

Since corpulence and leanness are the two great obstacles in introducing correct dress, a careful study of the natural laws which prevent and develop adipose tissue is important. Thinness and stoutness are not altogether accidental, but are the normal results of the conditions of life. As a rule, the lean are the hard workers and frugal livers; the stout are the indolent and well-fed. Physical development is to a great degree an inheritance, but this is the general law. The pounds avoirdupois usually give a clue to the temperament and avocation. The laboring class, whether the work be manual or mental, have a lean, hungry look. The denizens of society are fat and flourishing. They that dwell in king's houses wear the insignia of adipose tissue.

This is no new theory; it is as old as Lycurgus. We recognize the law in the lower orders of life. Poultry to be fatted for the market is shut up and corn-fed. The stalled ox tells its own story. The market furnishes us pate de fois gras at the expense of the Strasburg goose, overfed to the point of liver complaint. The Italians have a device for inducing the ortolan to sleep and eat constantly, and by this means the poor bird in three days becomes "a delicious ball of fat." Too many oats and too little exercise will make the best horse fat and lazy. A well-known New England educator has a favorite horse which he is in the habit of weighing every week in order to ascertain his physical status. At one of the customary weighings the scales showed a loss in the animal of one pound. The master gave orders immediately that the horse should not be driven for a week, until his normal weight was recovered. An establishment in Silesia advertises to make the thin fat. The gain guaranteed is a half pound daily. The patients are fed eight times per day. They eat slowly to the time of music. Exercise is regulated according to the need of the patient. Dr. Sargent says: "Appropriate exercise for the waist will soon reduce superfluous fat, and healthy muscle will take the place of the corset in supporting the bust and giving uprightness to the figure. One object of physical training is to keep down or reduce superfluous flesh." Obesity shows an unhealthful state of the system. It is a mild form of gout, and should be overcome by eradicating the disease. One may see the absurdity of attempting to reduce flesh by the use of the corset if one imagines a corpulent man attempting to lace in redundant fat. Excessive leanness also shows an impoverished, arid, unhealthy condition of the system. It is easier to reduce flesh than to acquire it, for the reason that self-denial, frugality, and exercise are possible to every one, while proper nourishment, physical ease, and mental repose are possible to few.

The first and vital step in physical culture is to overcome departures from nature in the common habits of life—breathing, standing, walking, and sitting. Women have, through weakness, indolence, acquired habit, and the restrictions of dress, assumed an unnatural carriage which has become second nature. Hardly one can be found who holds herself or walks in a natural or normal manner. An effort to assume the erect position and normal poise causes at first a feeling of constraint and awkwardness. The chest is sunken, the head dropped forward, the neck curved, the abdomen protruded, and the base of the back correspondingly depressed. In standing the weight of the body is thrown on the heels; in sitting the weight of the trunk is thrown on the spine. The breath is short and quick, seldom calling the diaphragm into action. These violations of the law of our being invite disease. The vital organs are thrown out of position, their altitude is lowered, their functions can go on only with abnormal strain, resulting in permanent weakness and disease.

Women must persistently, faithfully, systematically, begin a physical reformation. But how are these faults to be corrected? With a competent teacher one reaps the same advantage in this as in other departments. Opportunities for study are every day becoming more general, and physical culture will soon be summarily insisted on in our school as is now the "Three R's."

Edwin J. Houston

Edwin James Houston's "Destruction of the Forest" logically persuades of the many benefits of forest preservation. Houston, a member of the Pennsylvania Forestry Association and a professor of natural philosophy and physical geography at Philadelphia's Central High School, here combines a geographer's penchant for inventory with a philosopher's concern for situational ethics. Houston's main claim to fame was as the inventor of the arc-light generator which was sold by the Thomson-Houston Company that bore his name. Houston left the company in 1882 to focus on his environmental teaching, though he continued to publish books on electricity throughout the 1890s, including his most influential treatise *Electricity One Hundred Years Ago and To-day* (1894).

Simultaneously avoiding the role of doomsayer and Pollyanna, Houston's aim in the *Outlines of Forestry*, from which this passage is drawn, is not to argue against the economic use of forests and forest products by mankind, but to suggest methods by which such forests might be preserved or replaced through scientifically-guided reforestation. Houston's disarmingly simple observation that "regions best fitted for the growth of men are also best fitted for the growth of trees" suggests, on further consideration, antagonisms between man and nature, in a world of finite resources, that would only increase in the nineteenth and twentieth centuries. Houston's claims linking forestation with rainfall are interesting in light of the scientifically dubious "rain follows the plow" notion that spurred the rapid settlement of the Great Plains during the author's life. In the few years leading up to the publication of *Outlines of Forestry* in 1893, the Plains, in particular, continued to face perennially low precipitation and high wind erosion that reaffirmed, for some, the region's reputation as a "Great American Desert." Houston's ecological approach, grounded in physical geography, offers an unusually eclectic treatment of the far-ranging problems of ill-considered forestry.

Destruction of the Forest [65]

The removal of the forests from any considerable section of country, in the end, is invariably followed by some or all of the following results:

1. An increase in the frequency with which the rivers in that section of country overflow or inundate their banks.
2. An increase in the frequency and severity of droughts, as witnessed by a marked decrease in the amount of water in the river channels, and by an increase in the frequency with which the springs, in such section of country, either show a marked decrease in their flow or dry up altogether.
3. A rapid loss of the soil from such areas, resulting from the more rapid surface drainage of their surfaces.
4. A marked disturbance in the lower courses of the rivers, rising in or flowing through such section of country, produced by the filling up of their channels by sandbars or mudflats.
5. A decrease in the healthfulness of the district that borders on the lower courses of such rivers, that is, in those portions which lie in the lowlands near the rivers' months.
6. An increase in the number and severity of hailstorms, both over the areas themselves or in the countries bordering thereon.

When the forests are removed from any section of country, that part of the rainfall which formerly entered the ground, either by gradually sinking into the porous soil, or by running along the branches and trunks of the trees, and so entering and penetrating the more deeply-seated strata, now drains rapidly off the surface. Instead of reaching the river channel quietly and slowly through discharge from

---

[65] Edwin J. Houston, *Outlines of Forestry, or, The Elementary Principles Underlying the Science of Forestry: Being a Series of Primers of Forestry* (Philadelphia: J. B. Lippincott, 1893), 81–86.

the reservoirs of springs, it now rapidly drains directly off the surface into the river channel.

Instead of draining into the river channel continuously for a period of, say, three weeks, the rainwater now drains into the channel in often a period of as many hours. The channel rapidly fills, the river overflows its banks, and the floods so caused carry loss to the lowlands along the riverbanks, and, not infrequently, death to the inhabitants.

Not only are the riches of the rainfall thus squandered, to the loss of the inhabitants of the river valleys, from the excess of water immediately after a rainfall, but a still greater and more far reaching loss occurs from the failure of the rainfall to fill the reservoirs of the springs, the continuous discharge of which are necessary to maintain the proper flow of water in the river.

The springs, having their reservoirs but partly filled, are apt to fail shortly after the rainfall ceases, so that even limited droughts may cause them to dry up completely.

The damage, however, does not stop here. The soil in which the forest grew, being no longer held together either by the roots of the trees or under brush of the forest, or protected by a vegetable covering, is rapidly carried away by the water. The soil thus lost, resulted from the gradual disintegration of hard rocks, and contains as essential elements substances derived from the continued growth of former generations of plants, and probably required centuries for its production. Its removal in a few years is, therefore, a serious matter.

The soil, the wealth of the highlands, is now thrown into the river channel, and though some of it fertilizes the lowlands, over which it is spread during inundations, yet much collects in sandbars and mudflats on the lower courses of the river. These flats work injury because:

1. They hinder navigation, and thus interfere with the commerce between different parts of the country.
2. They become sources of contamination to the air of the lowlands, by breeding miasmatic and other diseases.

Besides the disturbances thus caused to the drainage of the region from which the forest has been removed, considerable changes are brought about in the rate at which the now bare soil receives the heat from the sun, and the rapidity with which it throws it off into the air.

Areas covered with forests both receive and part with their heat slowly, and are, therefore, not very apt to become very hot in summer, or very cold in winter.

Bare areas, or areas stripped of their vegetable covering, both receive and part with their heat rapidly, and are, therefore, apt to become very hot in summer and very cold in winter.

The presence of the forest, therefore, tends to prevent marked changes in the temperature of the air, while the removal of the forest tends to permit sudden changes in such temperature. These effects will be considered under the general head of climate.

The axe of the pioneer, so often regarded as the emblem of civilization, is more correctly to be regarded as an emblem of an entirely different character. The problem of the preservation and protection of the forest is one of extreme difficulty, for the following reasons:

The dense populations which now exist in most of the temperate regions of the earth could not continue to exist in the forest regions which once grew on large parts of their areas.

The regions best fitted for the growth of men are also best fitted for the growth of trees. Since civilized man cannot continue as a dweller in the forest, as the density of population increases, the forest must be cut down.

In removing the forest to make way for man, certain areas should be set aside in all sections for the purpose of perpetually maintaining trees thereon. The nature of such areas will, of course, depend on a variety of circumstances. In general, however, it can be shown that, on the slopes of mountain ranges, which form the natural places where rivers rise, forests should be especially maintained.

Laws should, therefore, be enacted providing for the replanting of trees on mountain slopes, either when they have been removed by the axe of the woodman, or by fire, or by any of the other enemies of the forest.

THOMAS SHAW

A Canadian employed by the Ontario Agricultural College Experimental Farm, Thomas Shaw considered himself, as did so many of the writers of the Golden Age of agriculture and conservation, kin to the practical farmer more than the theoretical scientist. As "The Prevalence of Weeds" makes clear, Shaw genuinely believed that, by force of will, a farmer could have a "perfectly clean" farm without the noxious weeds. In fact, he believed that farmers, working in concert the continent over, could render the most invasive weeds virtually obsolete. Shaw's own green thumb may have had something to do with his bravado, as he would publish numerous how-to manuals in the Progressive Era on subjects ranging from clover to grasses to animal breeding. Beyond his *Weeds and How to Eradicate Them*, which merited release in several editions, Shaw's most influential work was his 1911 book *Dry Land Farming*.

The essay that follows could only, it seems, have been written in an agrarian golden age. The subject matter, far from relevant to the era's rising class of urban professionals, was of urgent interest to the farmer battling a series of public enemies—Canadian thistle, wild mustard, and burdock, among others. Biblical in its passions and its allusions, Shaw's work personifies weeds as spoilsports and killjoys to be resisted with every ounce of the planter's energy. Weeds, "a foul reproach upon civilization" in the author's eyes, threatened the agricultural progress that defined Shaw's time. Evoking xenophobic analogies linking noxious weed species to the immigrants who unwittingly introduced them to North American soils, "The Prevalence of Weeds" occasionally troubles, though, on balance, it informs. Contemporary audiences unfamiliar with the scorn evoked by weeds in a pre-herbicide era will find this an enlightening if not paradoxical read, one that, in one sense, demonstrates a farmer's closeness to "nature" and, in another, epitomizes his blood oath to resist it.

The Prevalence of Weeds[66]

> *Cursed is the ground for thy sake; in sorrow shalt thou eat of it all the days of thy life; thorns also and thistles shall it bring forth to thee; and thou shalt eat the herb of the field.*

So reads the doom that was hurled down the centuries from the flaming gates of Eden, when man was ejected from a paradise lost, to earn his bread by the sweat of his brow. From that day to the present, weeds have followed in the footprints of man. He no sooner pitches his tent, or builds his more permanent home, than they entrench themselves around it. He no sooner commences to till the soil than they commence to dispute its possession with the plants that he sows, and thus they harass and perplex him, and complicate all his best devised methods for subduing the earth.

It is true, at the same time, that in lands that are as yet uninhabited, and which, therefore, were never tilled, we find some weeds, but they are indigenous, and the number of the weeds species is not only limited, but those which do exist seem unable to multiply to any great extent in the natural surroundings amid which they grow. On the other hand, in lands that have long been cultivated, we frequently find that foreign varieties of weeds are far more numerous and aggressive than the native species. Regions that have been settled with inhabitants drawn from different countries are peculiarly liable to be smitten with the various weeds which belong to the respective countries from which these inhabitants have come. The seeds of the weeds are imported along with the grain that is brought for sowing, and in various other ways are the foreign weed seeds introduced. Some of the varieties thus imported do not take kindly to the new conditions, but other sorts, like the people who have brought them, oftentimes find their new surroundings pre-eminently favorable to a greatly increased development.

---

[66] Thomas Shaw, *Weeds and How to Eradicate Them* (Toronto: J. E. Bryant Co., 1893), 1–8, 14–18.

The prevalence of weeds has respect, first, to the number of weed species found in any locality and, second, to the extent to which these various species are allowed to multiply.

With regard to the number of the various species of weed life which infest the several provinces and states of North America, we are as yet unable to speak with precision. The botany of not a few of these provinces and states is as yet unwritten. The story thereof, up to the present, has not half been told. It would therefore serve no good purpose, in the present condition of our knowledge, to enumerate the various species, or even to try to give an approximation thereto. Weeds are probably quite as numerous and varied now in the continent of America as in Europe, where it is well known that they have been continually increasing in number and variety with every passing century. But in addition to the noxious weeds of America that are indigenous to the continent, the major portion of those which have long harassed the inhabitants of Europe are now giving trouble to the inhabitants of America, and the weeds of the several provinces of Canada form no exception to this statement.

Our most troublesome and aggressive weeds are foreigners. The Canada thistle, which seems so completely at home in the central provinces of the Canada, was imported from Europe. The same is true of some varieties of the sow thistle. The wild oat, the ox-eye daisy, the ubiquitous burdock, the wild mustard, the wild flax, and indeed nearly all the various forms of weed life that are greatly troublesome to us, come from a foreign source. Foreign weeds in this country are far more numerous and characteristic than the people who brought them hither, and so they are likely to remain, for weeds, unlike nationalities, do not fuse and blend so as to lose their several individualities. For some of them, as, for instance, the Canada thistle, the new conditions have been found so favorable that they flourish to a greater extent than even in the lands whence they came.

Although the presence of weed life in any form is not desirable, some varieties, as, for instance, the dandelion, are not greatly harmful, while others, as the sow thistle, couch grass, and the Canada thistle, if given a chance, will soon render the growing of certain crops quite unprofitable. A large majority of the weeds found in this country may be kept in check by what may he termed good cultivation, that is to say, by such cultivation as is necessary to grow good crops, but, with reference to other varieties, some specific modes of treatment are required if, when the attempt is made to exterminate them, it is to prove successful. Happily, the number of the varieties of weeds which are really seriously harmful to crops, and difficult to eradicate, is not very large. In the present state of our knowledge of the subject, it would not be safe to name a definite number which would cover the entire list, nor would it be judicious to do so, as new varieties are coming forward all the time. Notwithstanding, it would probably be not incorrect to say that, at the present time, the varieties of really noxious weeds in Canada do not number much more than a score, and it is greatly encouraging to reflect that we seldom find more than half a dozen kinds entrenched in any one locality.

The extent to which certain varieties of noxious weeds have been allowed to multiply is simply alarming. Some of them are, in a sense, taking possession of the land. Notably is this true of wild mustard, the Canada thistle, and the ragweed in Ontario, and of the penny cress and the Russian thistle in Manitoba. In several sections of Ontario, the seeds of wild mustard are so numerous in the soil that, though no more seeds were allowed to ripen during the present generation, there would probably still be a few left to grow plants for the next generation to destroy. The penny cress and the Russian thistle have so entrenched themselves in some parts of the Northwest that fears have been expressed that, in consequence, the cultivation of the land there may yet have to be abandoned on account of them. Other varieties than those named are increasing with alarming rapidity. So that, unless some heroic measures are taken to destroy them, they will increase more and more, to the great injury of our agriculture. It is surely a stigma on the agriculture of any country, and a withering criticism on the defectiveness of the modes of cultivation that are practiced in it, when weeds increase rather than decrease. In the hope of doing something to stay the progress of the great tide of weed invasion and weed aggression, this book has been written. Hence the writer cherishes the hope that every interested reader will exert himself to the utmost to stay the progress of weed extension, by doing his best to utterly annihilate weeds in all their seriously noxious forms.

The Possibility of Destroying Weeds. The prevalence of noxious weeds in the United States and Canada is simply alarming. They abound on every hand.[67] In many sections, in one form or another, they flourish in every field, and luxuriate in every crop. Gardens, which, above all places on the farm, should be clean, are literally overrun with them. They occupy the sides of nearly every road throughout the whole continent. To so great an extent do they prevail everywhere that they form one great dark blot upon the boasted progress of the nineteenth century, and are a foul reproach upon civilization.

The extent to which weeds prevail in nearly all parts of this continent would lead one to suppose that the farmers had abandoned all efforts to destroy them, and were content to gather from their fields, in the form of crops, merely what weeds allowed to grow there without incurring so much labor and expense in doing so as to make the work unprofitable. As the matter presents itself to the writer, there is not a shadow of a hope that the weeds of this continent will ever be destroyed by the farmers thereof, so long as their complete eradication is looked upon as impossible, or so long as the belief is harbored that the outlay of labor and expense in completely eradicating them will not be repaid by the greater gains that will be obtained when once their destruction is effected.

Four propositions are now submitted which bear upon the subject of the complete eradication of weeds. So confident is the writer of the soundness of these propositions that he makes them as strongly affirmative as possible. They are as follows: 1) The noxious forms of weed life can be completely eradicated on every farm throughout the whole continent if the farmers of these farms resolve that so

---

[67] This sentence marks the beginning of Chapter III, pages 14–18, "The Possibility of Destroying Weeds," in the 1893 edition, represented here as a subheading of the same name.

it shall be. 2) Complete eradication can be effected without heavy outlay, if the work be done in the proper way. 3) When weeds are once eradicated, it will be easily possible, with but little outlay, to still keep them so. 4) The profits of farming will be, relatively, much larger where farms are kept entirely free from noxious weeds.

The writer is by no means unconscious of the fact that the assertion of these propositions will bring cold scorn to the lips of some who read them, but he finds comfort in the reflection that they will be distasteful to no one whose heart is really set upon the complete eradication of the noxious weeds that may exist on his own farm.

By the assertion that the more troublesome forms of weed life can be completely eradicated, it is meant that they can be so effectually exterminated that they will practically cease to interfere with any rotation that may be desired. More, it implies that they can be completely banished, root and branch, from every farm where the attempt is made, except insofar as their seeds are brought back again by natural or other agencies, and that with the necessary watchfulness the plants which grow from these can, in turn, be destroyed with but little difficulty. Many persons seem to hold the view that while weeds may be held at bay, and kept from seriously hindering the growth of crops, they cannot be wholly destroyed. They claim that while weeds may be thus far conquered, nevertheless they will come again, and therefore that the hope of eradicating them completely is not to be cherished. Those who hold this view shape their practice accordingly. They adopt some method of cleaning a field that proves fairly successful, and then during the years that immediately follow give the same field no further special attention. The consequence is that this field soon again requires to be put through some special cleaning process, owing to the increase of the weeds which were but partially eradicated by the previous cleaning process. If this practice were a good one, it would involve the correctness of the untenable theory that, in correcting error and uprooting evil, it is better to do it partially rather than wholly. So long as the belief is cherished by those who are most interested, that the complete eradication of noxious weeds is impossible, so long will weeds continue to prevail. To so great an extent is this belief indulged in that it would probably be found a greater task to correct it in the minds of many farmers than to uproot the weeds themselves from their fields.

To banish weeds completely from any farm will not only require the wise and diligent use of measures of a certain character, which will be described in succeeding chapters, but, when once they are gone, it will also require the most persistent watchfulness to still keep them away. With public sentiment on this subject as it is at present, it will be found impossible to get those who are most directly interested to act in concert in destroying weeds; hence the work of even materially reducing their numbers will necessarily be slow. The work of banishing weeds from any country would not of necessity extend over many years if all the farmers of the country would but act in concert. The spectacle would then be witnessed, for the first time since Edenic days, of an inhabited country without noxious weeds to harass and annoy the tiller of the soil. But because farmers, at present, cannot all be persuaded to put forth the effort to banish weeds, root and branch, from their premises, no one engaged in agriculture should refrain from doing all that he possibly can to bring about this result. Though our neighbors should not now believe in the possibility of being able to banish noxious weeds from their farms, if our own farms are made clean and kept clean, the evidence thus presented will in time have its due measure of influence.

FREDERICK JACKSON TURNER

Considered by some the single most influential piece of writing in American history, the reputation of Frederick Jackson Turner's "Frontier Thesis," as first articulated in "The Significance of the Frontier in American History," has suffered somewhat from a series of revisionist critiques. Indeed many contemporary scholars of American studies and of the American West have challenged nearly all of Turner's premises, including his most basic—the existence of a monolithic frontier.

The paper, delivered as a speech to the American Historical Association at the 1893 Chicago World's Fair, first appeared, per

Turner's footnote, in the 1920 publication of the book *The Frontier in American History*, in the Proceedings of the State Historical Society of Wisconsin in December of 1893, which, in turn, cited earlier publication in *The Aegis*, a publication of the University of Wisconsin student body in November of 1892. The chronological placement of the excerpt in this volume, consistent with its delivery in 1893, reflects a desire to place the work in its appropriate, historical milieu.

In the lengthy footnote written upon the publication of the expanded book-length version of *The Frontier in American History* in 1920, Turner consoles himself with the fact that Woodrow Wilson had since accepted many of Turner's original claims concerning the West as a preeminent factor in twentieth-century American history. At the time of its delivery, Turner's frontier thesis was largely ignored by conventional historians who believed America's social and political culture to be of European parentage. On the contrary, Turner insisted that America had evolved a unique culture by virtue of its singular geography and environmental history. While Turner, a professor of history at the University of Wisconsin, was certainly more an academic than an agrarian, in retrospect "The Significance of the American Frontier," excerpted briefly here, exists at the crux of a golden age of farm and conservation writing. Turner's reading of the evolution of American culture—from Indian, to trader, to pastoral rancher, to the "exploitation of the soil," to the "intensive culture of the denser farm settlement" to the manufacture and industry of the city—expresses, in a nutshell, the evolution on the land from the Gilded through the Golden Age. Moreover, his characterization of the American cultural traits as "that coarseness of strength ... that practical inventive turn of mind, quick to find expedients; that masterful grasp of material things ... that restless, nervous energy" serves as distillation of the predilections that determined, for good and for ill, the course of America's environmental history. Finally, Jackson's declaration of the closed frontier meant that America would have to accept, and better honor, the limitations of its land-wealth. Born in Portage, Wisconsin very near John Muir in time and place, Turner's theories are, in hindsight, as influential as Muir's.

The Significance of the Frontier in American History[68]

In a recent bulletin of the superintendent of the census for 1890 appear these significant words: "Up to and including 1880 the country had a frontier of settlement, but at present the unsettled area has been so broken into by isolated bodies of settlement that there can hardly be said to be a frontier line. In the discussion of its extent, its westward movement, et cetera, it can not, therefore, any longer have a place in the census reports." This brief official statement marks the closing of a great historic movement. Up to our own day American history has been in a large degree the history of the colonization of the great West. The existence of an area of free land, its continuous recession, and the advance of American settlement westward, explain American development.

Behind institutions, behind constitutional forms and modifications, lie the vital forces that call these organs into life and shape them to meet changing conditions. The peculiarity of American institutions is the fact that they have been compelled to adapt themselves to the changes of an expanding people—to the changes involved in crossing a continent, in winning a wilderness, and in developing at each area of this progress out of the primitive economic and political conditions of the frontier into the complexity of city life. Said Calhoun in 1817, "We are great, and rapidly—I was about to say fearfully—growing!"[69] So saying, he touched the distinguishing feature of American life. All peoples show development; the germ theory of politics has been sufficiently emphasized. In the case of most nations, however, the development has occurred in a limited area; and if the nation has expanded, it has met other growing peoples whom it has conquered. But in the case of the United States we have a different phenomenon. Limiting our attention to the Atlantic coast, we have the familiar

---

[68] Frederick Jackson Turner, *The Frontier in American History* (New York: Henry Holt, 1920), 2–5.
[69] The original footnote, numbered two, read "Abridgment of Debates of Congress, v, p. 706."

phenomenon of the evolution of institutions in a limited area, such as the rise of representative government; into complex organs; the progress from primitive industrial society, without division of labor, up to manufacturing civilization. But we have in addition to this a recurrence of the process of evolution in each western area reached in the process of expansion. Thus American development has exhibited not merely advance along a single line, but a return to primitive conditions on a continually advancing frontier line, and a new development for that area. American social development has been continually beginning over again on the frontier. This perennial rebirth, this fluidity of American life, this expansion westward with its new opportunities, its continuous touch with the simplicity of primitive society, furnish the forces dominating American character. The true point of view in the history of this nation is not the Atlantic coast, it is the great West. Even the slavery struggle, which is made so exclusive an object of attention by writers like Professor von Holst, occupies its important place in American history because of its relation to westward expansion.

In this advance, the frontier is the outer edge of the wave—the meeting point between savagery and civilization. Much has been written about the frontier from the point of view of border warfare and the chase, but, as a field for the serious study of the economist and the historian, it has been neglected.

The American frontier is sharply distinguished from the European frontier—a fortified boundary line running through dense populations. The most significant thing about the American frontier is that it lies at the hither edge of free land. In the census reports it is treated as the margin of that settlement which has a density of two or more to the square mile. The term is an elastic one, and for our purposes does not need sharp definition. We shall consider the whole frontier belt including the Indian country and the outer margin of the "settled area" of the census reports. This paper will make no attempt to treat the subject exhaustively; its aim is simply to call attention to the frontier as a fertile field for investigation, and to suggest some of the problems which arise in connection with it.

In the settlement of America we have to observe how European life entered the continent, and how America modified and developed that life and reacted on Europe. Our early history is the study of European germs developing in an American environment. Too exclusive attention has been paid by institutional students to the Germanic origins, too little to the American factors. The frontier is the line of most rapid and effective Americanization. The wilderness masters the colonist. It finds him a European in dress, industries, tools, modes of travel, and thought. It takes him from the railroad car and puts him in the birch canoe. It strips off the garments of civilization and arrays him in the hunting shirt and the moccasin. It puts him in the log cabin of the Cherokee and Iroquois and runs an Indian palisade around him. Before long he has gone to planting Indian corn and plowing with a sharp stick, he shouts the war cry and takes the scalp in orthodox, Indian fashion. In short, at the frontier the environment is at first too strong for the man. He must accept the conditions which it furnishes, or perish, and so he fits himself into the Indian clearings and follows the Indian trails. Little by little he transforms the wilderness, but the outcome is not the old Europe, not simply the development of Germanic germs, any more than the first phenomenon was a case of reversion to the Germanic mark. The fact is that here is a new product that is American. At first, the frontier was the Atlantic coast. It was the frontier of Europe in a very real sense. Moving westward, the frontier became more and more American. As successive terminal moraines result from successive glaciations, so each frontier leaves its traces behind it, and when it becomes a settled area the region still partakes of the frontier characteristics. Thus the advance of the frontier has meant a steady movement away from the influence of Europe, a steady growth of independence on American lines. And to study this advance, the men who grew up under these conditions, and the political, economic, and social results of it, is to study the really American part of our history.

JOHN MUIR

Wholly unlike any other nature writing of its day, John Muir's work in "Windstorm" and elsewhere approaches the ecstatic. Unabashed in his sentimentalities and energies, Muir here describes riding out a winter windstorm in the High Sierra. "Windstorm" serves as an ode both to the wind and to the trees that serve as the wind's "interpreters," allowing the naturalist a bird's-eye view of nature not as antagonist, but as divine orchestrator.

Though Muir makes passing reference in these pages to his more formal conservation fieldwork, his purpose here is more purely laudatory and narrative, as the windstorm teases from him deep memories of a childhood spent first in Scotland and later on a farm in Wisconsin. The essay forefronts not primarily Muir's technical expertise, but his willingness to engage nature on its terms, to describe rather than to prescribe. His writings about the Sierra, in particular, offer insights only available to a long-standing citizen of the woods, making an argument for purposeful locality as well as for careful witness. Muir's solidarity is especially evident when he writes: "We all travel the milky way together, trees and men; but it never occurred to me until this storm-day, while swinging in the wind." *The Mountains of California*, from which this passage is drawn, would, in its day, sell some ten thousand copies.

A Windstorm in the Forests[70]

The mountain winds, like the dew and rain, sunshine and snow, are measured and bestowed with love on the forests to develop their strength and beauty. However restricted the scope of other forest influences, that of the winds is universal. The snow bends and trims the upper forests every winter, the lightning strikes a single tree here and there, while avalanches mow down thousands at a swoop as a gardener trims out a bed of flowers. But the winds go to every tree, fingering every leaf and branch and furrowed bole; not one is forgotten; the mountain pine towering with outstretched arms on the rugged buttresses of the icy peaks, the lowliest and most retiring tenant of the

---

[70] John Muir, *The Mountains of California* (New York: The Century Co., 1894), 244–257.

dells; they seek and find them all, caressing them tenderly, bending them in lusty exercise, stimulating their growth, plucking off a leaf or limb as required, or removing an entire tree or grove, now whispering and cooing through the branches like a sleepy child, now roaring like the ocean; the winds blessing the forests, the forests the winds, with ineffable beauty and harmony as the sure result.

After one has seen pines six feet in diameter bending like grasses before a mountain gale, and ever and anon some giant falling with a crash that shakes the hills, it seems astonishing that any, save the lowest thickset trees, could ever have found a period sufficiently stormless to establish themselves; or, once established, that they should not, sooner or later, have been blown down. But when the storm is over, and we behold the same forests tranquil again, towering fresh and unscathed in erect majesty, and consider what centuries of storms have fallen upon them since they were first planted—hail, to break the tender seedlings; lightning, to scorch and shatter; snow, winds, and avalanches, to crush and overwhelm, while the manifest result of all this wild storm culture is the glorious perfection we behold; then faith in Nature's forestry is established, and we cease to deplore the violence of her most destructive gales, or of any other storm implement whatsoever.

There are two trees in the Sierra forests that are never blown down, so long as they continue in sound health. These are the juniper and the dwarf pine of the summit peaks. Their stiff, crooked roots grip the storm-beaten ledges like eagles' claws, while their lithe, cord-like branches bend round compliantly, offering but slight holds for winds, however violent. The other alpine conifers—the needle pine, mountain pine, two-leaved pine, and hemlock spruce—are never thinned out by this agent to any destructive extent, on account of their admirable toughness and the closeness of their growth. In general the same is true of the giants of the lower zones. The kingly sugar pine, towering aloft to a height of more than two hundred feet, offers a fine mark to storm winds; but it is not densely foliaged, and its long, horizontal arms swing round compliantly in the blast, like tresses of green, fluent algae in a brook; while the silver firs in most places keep their ranks well together in united strength. The yellow or silver pine is more frequently overturned than any other tree on the Sierra, because its leaves and branches form a larger mass in proportion to its height, while in many places it is planted sparsely, leaving open lanes through which storms may enter with full force. Furthermore, because it is distributed along the lower portion of the range, which was the first to be left bare on the breaking up of the ice-sheet at the close of the glacial winter, the soil it is growing upon has been longer exposed to post-glacial weathering, and consequently is in a more crumbling, decayed condition than the fresher soils farther up the range, and therefore offers a less secure anchorage for the roots.

While exploring the forest zones of Mount Shasta, I discovered the path of a hurricane strewn with thousands of pines of this species. Great and small had been uprooted or wrenched off by sheer force, making a clean gap, like that made by a snow avalanche. But hurricanes capable of doing this class of work are rare in the Sierra, and when we have explored the forests from one extremity of the range to the other, we are compelled to believe that they are the most beautiful on the face of the earth, however we may regard the agents that have made them so.

There is always something deeply exciting, not only in the sounds of winds in the woods, which exert more or less influence over every mind, but in their varied water-like flow as manifested by the movements of the trees, especially those of the conifers. By no other trees are they rendered so extensively and impressively visible, not even by the lordly tropic palms or tree ferns responsive to the gentlest breeze. The waving of a forest of the giant Sequoias is indescribably impressive and sublime, but the pines seem to me the best interpreters of winds. They are mighty waving goldenrods, ever in tune, singing and writing wind-music all their long century lives. Little, however, of this noble tree-waving and tree-music will you see or hear in the strictly alpine portion of the forests. The burly juniper, whose girth sometimes more than equals its height, is about as rigid as the rocks on which it grows. The slender lash-like sprays of the dwarf pine stream out in wavering ripples, but the tallest and slenderest are far too unyielding to wave even in the heaviest gales. They only shake in quick, short vibrations. The hemlock spruce, however, and the mountain pine, and some of the tallest thickets of the two-leaved species bow in storms with considerable scope and gracefulness. But it is only in the lower and middle zones that the meeting of winds and woods is to be seen in all its grandeur.

One of the most beautiful and exhilarating storms I ever enjoyed in the Sierra occurred in December, 1874, when I happened to be exploring one of the tributary valleys of the Yuba River. The sky and the ground and the trees had been thoroughly rain-washed and were dry again. The day was intensely pure, one of those incomparable bits of California winter, warm and balmy and full of white, sparkling sunshine, redolent of all the purest influences of the spring, and at the same time enlivened with one of the most bracing windstorms conceivable. Instead of camping out, as I usually do, I then chanced to be stopping at the house of a friend. But when the storm began to sound, I lost no time in pushing out into the woods to enjoy it. For on such occasions Nature has always something rare to show us, and the danger to life and limb is hardly greater than one would experience crouching deprecatingly beneath a roof.

It was still early morning when I found myself fairly adrift. Delicious sunshine came pouring over the hills, lighting the tops of the pines, and setting free a steam of summery fragrance that contrasted strangely with the wild tones of the storm. The air was mottled with pine tassels and bright green plumes that went flashing past in the sunlight like birds pursued. But there was not the slightest dustiness, nothing less pure than leaves, and ripe pollen, and flecks of withered bracken and moss. I heard trees falling for hours at the rate of one every two or three minutes; some uprooted, partly on account of the loose, water-soaked condition of the ground; others broken straight across, where some weakness caused by fire had determined the spot. The gestures of the various trees made a delightful study. Young sugar pines, light and feathery as squirrel-tails, were bowing almost to the ground, while the grand old patriarchs, whose massive boles had been tried in a hundred storms, waved solemnly above them, their long, arching branches streaming fluently on the gale, and every needle thrilling and ringing and shedding off keen lances of light like a diamond. The Douglas spruces, with long sprays drawn out in level tresses, and needles massed in a gray, shimmering glow, presented a most striking appearance as they stood in bold relief along the hilltops. The madronños in the dells, with their red bark and large glossy leaves tilted every way, reflected the sunshine in throbbing spangles like those one so often sees on the rippled surface of a glacier lake. But the silver pines were now the most impressively beautiful of all. Colossal spires two hundred feet in height waved like supple goldenrods chanting and bowing low as if in worship, while the whole mass of their long, tremulous foliage was kindled into one continuous blaze of white sun-fire. The force of the gale was such that the most steadfast monarch of them all rocked down to its roots with a motion plainly perceptible when one leaned against it. Nature was holding high festival, and every fiber of the most rigid giants thrilled with glad excitement.

 I drifted on through the midst of this passionate music and motion, across many a glen, from ridge to ridge, often halting in the lee of a rock for shelter, or to gaze and listen. Even when the grand anthem had swelled to its highest pitch, I could distinctly hear the varying tones of individual trees—spruce, and fir, and pine, and leafless oak—and even the infinitely gentle rustle of the withered grasses at my feet. Each was expressing itself in its own way—singing its own song, and making its own peculiar gestures—manifesting a richness of variety to be found in no other forest I have yet seen. The coniferous woods of Canada, and the Carolinas, and Florida, are made up of trees that resemble one another about as nearly as blades of grass, and grow close together in much the same way. Coniferous trees, in general, seldom possess individual character, such as is manifest among oaks and elms. But the California forests are made up of a greater number of distinct species than any other in the world. And in them we find, not only a marked differentiation into special groups, but also a marked individuality in almost every tree, giving rise to storm effects indescribably glorious.

Toward midday, after a long, tingling scramble through copses of hazel and ceanothus, I gained the summit of the highest ridge in the neighborhood; and then it occurred to me that it would be a fine thing to climb one of the trees to obtain a wider outlook and get my ear close to the Aeolian music of its topmost needles. But under the circumstances the choice of a tree was a serious matter. One whose instep was not very strong seemed in danger of being blown down, or of being struck by others in case they should fall; another was branchless to a considerable height above the ground, and at the same time, too large to be grasped with arms and legs in climbing; while others were not favorably situated for clear views. After cautiously casting about, I made choice of the tallest of a group of Douglas spruces that were growing close together like a tuft of grass, no one of which seemed likely to fall unless all the rest fell with it. Though comparatively young, they were about one hundred feet high, and their lithe, brushy tops were rocking and swirling in wild ecstasy. Being accustomed to climb trees in making botanical studies, I experienced no difficulty in reaching the top of this one, and never before did I enjoy so noble an exhilaration of motion. The slender tops fairly flapped and swished in the passionate torrent, bending and swirling backward and forward, round and round, tracing indescribable combinations of vertical and horizontal curves, while I clung with muscles firm braced, like a bobolink on a reed.

In its widest sweeps my treetop described an arc of from twenty to thirty degrees, but I felt sure of its elastic temper, having seen others of the same species still more severely tried—bent almost to the ground indeed, in heavy snows—without breaking a fiber. I was therefore safe, and free to take the wind into my pulses and enjoy the excited forest from my superb outlook. The view from here must be extremely beautiful in any weather. Now my eye roved over the piney hills and dales as over fields of waving grain, and felt the light running in ripples and broad swelling undulations across the valleys from ridge to ridge, as the shining foliage was stirred by corresponding waves of air. Oftentimes these waves of reflected light would break up suddenly into a kind of beaten foam, and again, after chasing one another in regular order, they would seem to bend forward in concentric curves, and disappear on some hillside, like sea waves on a shelving shore. The quantity of light reflected from the bent needles was so great as to make whole groves appear as if covered with snow, while the black shadows beneath the trees greatly enhanced the effect of the silvery splendor.

Excepting only the shadows there was nothing somber in all this wild sea of pines. On the contrary, notwithstanding this was the winter season, the colors were remarkably beautiful. The shafts of the pine and *Libocedrus* were brown and purple, and most of the foliage was well tinged with yellow; the laurel groves, with the pale undersides of their leaves turned upward, made masses of gray; and then there was many a dash of chocolate color from clumps of manzanita, and jet of vivid crimson from the bark of the madronnos, while the ground on the hillsides, appearing here and there through openings between the groves, displayed masses of pale purple and brown.

The sounds of the storm corresponded gloriously with this wild exuberance of light and motion. The profound bass of the naked branches and boles booming like waterfalls; the quick, tense vibrations of the pine needles, now rising to a shrill, whistling hiss, now falling to a silky murmur; the rustling of laurel groves in the dells, and the keen metallic click of leaf on leaf—all this was heard in easy analysis when the attention was calmly bent.

The varied gestures of the multitude were seen to fine advantage, so that one could recognize the different species at a distance of several miles by this means alone, as well as by their forms and colors, and the way they reflected the light. All seemed strong and comfortable, as if really enjoying the storm, while responding to its most enthusiastic greetings. We hear much nowadays concerning the universal struggle for existence, but no struggle in the common meaning of the word was manifest here; no recognition of danger by any tree; no deprecation; but rather an invincible gladness as remote from exultation as from fear.

I kept my lofty perch for hours, frequently closing my eyes to enjoy the music by itself, or to feast quietly on the delicious fragrance that was streaming past. The fragrance of the woods was less marked than that produced during warm rain, when so many balsamic buds and leaves are steeped like tea, but, from the chafing of resiny branches against each other, and the incessant attrition of myriads of needles, the gale was spiced to a very tonic degree. And besides the fragrance from these local sources there were traces of scents brought from afar. For this wind came first from the sea, rubbing against its fresh, briny waves, then distilled through the redwoods, threading rich ferny gulches, and spreading itself in broad undulating currents over many a flower-enameled ridge of the coast mountains, then across the golden plains, up the purple foothills, and into these piney woods with the varied incense gathered by the way.

Winds are advertisements of all they touch, however much or little we may be able to read them, telling their wanderings even by their scents alone. Mariners detect the flowery perfume of land-winds far at sea, and sea-winds carry the fragrance of dulse and tangle far inland, where it is quickly recognized, though mingled with the scents of a thousand land-flowers. As an illustration of this, I may tell here that I breathed sea air on the Firth of Forth, in Scotland, while a boy; then was taken to Wisconsin, where I remained nineteen years; then, without in all this time having breathed one breath of the sea, I walked quietly, alone, from the middle of the Mississippi Valley to the Gulf of Mexico, on a botanical excursion, and while in Florida, far from the coast, my attention wholly bent on the splendid tropical vegetation about me, I suddenly recognized a sea-breeze, as it came sifting through the palmettos and blooming vine tangles, which at once awakened and set free a thousand dormant associations, and made me a boy again in Scotland, as if all the intervening years had been annihilated.

Most people like to look at mountain rivers, and bear them in mind; but few care to look at the winds, though far more beautiful and sublime, and though they become at times about as visible as flowing water. When the north winds in winter are making upward sweeps over the curving summits of the High Sierra, the fact is sometimes published with flying snow banners a mile long. Those portions of the winds thus embodied can scarce be wholly invisible, even to the darkest imagination. And when we look around over an agitated forest, we may see something of the wind that stirs it, by its effects upon the trees. Yonder it descends in a rush of water-like ripples, and sweeps over the bending pines from hill to hill. Nearer, we see detached plumes and leaves, now speeding by on level currents, now whirling in eddies, or, escaping over the edges of the whirls, soaring aloft on grand, upswelling domes of air, or tossing on flame-like crests. Smooth, deep currents, cascades, falls, and swirling eddies sing around every tree and leaf, and over all the varied topography of the region with telling changes of form, like mountain rivers conforming to the features of their channels.

After tracing the Sierra streams from their fountains to the plains, marking where they bloom white in falls, glide in crystal plumes, surge gray and foam-filled in boulder-choked gorges, and slip through the woods in long, tranquil reaches—after thus learning their language and forms in detail, we may at length hear them chanting all together in one grand anthem, and comprehend them all in clear inner vision, covering the range like lace. But even this spectacle is far less sublime and not a whit more substantial than what we may behold of these storm-streams of air in the mountain woods.

We all travel the milky way together, trees and men, but it never occurred to me until this storm-day, while swinging in the wind, that trees are travelers, in the ordinary sense. They make many journeys, not extensive ones, it is true; but our own little journeys, away and back again, are only little more than tree-wavings—many of them not so much.

When the storm began to abate, I dismounted and sauntered down through the calming woods. The storm-tones died away, and, turning toward the east, I beheld the countless hosts of the forests hushed and tranquil, towering above one another on the slopes of the hills like a devout audience. The setting sun filled them with amber light, and seemed to say, while they listened, "My peace I give unto you."

As I gazed on the impressive scene, all the so-called ruin of the storm was forgotten, and never before did these noble woods appear so fresh, so joyous, so immortal.

Lucius "Lute" Wilcox

Longtime Denver publisher and editor of the periodical *Field and Farm* Lucius "Lute" Merle Wilcox crops up in horticultural programs and publications throughout the Progressive Era, more often than not advocating irrigation for the Plains. Drawing on twenty years of experience in agriculture in "The Advantages of Irrigation," Wilcox rightly notes that large-scale irrigation in America in the 1890s was in its "infancy." He views irrigation as a kind of agricultural panacea, as "simple as child's play." An understanding of the relative frequency of drought in Colorado and surrounding states in Wilcox's time leads the reader to a more complete understanding of the author's sometimes desperate faith in a practice that caused the desert to "blossom" without the fears of soil exhaustion. In an 1896 issue of *Garden and Forest*, the proceedings of the Nebraska State Horticultural Society described consecutive seasons of drought and the financial depression they helped precipitate. Thus, when Lute Wilcox later spoke to the same Nebraska historical group, he had a curious if not eager audience, as he did in most of his stops in the Plains.

The ambitiousness of Wilcox's claims in "The Advantages of Irrigation," supported by a number of hydraulic engineers and scientists of the time, were given a sympathetic, even wishful, read by many reviewers, including *Garden and Forest*, which declared "no tiller of the soil can read these chapters without having his views broadened and gaining ideas which he can put to practical use on his own land, wherever it may be." Indeed, a unique aspect of *Irrigation Farming* was its appeal to the farmers in the "rain belt" enduring the boom-bust cycles of what Wilcox calls "rain farming." Touting irrigation-increased growth of orchards, vineyards, and gardens as well as of crops as varied as potatoes, hops, tobacco, and cotton, Wilcox exhorts a variety of farmers with the mantra "till and keep tilling," while condemning what he calls "the wanton waste of water" by "shiftless" cultivators. A predictor of Western land reclamation policies in the decades that followed, Wilcox's argument here is founded on the availability of clean, minerally-rich water from springs and streams and the resiliency of soils—both premises later questioned.

The Advantages of Irrigation[71]

Someone has spoken of irrigation as the "wedding of the sunshine and the rain." A great many people hearing the word irrigation experience the same sensations that they do when Madagascar or Wiju is spoken of. They have a feeling that it is something a great distance off—hard to reach—intangible. They read about it as they like to read *Arabian Nights* or Hans Andersen's *Fairy Stories*, and it leaves on their minds about the same impressions of wonder, magnificence and untruth as do the stories named. To them the very word "irrigation" puts their reasoning to flight, and they imagine that the art of applying water to cultivated lands is some complicated and wonderfully intricate process, not easily understood or attained by mortal man. The fact of the matter is, as the author proposes to show in the succeeding chapters, that irrigation is as simple as child's play and may be accomplished by the most commonplace day laborer in the fields. In enumerating a few of the advantages attendant upon irrigating methods, we will cite the facts that irrigation reclaims arid wastes; makes a prosperous country; causes the desert to blossom and overcomes the destructive effects of the parching southern winds; insures full crops every season; improves land at each submergence, and consequently does not wear out the soil; produces support for dense population; multiplies the productive capacity of soils; destroys insects and worms and produces perfect fruit; creates wealth from water, sunshine and soil; makes the farmer independent of the rainfall, will redeem 100,000,000 acres of desert lands in the United States alone; yields large returns to investors; adds constantly to the security of investments; will yield support for 50,000,000 of increased population in

---

[71] Lucius Wilcox, *Irrigation Farming: A Handbook for the Practical Application of Water in the Production of Crops* (New York: Orange Judd Co., 1895), 11–18.

America; makes the production of choicest fruits possible, and prolongs the harvest period of various crops if so desired; affords a sure foundation for the creation of wealth; lessens the danger of floods; utilizes the virgin soil of the mountain regions; is now employing more than $1,000,000,000 of capital; insures two or more crops annually in the lower latitudes; will increase threefold the value of lands having rainfall; keeps off the early approach of Jack Frost; improves the quality and increases fully one-eighth and oftentimes one-fourth the size of fruits, vegetables and grains; makes farming profitable in waste places and forever forestalls the inroads caused by the ghost of drought; and will finally solve the great labor question and fortify against the alarming increase of city populations.

The farmer who has a soil containing an abundance of all the needed elements, in a proper state of fineness, cannot but deem himself happy if he have always ready at hand the means of readily and cheaply supplying all the water needed by his soil and growing crops, just when and in just such quantities as are needed. Happier still may he be if, besides fearing no drought, he has no rainfall to interrupt his labors or to injure his growing or harvested crops. And happier still may he be when he realizes that he need have no "off years," and he knows that the waters he admits to his fields at will are freighted with rich fertilizing elements usually far more valuable to the growing crop than any that he can purchase and apply at a costly rate—a cost that makes serious inroads upon the profits of the majority of farmers cultivating the worn-out or deteriorated soils in the older states year by year. Fertilizers are already needed for the most profitable culture on many farms in Iowa, Minnesota, eastern Kansas and Nebraska, in Missouri and in all states east of those named.

In proof of this assertion the writer can best be qualified in his statement by mentioning the fact that there is an oat field in Saguache County, Colorado, that up to 1894 had produced twenty-three consecutive crops, each of which averaged forty bushels to the acre through all the years. The yield of the twenty-third crop averaged sixty bushels, which would indicate that the fertility of that field was keeping up remarkably well without rest or rotation. This unusual result was made possible by means of irrigation alone, and there is no doubt much truth in the theory that the irrigating waters from the mountains contain great quantities of mineral fertilizing element in solution. Even by shallow plowing and the most shiftless methods of land preparation, a Mexican farmer named E. Valdez, of Chromo, Colorado, produced twenty-five consecutive crops of wheat on the same soil, and without manure or change of seed in the interim. This peculiar result was made possible only by the use of irrigating waters, applied as they were regardless of scientific principles or any defined method whatever. The yield the last season was forty-five bushels to the acre, as heavy as any throughout the quarter of a century of constant croppage.

Irrigation farming has peculiar characteristics. It is a higher and more scientific industry than rain farming; it succeeds best by what is known as intensive culture, or what is better described as scientific culture. The soil to be at its best should be carefully prepared, and cultivation ought to be minute and thorough. To make such agriculture pay such crops must be raised as will yield the greatest value to the acre. The irrigated lands are better adapted to the growth of orchards, vineyards, gardens, potato fields, hop yards, tobacco and cotton plantations, and whatever extra work may be required to cover the land with water will be repaid tenfold from the first crop that is taken off. In traveling in the far West over long stretches of parched and dusty plains or through mountain gorges, the writer has often seen fields, orchards, vineyards and gardens all dressed in living green. The life, vigor and fruitfulness were in surpassing contrast to the general aspect. And why this contrast? Because of the tapping of mountain streams, fed by crystal springs or banks of perpetual snow, and turning a portion of their waters upon the lands. From great eminences the course of these life-giving waterways made by the hands of man could be traced by the eye, until they were lost in the dimness of distance. There was no need of being told where were the irrigating ditches. The eye of a novice could mark them with accuracy as they wound about the foothill slopes, dotting the landscape with patches of emerald, where lone settlers and busy towns were located.

It is in the horticultural pursuits that the highest degree of perfection as the patrimony of modern irrigation is to be realized. Under any system of irrigation where a constant supply of water is to be had, the horticulturist can plant with almost a certainty of gathering a crop. Untimely frosts, insects and fungous diseases are often to be contended with, but it is a great consolation to feel sure that drought cannot prevent the starting of trees, plants and seeds in springtime, or cut short a growing crop. Neither are floods likely to overflow, except on low bottoms, and these are not the best places for the most profitable orchards. One field or a small portion of it can be watered without the rest being deluged or even sprinkled, if desired.

It is the writer's desire at this time to direct the attention of horticulturists and farmers generally in the "rain belt" to the benefits to be derived from an artificial supply of water to their crops. Some may scout the idea and say it is not practicable—that it will not pay to go to so much expense for the little use to be made of the water; but in all seriousness it may be said that it will pay, and there are many places east of the arid regions where irrigation is now considered by those who have long tried it as almost indispensable. There is scarcely an acre of ground under cultivation in North America that would not produce more and better crops if there was at hand an abundant water supply. There are seasons now and then in which the rains come just right and irrigation might not be needed even once, but they are rare. Usually there are several dry spells during each year that cause serious injury to the crops, and were irrigation possible all harm from this source might be prevented. A very little water at the right time would make all the difference with the crop and turn into success what otherwise would have been a partial or total failure. The work already put upon the land would be saved, as well as seeds and plants. Satisfaction and plenty would take the place of disappointment and scarcity. If eastern pomologists would only adopt irrigation, there would be no good cause for having weakly plants and trees or for the premature dropping of leaves. The buds would develop early, and be plump and vigorous. There would be no winter-killing of trees and plants because of their feeble condition. Many things are considered tender that are so in some places only because of their inability to make sufficient growth to fortify against the evaporating influences of the winter. It would not be reasonable to expect that any of the many systems of irrigation can be applied to all sections of our country, or to every farm in any section. Neither is it always practicable that all of a large farm should be placed under irrigation, except in rare cases. But where there is now, or may be created, a supply of water that can be drawn upon in time of need for at least a small part of the farm, it is a great mistake not to make use of its benefits. There are special crops, such as asparagus, celery, and the strawberry, which need an amount of water that is not required by most others, and which could be grown much more cheaply than at present if aided by irrigation. In this connection it might be well to add that statistics show that in all rainy countries—that is, where the farmers depend upon the rains to make their crops—the seasons of drought and the seasons of too much rain constitute three out of every five, giving the farmer three bad crops to every two good ones. As a matter of fact, the intrinsic advantages of irrigation concern and are within reach of the farmer of the humid region quite as much as his fellow in the arid climate, and in many, if not in most, cases his water supply will cost him less, and when once applied, will never be given up. There can be no doubt that when the available waters of the humid region are examined in regard to the supplies of plant food they are capable of giving to lands irrigated with them, they will be found to be nearly, if not quite, as valuable in this respect as those of the arid region.

Another suggestion along this line presents itself right here: as there is no material difference in the cost of cultivation of an acre yielding ten bushels of wheat and another acre yielding sixty bushels, it must be evident that the man who gets only ten bushels pays six times as much as does the man who produces sixty bushels. The profits to be derived from "the new agriculture," as irrigation has aptly been called, comes not alone from the annual return from the watered acres, but from the constantly increasing valuation of the land itself. Many individual instances could be cited, especially in regions devoted to fruit culture, where the returns are almost fabulous. Lands which were worth from two to ten dollars an acre have by the expenditure of from ten to twenty dollars an acre in the construction of irrigation works become worth $300 an acre and upward. The same lands, set out with suitable varieties of trees and vines, have sold within five years of planting at $1,000 or more an acre. So valuable are irrigated lands in Spain that they sell for $720 to $880 an acre, which is ten times the price of the unirrigated, and the same ratio of values prevails elsewhere.

In summarizing the manifold advantages that the irrigation blessing has brought to humanity through all the ages of persevering man, and anticipating those benefits that are to be commanded by "the nations yet to be," we may conclude that irrigation means better economic conditions; means small farms, orchards and vineyards; more homes and greater comfort for men of moderate means. It means more intelligence and knowledge applied to farming, more profit from crops, more freight and more commerce—because special products of higher grade and better market value will be enhanced. It means association in urban life instead of isolated farms. It means the occupation of small holdings. It means more telephones, telegraphs, good roads and swift motors; fruit and garden growths everywhere; schools in closer proximity; villages on every hand, and such general prosperity as can hardly be dreamed of by those who are not familiar with the results of even the present infancy of irrigation in America. It can hardly be doubted that in time the lessons conveyed by history, as well as by the daily practice and results of irrigation in the arid region, will induce the dwellers in the regions of summer rains to procure for themselves at least a part of the advantages which are equally within their reach, putting an end to the dreadful seasons when "the skies are as brass and the earth as a stone" and the labors of the husbandman are in vain.

ISAAC PHILLIPS ROBERTS

Liberty Hyde Bailey called *The Fertility of the Land* "the ripened judgment of the wisest farmer whom I have ever known." Written by Isaac Roberts at Bailey's urging, *The Fertility of the Land* demonstrates what Bailey found so valuable in his Cornell University colleague—a rare combination of professor, practicing farmer, and able philosopher. Raised on a farm near Ithaca, Roberts labored in his young adulthood as a carpenter and occasional teacher, working his way westward, where he would marry and settle on a farm in Mt. Pleasant, Iowa in 1862. After Iowa accepted the terms of the Morrill Land-Grant College Act of that same year, it began a search for a superintendent of the farm and secretary of the board of trustees of the Iowa Agricultural College at Ames. Roberts fit the bill, and assumed his new position in 1869 by cleaning up the grounds of the college dramatically while interested students followed him into the field. By serendipity, Roberts and his tag-alongs arrived at what would become the hands-on, pedagogical method favored thereafter by agricultural education and extension. When another newly authorized land-grant, Cornell University in Ithaca, learned of the success of an Ithaca native in Iowa, they lured Roberts—who had never been to college—back home by naming him dean of faculty of agriculture and director of the experiment station.

In "The Fertility of the Land: A Chat with a Young Farmer," Roberts moves easily from historical survey to biting philosophy to bracing commentary, all for the benefit of the young farmer Professor Roberts imagines as his interlocutor. While sometimes presumptuous in its omniscience, Roberts' innate understanding of farm boys—having been one himself—seems startlingly accurate even to contemporary ears. As with Bailey, Roberts proves unafraid of abandoning agricultural history and science for homespun lyricism, as when he writes of the farmer's son, "So intently have you longed to have some great corporation brand and number you, when the tasks at home were hard, that you have even planned to slip down those huge porch posts at night with your little bundle on a stick." And yet it is the farm boy's love of the land, a love shared by the author, that causes the young farmer inevitably to take up his father's trade, with new training, despite the steepest odds. "The city may need you later," Roberts writes to his young audience, "but sidewalks are hot and hard, while the country roads are soft and cool." The breadth of Robert's agrarian vision is implicit in his book's lengthy subtitle, "a summary sketch of the relationship of farm-practice to the maintaining and increasing of the productivity of the soil," a hodgepodge of concerns that coalesces with Roberts' injunction "be honest with the soil and with yourself." In his thirty years at Cornell, Roberts wrote an estimated fourteen hundred short articles, published four books, and served as the associate editor of the periodical *The Country Gentleman*.

The Fertility of the Land: A Chat with a Young Farmer[72]

In the hurry and unrest of a new country, few have time or inclination to become familiar with plant and animal life as seen in the field and wood, and fewer still have looked upon the surface of the earth as anything but a mass of dirt, the particles of which are to be avoided or removed whenever they offend the sight or interfere with comfort.

Fill a flower pot with the soft, dark earth and mold from the border of the wood, and carry it to the student of entomology, and see if he can name one half of the living forms of this little kingdom of life; or hand it to the botanist, well trained in the lower orders of plants, and see how many of the living forms which these few handfuls of dirt contain he can classify. Present this miniature farm to the chemist and the physicist, and let them puzzle over it. Call in the farmer, and ask him what plants will thrive best in it; or keep the soil warm and moist for a time, and have the gardener say of the tiny plants that appear as by magic which are good and which are bad. Mark well what all these experts have said, and call in the orchardist to tell you how to change dead, lifeless, despised earth into fruit; ask the physiologist to explain how sodden earth is transformed into nerve and brain. With this extended little field in view, choose the profession of agriculture if you love rural pursuits, but comprehend fully that in doing so you are entering upon the most difficult of all pursuits: difficult in ordinary times, doubly so under the present conditions, which have come about so rapidly that they are almost incomprehensible.

The American inherits from his European ancestors an inordinate desire for landed estates. In earlier days, many farmers acquired land

---

[72] Isaac Phillips Roberts, *The Fertility of the Land: A Summary Sketch of the Relationship of Farm-practice to the Maintaining and Increasing of the Productivity of the Soil* (New York: Macmillan, 1897), 1–8.

by the square mile, and all secured more than they could farm well. The federal government sold at nominal prices, gave away and indirectly forced land upon all comers, not even reserving the hilly timber lands which, if they had been reserved, would have tempered the climate and have been an ever-present source of wealth. The Homestead Act has not brought unmixed blessings. The whole course of our federal policy towards public lands has tended to produce soil-robbers, not farmers. Transportation by steam power has made the products of vast inland areas salable, giving value to lands which were valueless, but the same power has also brought the products of Asia, Africa and South America into competition in the markets of the world.

From 1861 to 1865, vast numbers of men were transferred from the producing to the consuming class, and the prices of farm products became abnormally high when measured by an inflated currency. These conditions could not fail to mislead and disappoint many when the population and the currency were restored to normal conditions. At the close of the Civil War, in addition to a vast influx of foreigners, there were added to the farming community many soldiers who, in the high prices, saw quick and large returns from the rich lands which had by this time been opened to settlers by the construction of extended systems of railway. During the third quarter of the century inventive genius so improved the appliances of agriculture as to quadruple the productive power of each farmer. From 1870 to 1880, the percentage increase of new farms was 50.71 percent, while the percentage of increase of population was 30.8 percent. From 1880 to 1890, the increase was but 13.86 percent, but the increase of population was 24.86 percent. This shows that, for a time, the percentage increase of farms vastly outran the percentage increase of population. It also shows that the conditions prevailing before 1880 are being so rapidly reversed that the percentage increase in population may out run that of farms far enough to greatly improve the home markets of many farm products in the near future. Be this as it may, the farmer is wise who adjusts himself quickly to present conditions, so unlike those of his father. To do this, he must see clearly and think straight; he must have good executive ability, as well as training and practice in well defined business methods. To see clearly, the eye must be trained to take in a multitude of objects quickly, to sort, compare and photograph on the sensitive brain those which are worth preserving. To think straight, many scientific facts, or items of knowledge, arranged in order, must be acquired, and these can be secured only by long, painstaking effort.

But to know is not enough; the ability to execute must be joined to knowledge, and executive skill is acquired in its highest form only by the direction and management of large affairs. It can not be learned in the classroom, nor formulated in a textbook, and it is seldom learned by the farm boy because of want of opportunity; hence the lessons of the beginner are usually manifold, the tuition for the first term high, and the whole is paid for from his own resources, while young teachers and professional and business men get free tuition because they learn at the expense of their employers. What has been said of executive ability applies with nearly equal force to business ability, the lack of which in city and in country is evidenced in the newspapers by the word *assignment*.

To the clear eye, to the intellectual equipment, to executive ability and trained business methods, must be added manual dexterity. Until recently the untold, fertile acres, the favorable conditions, and the simple wants of the people, have arrested, in agriculture, the operation of that great law the survival of the fittest. It has been said that anybody can farm. That was, but is not true. From this time on the struggle in farming will be such as it has been in mercantile affairs for some time. The unfitted in agriculture will have to yield for the same reason that many little factories, located off the lines of transportation, furnished with inadequate power, machinery and brains, have been abandoned. Many hillsides will be left to cover their nakedness with a new growth of hardy vegetation. It will thus be seen how well-equipped the farmer should be, how fertile in brain, in imagination and in resources; how full of wisdom, of enthusiasm, of faith; how quick to see, how prompt to execute, how patient to endure under difficulties, if the fertility of his land is to be transformed into abundant and perfect fruits and flowers.

This is dedicated to the young farmers of America. I am well acquainted with you all, though you are not acquainted with me, and being acquainted and older than you are, I cannot forbear entering into a little familiar chat. I know your thoughts, your toils and sorrows and discouragements; your aspirations, hopes and joys. I know, too, what fiber, endurance and patience farm work gives to the boys who make the most of what an outdoor life with nature has to offer. I know how hot it is in August under the peak of the flat-roofed barn, how large the forkfuls are that the stalwart pitcher thrusts into the only hole where light and air can enter. I know how high the thistles grow, and how far the rows of corn stretch out. I know, too, the freedom, fun and work of the old farm that make one expand, enjoy and grow, and leave no bitter memories. I know you well, my boy—how green and brown you feel when you come to the noisy city, and how you would like to be free and cool again!

Yon have seen for the thousandth time the long, wavy line of smoke as the train goes swinging by, winding in and out among the hills. Then you have longed to drive that mighty iron horse, feed him on fire, and make him leap away in wild freedom. Or, perhaps you do not aspire so high, and would be content to run a streetcar. You have even admired the bright letters on the caps of the motormen, and you would exchange your freedom for the blue coats and shiny buttons. So intently have you longed to have some great corporation brand and number you, when the tasks at home were hard, that you have even planned to slip down those huge porch posts at night with your little bundle on a stick. But when night came you fell asleep, and the morning sun found you with thickened blood, temptation gone and courage for another day. Love and inherited pluck saved you. You were not ready for the city; you lacked knowledge, seasoned fiber and judgment. We never send colts to the city; they lose their heads and get stove up by rapid pace and rough, hard streets. The city may need you later, but sidewalks are hot and hard, while the country roads are soft and cool.

My scientific reader is getting anxious to know what manner of book this is, and in his heart he thinks I would better have been telling you how energy is changed into heat, heat into motion, and then back into potential energy again. But let him wait; you and I are to finish our chat before we sit down to hard study.

All are greatly interested in yon, my boy! We cannot see how to get along without you, and yet no one cares very much where you were born, where you live or how low you start, how high you climb or what you do, so long as you do right and lead a useful life. The world cares how you work, and it is interested in the progress of civilization. It asks that every one of you start from just where you are, without grumbling and with courage, and climb faithfully, honestly and in harmony with natures modes of action, and the bars which guard the wealth of soil and the accumulation of man's toil will then fly back at your bidding. But wealth should be sought not for the pleasure of securing and possessing it, but as a means to higher ends. When rightly used, it relieves its possessor from a too severe struggle for mere existence, and gives time and opportunity for acquiring useful and pleasurable knowledge, which in turn naturally leads to a fuller comprehension of the real and enduring verities which, though unseen by the natural eye, are all that remain at the close of life. Financial reserves and mental training are the two great stepping stones by which mankind may reach a higher plane of existence. On this higher plane, the environment is so broad and grand, the air so pure and thoughts so lofty, that all work, however menial, becomes inspiring, and study, however hard, is pleasant and ennobling.

So, my young farmers—who should be the pride of the nation and the anchor which holds the thoughtless from drifting towards anarchy—be honest with the soil and with yourself. As you acquire health, fiber, purpose, and courage in mounting the first step, do not stop at the second or the third. Aim high, for it has been written: aim at the sun, and you may not reach it, but your arrow will fly far higher than if aimed at an object on a level with yourself. In the hurry of this intensely utilitarian age, not only may health and life be curtailed, but the better and loftier sides of our nature are in danger of becoming dwarfed.

Frances Elizabeth Willard

Having spent her girlhood on a farm in what was then the territory of Wisconsin, Frances Elizabeth Willard here advocates for a controversial career for an independent woman in the late nineteenth century: farming.

Her essay "Women as Farmers," best classed as a kind of opinion journalism, corrects the misconception then held by many urban women of farm work as "drudgery." For the farm wife who fears she may have to abandon rural living for work in the city—a thoroughly contemporary dilemma—Willard cites as case studies several women farmers who, by recognizing the productive capacity of the land around them, kept both farm and sanity. Likewise, Willard's belief in the balm of hard work is expressed in her famous aphorism, "Let us have plain living and high thinking." The Dean of Women of the Woman's College of Northwestern and later the national president of Women's Christian Temperance Union (WCTU), Willard was known for her charismatic and eloquent oratory as well as her temperance work among the disenfranchised in Chicago, the city with which she is most closely associated.

In "Women as Farmers" Willard imagines a kind of feminist ecotopia consisting of a series of closely networked farms predicated on close cooperation between women tillers. While overzealous in its claim that "starvation and pauperism" would be impossible for the farmer—as the farm Depressions of the 1890s and 1930 would prove—Willard's short chapter evidences not only the growing strength of the women's suffrage and temperance movements but also the gender role inversion that would be precipitated by the World Wars to follow. While the idea of woman as primary tiller rather than tiller's assistant would never predominate in American agriculture, the can-do sentiments expressed here are echoed by popular twenty-first century women ranchers and writers such as Linda Hasselstrom.

Women as Farmers[73]

One both smiles and sighs when she hears a woman who toils from fourteen to sixteen hours a day with her needle, earning, perhaps, from seven to nine dollars a week, or a clerk who every day stands

---

[73] Frances Elizabeth Willard, *Occupations for Women: A Book of Practical Suggestions for the Material Advancement, the Mental and Physical Development, and the Moral and Spiritual Uplift of Women* (New York: Success Co., 1897), 102–107.

from eight to six o'clock in a stuffy corner of a stuffy store retailing various cheap articles, with a salary of from three to five dollars a week, speaking of farm work as "drudgery."

"Americans do everything but think," someone remarked not long since—an extreme assertion, for surely many Americans think to splendid purpose, but when one realizes how much multitudes bear for want of a few hours thought and a modicum of energy and decision, the remark does not seem wholly unjustifiable. Most needlewomen and store employees could hardly work under more distressing conditions, and, through a lull in their employment, might starve or become paupers. As farmers, starvation and pauperism would be impossible.

If it is objected that many girls are too delicate for outdoor employment, it may be answered that in numerous cases these girls are too delicate for anything else. Sunshine, air, and exercise are three of their most vital needs. Many a consumptively inclined person has become healthy and happy by close daily contact with the soil, the facing of free winds, and plenty of outdoor employment.

Of course the rule holds good here as it does regarding other kinds of employment. No one should adopt farming as an occupation who does not love outdoor pursuits and farm belongings. To any other it would surely mean drudgery, and slavery as well. But there are thousands who love "all outdoors," and any occupation which had to do with country wideness, and green, growing things would be their delight. If these could be weeded out from the city workers much sorely needed relief would be afforded to thousands of other workers as well as to themselves.

The woman farmer is no longer sufficiently unique to be wondered at, sneered at, or smiled at. She is found in many parts of the country, and is, if one may judge from the facts brought to light, as successful in her chosen work as is her brother tiller of the soil.

It will seem surprising if in the near future we do not see communities of girl farmers located near enough together to be helpers and companions to each other. Cooperation would lighten the heaviest toil, and the recreation and relaxation which such a neighborhood would make possible would do away with that which is usually a farm's most objectionable feature—its loneliness.

One can begin her agricultural pursuits with very little land if necessary. A writer on this subject says:

> Americans are only beginning to understand that a small patch of land may be cultivated with great profit. The Japanese immigrants who have settled in California within the last few years have aroused the interests of horticulturists to their method of tillage which has prevailed for ages in Japan. They understand the art of getting a bountiful supply from every inch of soil. With three or four acres the Japanese farmer satisfies his every want, keeps clear of debt, and lays up money. With one acre in vegetables he is independent.
>
> Many a woman has a home with a bit of ground attached, which hardly pays the taxes. She is fretting and struggling to make a little money to live on. The only way she can think of is to sew or teach or find something to do for which she will be paid, however small a sum. Her bit of ground can be made to pay like a bank, if she goes at it right. Let her buy a good book on market gardening, study it, and set to work to get the most out of her bit of ground; *Onions for Profit*, published by a Philadelphia publisher, will give her instructions on that profitable specialty. *Market Gardening and Farm Notes*, by Burnet Landreth, one of the foremost practical and scientific horticulturists in the United States, will be as good an education in gardening as can be had from a book.

A Chicago paper is responsible for the following story concerning an Illinois widow:

> Her capital consisted of a comfortable house located in a large barren village lot, a stable, and one cow. She had three dependent children, and no income. After due consideration and preparation, she had the lot plowed in early

spring, and converted it into one large strawberry bed, while around its sides were planted blackcap raspberries. She selected standard reliable varieties, and gave her plants good and thorough cultivation. The next spring her plants were strong and thrifty, and in good bearing condition. A compact was made with her grocer, who undertook the sale of the entire crop. When the season was over and settlements made, the widow felt well-repaid for all her work and anxiety, for her berries had returned sufficient over expenses to provide for all the needs of herself and children till the next spring. Then she secured an adjacent vacant lot on a long lease, at a low rent, and filled it with the increase of plants from her original patch. The question of support was settled. There was no need for her to leave her home to labor, and last but by no means least, she was able to interest and employ her children, to teach them the lesson of self-help and mutual help, and to keep them under her care. In tilling the soil on a large scale, women seem to be as successful as in the berry patch.

The success of Kate Sanborn as a farmer has been too widely and interestingly heralded to need more than passing mention here.

About seven years ago there moved about the town of Uxbridge, Massachusetts, a young girl named Sarah A. Taft, to whom life had offered no occupation which was at all congenial to her tastes. Her friends, noting her slim figure, pale face, and the tiny hands which mated feet which number two shoes covered, shook their heads and smiled when she declared that she wanted a farm. After a time she managed to gain her heart's desire in a farm located two or three miles from the town of Uxbridge. It was pretty discouraging for the first two years, but knowledge and experience were being gained, and the third year some profit was realized. Then came numerous evils to the young farmer. Her barn, hay and entire stock were burned, she broke her arm, money was stolen from her desk, and her hired help seemed determined to give her all the trouble they possibly could. But she went straight on, rebuilding, reconstructing, learning by every present failure how to make a future success. One who is so fortunate as to visit Beechwood today is driven from the station by a healthy-looking young woman whose small, strong hands guide a pair of handsome grays which are harnessed to a luxuriously upholstered double carriage. After drinking a glass of milk which makes one wish that her hostess might become her milk provider, the visitor is shown over a neat farm now mostly given up to hay, small fruits and poultry. Miss Taft has just built a poultry shed 144 feet long, and expects, with her experience with hens, to reap a good profit from poultry culture. She has done well with small fruits and milk, all her wares being disposed of in the town. Each succeeding year, since she recovered from the effects of her disasters, has brought her more gain, and considering the time of her trial, she may be considered one of the most successful farmers "in all the region round about."

A newspaper correspondent tells the story of a southern woman who found sheep-raising profitable:

> If one has decided to try the sheep venture, as did a southern woman on the same line, let any priestess of an abandoned New England farm, or a Virginia plantation, or an old Pennsylvania homestead, buy her livestock from some reputable farmer or drover, and pay not more than $3.00 apiece for her ewes. If a small flock of sturdy animals are purchased in September, and turned to grass at once, they will feed themselves and ask no care till the stress of winter comes. Somewhere on the bookshelf should be kept a volume of commonsense advice on sheep-raising, and when in doubt as to what is best to be done, counsel should be taken with the author. Under fairly good conditions the drove of eight or ten ewes between January and March ought to be increased to a respectable flock of fifteen or eighteen lambs, and lambs born in January sell in the spring for $7.00 and $10.00 apiece in good markets.
>
> Because her pasture was not large enough, and because she taught school for a living, and so had no great amount of time to give to her flock, the southern

woman did not let her number increase beyond sixty ewes, but some years she drew as much as $500 from her sheep.

At Greenwich, Connecticut, Miss Churchill owns and manages a large dairy farm, making a good profit by sending her milk and cream into the country to supply customers.

The three daughters of the late J. D. Gillett, of Logan County, Illinois, manage three farms whose acres aggregate over four thousand acres. These three young women, who are finely educated, speak French, and have a taste for art, literature and music, are enthusiastic over farming as a profession for women. The farms now yield four times as much as they did when managed by Mr. Gillett. They are divided into small sections which are tilled by tenants with whom the crops are divided. A lake on this land was drained by digging a ditch a mile and a half long. These women often ride thirty or forty miles a day on their tours of inspection.

Mrs. Taber Willett, the woman who so successfully manages a farm of 250 acres at Roslyn, L. I., is described as a small, lithe person, with winning manners, a sweet face, and fine mind.

"I was born a farmer," she declares. "Farmers are born, not made."

"You speak of a new woman farmer, a new woman this, and a new woman that," said Mrs. Willett to a newspaperwoman. "There are no new women, but there are new men; for they are beginning to recognize the worth of women, and to acknowledge it. Women are the same as they always have been, only the sudden opening of the world's eyes to their power has given them courage to strike out and conquer new fields.

"These are my farmer friends," she continued, as she tapped on the glass doors of an immense bookcase, assuring her caller that every reliable work on farming was there, as being acquainted with scientific methods was the only way to farm with profit. On being asked if there was really any profit in farming, she replied emphatically:

> There is just as much profit in farming as ever, and even more, for modern machinery and implements

have reduced the work to a minimum. The farm of today is just like a great factory, and instead of requiring competent hands to turn out hard work, in many cases it only requires raw hands to see that the wheels go round. About a year ago I had about the largest yard of thoroughbred Guernsey cattle in the state, and I used to make all the butter, and attend to a large share of the milking. There were over fifty of them.

In reply to the inquiry if she believed that women were as capable of managing farms as were men, Mrs. Willett replied:

Indeed, I do. Sex makes no difference. Women who work on farms become as healthy and rugged as men. Then they have more patience, and the power to adapt themselves more readily, and their dispositions are such that they grow to love their work in the fields because it brings them nearer to nature, and their work is a constant reminder of the goodness of their Maker. I have done everything that can be done upon a farm, from hoeing potatoes to stacking hay, and there was no task, however heavy, but was lightened by the thought of His touch having been there before.

Of course, there are plenty of women who could not be successful farmers, as there are plenty of men. If a woman loves farming well enough to make a success of it, she'll manage to get a farm somehow, and when she does get it you may be sure she'll make it pay.

EDWARD FRANCIS ADAMS

By 1899, when *The Modern Farmer and his Business Relations* was published, Edward Francis Adams had retired to his farm from a twenty-five year career in business, where he described his role as "agricultural editor of a daily journal." Self consciousness about his lack of formal training in agricultural economics as well as a three-decade hiatus from conventional farm work prompt Adams to declare: "My pecuniary interests and my sympathies are with the farmer." In fact, Adams' deference to the new class of agricultural scientists caused him to solicit a chapter from L. A. Clinton, an assistant agriculturalist in the College of Agriculture at Cornell University, on the subject of the farmer as a man of science—a topic Adams claimed to know little about. Such modesty was unusual for Adams, whose most well-known book was titled *The Inhumanity of Socialism: The Case Against Socialism and A Critique of Socialism*, a treatise based on an inflammatory address Adams delivered to a group of ardent socialists in Oakland, California. The transcript of the speech was lost in the San Francisco fire and earthquake that followed but republished in a revised and expanded edition in 1913.

In "The Agricultural Colleges" Adams provides an important historical record of the late nineteenth-century agricultural college debate as it raged in the years following the passage of the Hatch Act of 1887. In particular, the author summarizes common complaints farmers leveled at the new colleges: their tendency to sway farm children away from yeoman careers and their lack of qualified faculty who practiced what they preached, agriculturally speaking. An advocate of agricultural specialization, on one hand, and a broad liberal arts education on the other, Adams' thinking sometimes proves needlessly dichotomous, as when he writes that a faculty member cannot be an equally accomplished farmer and scholar any more than a student at an agricultural college could be both an able pupil and tireless fieldworker, a myth debunked in the recent anthology *Black Earth and Ivory Tower: New American Essays from Farm and Classroom*. As *Black Earth and Ivory Tower* makes clear, farmer-teachers and farmer-legislators, to name two twin callings practiced successfully and widely during the Golden Age and since, fly in the face of Adams' claims of mutual exclusivity.

In the title section, "The Discontentment of the Farmer," Adams, writing at the end of a period of agricultural depression in the 1890s, takes aim at the farmers' unions and populist parties that had proliferated in lean times. Indeed, Adams' observation that the most economically marginal farmers were also the loudest complainers echoes his critique of socialism, which he claimed "lulls to indolence those who must struggle to survive; because the theories of good men who are enthralled by its delusions are made the excuse of the wicked who would rather plunder than work; because it stops enterprise, promotes laziness, exalts inefficiency, inspires hatred, checks production, and assures waste." While Adams takes care to present all sides in *The Modern Farmer and His Business Relations*, he clearly falls victim to age-old notions of agrarian superiority in arguing "the farmer is the only necessary man." In sum, Adams' work represents an important cataloguing of the dissatisfaction of some American farmers.

The Agricultural Colleges[74]

As is well known, the agricultural colleges of the country received their first endowment by a donation of lands from the United States (see Morrill Acts and Hatch Act). The sale of these lands and the application of the funds were left to the states, some of which have conserved them and have large endowments, while others permitted them to be sold to speculators at low rates, and now suffer for their folly. In addition to this endowment, these colleges now receive an annual money appropriation from the general government, whose expenditure is in some degree supervised by national authority. All these are colleges of mechanics as well as of agriculture, and while called "agricultural" colleges, are as much bound to develop mechanical as agricultural science. In most states there is now a "state university." In some states the national endowment for the agricultural college has been turned over to the state university, in which case the institution is usually known as the "Agricultural College of ..." In other cases the management has been kept separate, and the location made in some other place than the university town. This has frequently been determined simply by the relative strength in the legislature of rival places seeking the advantage of the school and the trade it would bring. For

---

[74] Edward Francis Adams, *The Modern Farmer in his Business Relations: A Study of Some of the Principles Underlying the Art of Profitable Farming and Marketing, and of the Interests of Farmers as Affected by Modern Social and Economic Conditions and Forces* (San Francisco: N. J. Stone Company, 1899), 39–50.

the most part, the so-called "purely agricultural" colleges are weak for the reason that they have not the means to support a sufficient number of instructors, or provide suitable laboratories and shops. To make their money go as far as it will, they are compelled to pay lower salaries than the stronger colleges, and so are unable to attract the strongest men or retain such strong men as they develop. Certainly, I do not say this of all separate agricultural colleges, but this is the tendency. An agricultural college attached to a strong university does not have to maintain professors of botany, mathematics, chemistry, physics, languages, and the like, which are part of the equipment of all great universities, but can expend all its means in the applications of those sciences to agriculture. Higher education is a very costly thing. Mechanics includes machinery, and can not be taught without machinery, and the maintenance of the necessary engineering plant is too great a strain upon the resources of most of the separate agricultural colleges. The inevitable result is that many students are drawn from the colleges of their own states to the larger and better-equipped institutions of other states. In most cases, whether the colleges are separate or included in the state university, they receive aid from the treasury of the state. A great university requires an income of at least $1,000 per day, with a tendency to increase. A first-rate agricultural college alone will need at least $500 per day. It is not believed that any of our strictly agricultural and mechanical colleges have anything like this income, but a first-rate agricultural college can be maintained in connection with a great university upon an income of from $200 to $300 per day.

The proper office of the agricultural college is not well understood. The popular impression of it is as a place where young persons are educated to become farmers. This impression is wholly erroneous. Such was never the intention of an agricultural college, nor is any such function possible to it. We can not afford to spend years in learning one trade, only to forsake it and practice another. Nor is it done. Those who have been sent to agricultural colleges with the expectation of becoming farmers have almost invariably ended by being something else. This has led to the common complaint that agricultural colleges "educate the boys away from the farm." Of course they do. That is what they are for. The error is not in the agricultural colleges but in the false popular impression of their nature and function. A rich man's son, who may expect to become the owner of a large, landed estate, or the poor man's son, who may hope for the superintendency of such a property, will do well to graduate from an agricultural college. As a preparation for the life of an ordinary farmer, a course at an agricultural college is a foolish waste. The reason is, as we shall see, that nine-tenths of the time of the student will be expended in learning what he will have no occasion to apply, and will soon forget.

The office of the agricultural college is to investigate phenomena connected with farm life and the operations of farming, and disseminate the information gathered. Of its graduates, the functions of some must be to continue original investigations, and others to convey what is learned to the people at large. Occasionally, as stated, they will serve as managers of large properties. For this, however, vigor, executive ability, and common sense are essential. An individual possessing these qualities will be made more valuable by the education to be obtained in an agricultural college. But if the knowledge gained is to be applied only to an ordinary farm, the cost of obtaining it would be as unwise an expenditure as the purchase of a combined harvester by the farmer who raises only twenty acres of wheat. Besides, eight years spent in sedentary employment when a young man is almost sure to cause a dislike for the physical work of a farm. Habits are not easy to change.

The graduate of an agricultural college should be able to analyze soils and foods. The farmer does not need to do this, nor could he usually have the laboratory and appliances for accurate work; neither, unless constantly engaged in it, would he be able to retain his skill. Nowadays, when a farmer desires an analysis of his soil, or of any special food, he gets it done for nothing, by experts, by sending it to his agricultural college. Why spend much money to prepare one's self to do poorly that which he can get well done for nothing? The agricultural graduate should be a bacteriologist. That is, he should be familiar with known forms of bacteria, and be able to investigate, isolate, and propagate new forms. It is never necessary for the farmer to do this, nor would he have the appliances to do it, or the time to devote to it. He does not even need to know how they look, for he can never see them without an expensive microscope, nor, if he should see them, would he be able to distinguish one from the other, unless constantly engaged in observing them. It requires a long time to become a bacteriologist, and constant practice to retain the art. It does not pay to teach such a difficult thing to one who can never make use of it. If the farmer suspects the presence of malignant bacteria in a plant or a product, his agricultural college will make the investigation for him. Such accurate knowledge of the anatomy and physiology of plants and animals as will enable one to trace and remove the causes of disease is gained only by dissection and microscopic examination of living organisms, or those which have lived. This art, also, could not be practiced by the farmer if he knew it, for lack of time, practice, and appliances. Why, then, learn it? Most of what he needs to know, in this and other lines, has already been discovered at great cost. It is silly to spend money to do over what has already been well done. His agricultural college and its graduates will tell the farmer what is now known that concerns him, and will undertake the investigation of all new phenomena.

The phenomena of nature are infinite in variety, and new occasions for investigation are constantly arising. Upon these problems able men in all civilized countries are engaged, and the results of their studies are embodied in the special literature of their own languages. To keep abreast of these discoveries, and therefore save the cost of unnecessary duplicating work in investigation, the ability to read other languages than one's own is required. For this reason some part of the time spent in an agricultural college is, or should be, devoted to the acquiring of a reading knowledge of one or more modern languages, usually the German and French, which are rich in accumulated information not accessible in English. It is seldom that any real mastery of these languages is attained by students, but they do become qualified to discover and dig out what they need from time to time as they need it. It is evidently a waste of effort to devote time and money to learning how to imperfectly read what one will never see, when it is certain that the imperfect knowledge acquired will vanish in a year or two without constant use.

The time required to become a graduate of a first-class agricultural college is seven or eight years, of which three or four are spent in preparatory work, and four in the college. The cost can not well be less than $300 per year, or $2,400, besides eight years of time. This sum, and time, after deducting the cost of a reasonable and well-directed education after leaving the common schools, would go far to buy and stock a small farm. If applied entirely to an education, it should be considered a capital sufficient to assure its possessor a means of livelihood for his life without the additional expense of buying a farm.

It should be plain, then, that the agricultural college is an essential factor in assuring agricultural prosperity, but that its graduates are not necessarily to be working farmers. I do not say that the working farmer will not be the better for a complete education, provided he will work on his farm, but after spending time and money to fit himself for other work, he is not likely to do so and seldom does. A person, however, who can not or will not work on his farm ought never to own one. It is useless to know how to do things if we do not do them, and if an agricultural education is worth anything, it must qualify its possessor to do what the uneducated man can not do. The educated farmer who does not work on his farm may have the barren satisfaction of knowing somewhat more definitely than others just where he loses money, but this knowledge not followed up is hardly worth the cost of it.

The agricultural graduate is a professional man, who will usually, like the doctor, the lawyer, and the engineer, gain a living by the sale of his professional knowledge. There is no more need, and need be no more intent, that the graduate in agriculture should own a farm than that a mechanical engineer should own a factory. He is equipped to give professional advice, and, if desired, superintendence. Until lately there has been almost no demand in America for the services of agricultural graduates, and their services are never likely to be so well rewarded as those of other professions. As a consequence, few students are found to take the full agricultural course, and of such as have done so the majority have drifted into other occupations. Nevertheless, as the demand for instruction in the branches of science bearing upon agriculture increases, the attendance at agricultural colleges will increase. It is increasing now. But most students are poor, struggling painfully to acquire knowledge and training by whose sale they can live. They have little use for education which will not sell. General culture is a beautiful thing, but it is a luxury, like a greenhouse or a yacht. A practical education is an education which will sell.

This, of course, is the purely commercial view of the subject. It is the view, so far as my observation goes, which the majority of university students are compelled to take. In the older and richer communities, from which I have been absent for many years, there are doubtless many who, having graduated from an agricultural college, return to the manual labor of the farm. Whether they do or not, however, they have acquired true conceptions of the dignity of agricultural life; of the relationship of agriculture to the other arts; of the claims of agriculture upon the state for material aid, and of the nature of the aid which should be given; and they are prepared, as well-rounded and useful citizens, to become centers of powerful influence in the communities in which their lot may be cast.

For there is another view of higher education, far more comforting and inspiring than the commercial aspect in which I am presenting it, and with which, although I do not make it prominent in these pages, I can not permit myself to be thought unimpressed. It is well to be a good farmer, but it is far nobler to be a good citizen. While from the standpoint of the poor student the mercantile view may of necessity be almost overpowering, yet from the standpoints of the wise statesman who plans public aid for higher education, the strong teachers who rise to the direction of its effort, or the mature citizen who is a lover of his race, the true end of education is not that the educated may be housed, and clothed, and fed, but that with a mind trained to think, and a heart inspired by daily contact with great souls, living and dead, he shall go forth in the fullness of his youthful strength, a good citizen and a noble man, to carry the light which he has received to all who may cross his path.

The Experiment Stations. The original idea of the agricultural college included an experimental farm, usually with the collateral idea that students might earn part of their support by working upon the farm, and that the farm would be some source of income.[75] None of these things were found practical. In the first place, it is not possible to endure the strain of securing a modern education and at the same time to do any considerable amount of physical labor. At least it is possible only to persons of far more than ordinary strength. An old-fashioned education such as answered very well a half century since could be obtained while performing a good deal of physical labor, provided the student had ordinary strength. So much, however, is now required to equip students to compete with others that the strength of the strongest is fully taxed. In fact, no good student can begin to do what he sees he needs to do, and it is only a question of what he shall neglect least or most. Students so engaged certainly need exercise and change, but hoeing potatoes is not found to answer the purpose. It distinctly lacks the element of recreation which all young (and old) people desire and require, and, besides, the potatoes must be hoed when they need it, especially if there is a rain coming up, and the farm necessities were found to continually clash with the requirements of the classroom. The work done under such circumstances had no heart in it, and so was not well-done; a portion only of the students engaged in it, and these could not keep pace with the others in the classes, and the practice has been abandoned nearly everywhere.

Besides the student difficulties there was the trouble with the instructors who were supposed to supervise the operations. It was discovered that there were first-rate professors of agricultural science who were horrible farmers. An accomplished entomologist might not know wheat from barley or a sulky plow from a hay rake. He might know all about bugs, and have a rare faculty of imparting instruction, and so be invaluable in the classroom, and yet be perfectly helpless if

---

[75] This sentence marks the beginning of Chapter III, page 46, "The Experiment Stations," in the 1899 edition, represented here as a subheading of the same name.

called upon to deal with a piece of boggy ground. So it finally came to be seen of all men that few men could be good farmers and good scientists at the same time—not even agricultural professors. A good farmer must be a good executive man, and many good agricultural professors are not such men, and while there are always attached to agricultural colleges some who are both good instructors and good farmers, the two arts can not usually be practiced by the same man at the same time.

Neither was there any chance of an income from an experimental farm. Experiments are costly, and the most of them fail. New plants are tried in order to see whether or not they can be cultivated at a profit; new methods of culture or of feeding are tried, for the same reason. If the plant or the method is not profitable, it is worthwhile to know that, and it is better to have one experiment tried at the public expense, and the result widely published, than for hundreds of individuals to try the same thing, fail in it, and say nothing. These things are not always well understood by farmers, and as a farm which continually tries things which do not work is sure to get a bad name, the college farms, however unjustly, became the subject of derision in their communities to an extent which largely accounts for the general prejudice against agricultural colleges which so long existed among farmers. It was also the fact that it is not always possible, when starting an agricultural college, to secure the services of a full staff of clear-headed, sensible men, with the requisite knowledge and executive ability. Political methods have often controlled appointments, and there have been some weak men connected with the colleges and some foolish things done, such as could hardly occur now. Men have found their places.

But with the abandonment of actual farming in connection with the agricultural colleges, the necessity became more urgent for actual experimental work for the public benefit. To meet this necessity the United States government appropriates $15,000 per annum to each state to be expended exclusively in experiments for the benefit of agriculture. This sum is, I believe in all cases, placed in the hands of someone connected with the agricultural college, and usually the head of it. This person is known as the director of the experiment station, and is held strictly accountable to the United States Treasury for the expenditure of the funds. He is also obliged to publish at least four bulletins in each year, in which the work done at the station is described, with the results.

Some of the most valuable work ever done in the agricultural interest has been performed by these stations, and the work is likely to go on forever, continually increasing in value as more experience is gained. These bulletins, issued by these stations, have come to be the principal sources of exact information in agricultural matters. Experimenting is an art, and it is by no means everyone who can experiment in any such way as to secure valuable results. It is also very costly, and entirely beyond the means of the ordinary farmer.

In a general way all intelligent farmers are now familiar with the work of these stations, and yet their educational value is not well understood. Perhaps their greatest value is in teaching what to avoid. An experiment which is successful on a small scale and with constant watchfulness and care may be wholly unsuccessful under the ordinary conditions of farm life, but a culture which will not succeed under station conditions should not, as a rule, be attempted under ordinary conditions.

Experiments in feeding and digestion have been very valuable. An animal can only grow or yield work by the assimilation of food. When food was abundant and cheap, and the market known, almost any kind of feeding might show a profit. But when food is dear, or prices of animals very low, it becomes an object to know, as exactly as possible, the relative feeding value of all feedstuffs. It is not what an animal eats but what it digests that counts, and if an increase of muscle is desired, it is a waste to feed fat-forming foods in excess, or if work is desired, an excess of flesh-forming foods. If the weight of an animal be taken, and all it eats and drinks for a period be weighed, and at the end of the period the animal be weighed again, it can be determined exactly how much the animal has assimilated from its food. If, then, the food be analyzed, and also all the excreta of the animal, it can be determined just how much of the flesh-forming and heat-giving elements the food contained, how much was digested, and how much voided unused. This, of course, will give the facts only for that particular animal under the particular conditions obtaining, but in the course of time a large number of such experiments have been made at different places and with different foods and animals, from all of which an average can be had which should show very nearly the feeding value of the food in question. At this point exact knowledge stops, and the result is ready to be turned over to the farmer to be used in the light of his own observation and common sense with his own stock. The result is found to be a large saving in the cost of feed per pound of weight gained, or per horsepower of work done. Those who employ these methods can do work or produce meat or milk or wool cheaper than those who do not use them. In like manner all farm operations are experimented upon and tested by exact methods for the benefit of the farmer. Forage and other plants are tested for their food or other value, and new plants as to their adaptability to soil and climate. The effects of fertilizers are also closely tested and different methods of intensive culture.

The experiment stations are among the most valuable of the educational agencies which the public puts at the disposal of the farmer. Their means are limited, and no station can do all the desirable things at once, but gradually an immense fund of accurate information is being gathered by the different stations in this country and Europe, all of which is made available to the agricultural world. Most stations send their bulletins freely to all applicants, and all stations do so to applicants within their own states. The farmer, however, must apply for the bulletins, and, to be benefited, must study them after he gets them and make use of their lessons. The number who as yet do this is extremely small as compared with the whole number of farmers but is rapidly increasing. Those farmers who live near enough to their stations to make them an occasional visit, can benefit still more from them if they will only ask questions. What they will see there will be mostly new to them, for the stations do not spend money in finding out what is known already. A visitor, therefore, who merely passes through without inquiry is not likely to learn much. He will not understand what he sees, and possibly will imagine that the station men are also working in the dark, in which he will be wholly wrong. I do not think the station work is usually well appreciated by farmers living nearest the stations.

Akin to the bulletins of the experiment stations are the publications of the Department of Agriculture in Washington. The greater part of these are prepared in a popular form for general use, and are distributed free upon application to the Secretary of Agriculture at Washington. Some, however, are sold at the cost, as fixed by the public printer, and are obtained by enclosing the price to the Superintendent of Public Documents. None of the documents are sent regularly, as they appear, to any address, except a monthly list of publications, which is mailed regularly to all who request it of the Secretary of Agriculture. From this list the farmer can see what the nation has published for his benefit, and by application can obtain what he desires.

These publications of the experiment stations and the department are among the most available sources of information for farmers. They are supplemented in many, and perhaps all states, by the publications of state boards of agriculture, horticulture, dairying, and similar official bodies, all of which are always mailed free to residents of the state. It is the part of a live modern farmer to know the exact places from which this information is to be had as it appears, and to apply for it and make use of it. Great sums of public money are spent yearly for the benefit of the farmers. The information supplied is authentic and useful. No other industry receives any such assistance from the public, and thus far the hardest task of all has been to get the mass of farmers to take and use the information which is supplied to them without price.

The Discontent of the Farmer[76]. The farmer mingles less than others with other men, and so tends to become introspective—to think about himself.[77] This, when carried to extremes, always produces a morbid condition of mind, as is shown in criminals condemned

---

[76] The distinct chapter "The Discontentment of the Farmer," originally pages 120–129 of *The Modern Farmer in his Business Relations*, is presented here as a subheading under the title "The Agricultural Colleges" by consequence of its sustained discussion of the farmer's intellectual life and alleged resistance to change.

[77] The following sentence has been removed from the original text for contextual purposes: "The subject of this chapter includes the relation of the farmer to himself and to mankind."

to solitary confinement, who tend to become insane. There is no doubt that the comparatively isolated life of the farmer often produces an unhealthy condition of mind, of which the leading symptom is an unreasonable discontent with his own lot. I do not mean by this that all discontent of the farmer is unreasonable, but that some of it is, and that it is often hard to distinguish between that which is and that which is not reasonable.

In discussing the relation of the farmer to the politicians, I have had occasion to point out the low esteem in which the intelligence and astuteness of the farming class is held by those whose trade it is to induce masses of men to vote as they desire, and the same feeling is prevalent to a considerable degree among all classes of men and with regard to all classes of subjects. The business world does not believe in the mental competence of the farmer.

I do not think this popular judgment upon the farmer's capacity well-founded. I think the farming class contains as large a proportion of vigorous although untrained minds as any other class, but they are, for the most part, inarticulate. The habit and environment of the farmer are not favorable to the practice of oral or written speech, nor has he at command the library facilities wherewith to fortify himself for public utterance. It is also true that the majority of farmers, like the majority of other classes, are occupied in transacting business rather than in thinking upon its principles. Finally, they suffer in the estimation of mankind from the utterances of their self-constituted champions. As with working men, the mercantile class, and even lawyers, the capacity of farmers for public speech is often in an inverse ratio to their practical wisdom, and the desire to engage in it is very apt to be in such ratio. The deliverances of men and organizations in whose judgment the substantial farmers themselves have little confidence are assumed to be the opinions of the great body of farmers, from whom, in fact, we really hear very little.

So far as the farmer is discontented it is because he is unable to obtain from the products of his farm the satisfactions which he desires. This covers the entire subject, and so far as the mere statement of the causes of his discontent is concerned, this chapter might well end here. When I have said all that I shall say, this will be the final summing up. But in this the farmer is not peculiar. The majority of mankind are in the same condition. The workman skilled and unskilled is also discontented because the sum of his possible sacrifices will not secure the aggregate of his desired satisfactions. So is the merchant, the lawyer, the doctor, and the money lender.

If it be said here that I should confine myself to such causes of discontent as affect only the farmer, I may say that if I am to deal with real causes, and not their mere operation or manifestation, I know of none which do not operate upon all classes alike, nor do I believe there are such. One law seems to me to govern mankind. Its violation always, and sometimes its fulfillment, causes suffering, and suffering, discontent. Through all nature runs the grim story that the strong survive and the weak perish, and that discontent lies mostly with the weaker. I do not find that well-to-do farmers, or shopkeepers, or working men suffer very much, or that they are possessed of any but that rational discontent which we usually call ambition, and which is the mainspring of progress.

It is then with the poorer farmers that we have mostly to deal in considering the causes of such discontent as is the result of real suffering from economic causes. Almost invariably these will be found in debt beyond their ability to pay. The actual owner of five acres of land from which he gets a living for himself and family is a small capitalist who proves by keeping out of debt that he has adapted himself to his environment and is thus in a way to survive. If he has irrational discontent, the causes thereof are incident to his personal character, and demand no special thought. It is the indebted farmer who complains aloud, and the causes of whose discontent deserve investigation. And yet, putting aside the question of his weakness, which we can not mend, we can not reach the real sources of his discontent without considering the situation not only of all other farmers, but of all other classes.

The causes which are usually assigned as the sources of the farmer's discontent are appreciation of money, inequality of taxation, monopolistic combinations, excessive transportation charges, donations of public lands to be worked in competition with land which farmers have paid for, the exploitation of new lands and inferior races, and the competition of expensive machinery which the small farmer can not own. Most of these subjects are considered in detail in other chapters of this book. In this place it is only necessary to call attention to the fact that the operation of these causes is not uniform even upon farmers. They help some while injuring others. The Nebraska farmer who owes an old debt is hurt by the appreciation of the purchasing power of money, while the Connecticut farmer who holds the little mortgage is greatly comforted thereby. The farmer who hires help in fruit harvest desires cheap labor. The one who, with his family, works in the cannery after his grain is cut, desires wages to be high. All these are farmers. The farmer desires cheap cloth, cheap machinery, and cheap transportation, but the owners and operatives who produce these commodities all desire them to sell high. Here the farmer clashes with other classes. By as much as he is satisfied in these particulars, by so much some other class tends to discontent. We are on boggy ground unless we can hit upon some law which operates uniformly on all.

And this brings us face to face with a question whose determination is an essential prerequisite to any intelligent or useful discussion of the causes of discontent in any class. What ought the farmer to get from his farm? Is his discontent reasonable or unreasonable? If the latter, we need pay no attention to it; if the former, there must be a remedy. But certainly before we can consider a remedy we need to know the exact nature and extent of the injustice. Otherwise we at once sink in a quagmire of muddy thought. The farmer on rich bottomland is discontented because his smaller neighbor will not sell out to him. Ought his farm to yield him the means of offering a temptation which his neighbor can not resist? The farmer on some poor hillside is discontented because his farm will not pay off a debt improvidently incurred for unrenumerative improvements, Ought the poor farm, or society at large, to be held responsible for the unwisdom of its owner? What is a reasonable standard of life for the farmer? What ought to be the reward of the thrift by which means to buy and stock the farm were provided, and the physical and mental capacity required for its successful management? And since the satisfactions of the farmer must largely come from the sacrifices of other classes, we can not determine the measure of the injustice, if any, which the farmer endures, without considering what is due also to others. And if we do not know the exact injustice we can not know the real cause of his just discontent, much less intelligently discuss a remedy. What is the rational standard of life for all of us? What ought we to have the means to procure? Ought the standard to be uniform for all? If not, what ought each class to get, and individuals, according to their ability and thrift, within their class? Can we improve on nature's method of letting us all fight it out among ourselves? It is easy to make a list of things which farmers do not like, and call them the causes of their discontent; but are they causes, or are they mere manifestations of some deeper cause?

I think the latter, and that in this country the one cause of unreasonable discontent among farmers is the absence of any well-defined ideal of a standard of life such as most farmers may reasonably hope to reach and maintain. Our experience in the development of a new country has been such that boundless opportunities seemed open to all; all have had, consequently, boundless ambition, and boundless ambitions are never satisfied. Of course we may go still deeper, if we desire, and inquire whether unreasonable discontent is not part of man's nature, and irremediable except by a slow evolutionary process which shall kill off the discontented, if indeed it is not that class which is likely to survive. But it is bad enough to get out of economics into ethics without leaving that perhaps rather boggy ground for the wide ocean of psychology and anthropology and evolution, wherein no man may touch bottom. And yet, when we approach an economic subject, we are compelled to recognize that beneath every problem in economics there lies a deeper problem in ethics which can not be ignored if there is to be intelligent discussion. The ethical problem which underlies the question of the causes of the farmer's discontent, is that of a rational standard of the farmer's life, and as the farmer is not alone in the world, of the rational standard of all life.

In attacking this subject I am compelled to break what, so far as I can discover, is new ground. So far as this subject has been considered at all it has been approached from the standpoint of the artisan or the urban resident. It seems to me that the first class to be considered is the farmers, because this is the only class essential to the existence of the race. Other classes exist for its comfort, and are entitled to due consideration, but from an economic standpoint the farmer is the only necessary man. In addition to that, his manner of life is in a great measure fixed by external causes entirely beyond his control, which is true of hardly any other class except sailors. Early in the morning and late at night the farmer must minister to the animals which serve him. When the ground is right, he must cultivate it, sometimes as long as he can see. When the crop is ripe, he must work hard and for long hours to gather it. There can be no eight-hour day for him. Why should he work twelve hours that those who make his shoes or build his houses should work only eight hours? And the farmer's wife, how many hours must she work? And the workman's wife? Who ever heard of an eight-hour day for them. What is just for one is just for all, men and women. And the necessary standard of the farmer must be the basis for other standards.

It seems to me that the farmer and his wife must expect to work twelve hours, on the average, every day, some of the time at light work. I believe they are perfectly healthy in so doing, and happy when they get reasonable satisfactions in exchange for their sacrifices. I think a reasonable satisfaction for a farmer is a comfortable but modest home, abundant but plain food, plenty of stout work clothing, and a good suit for Sunday, a comfortable conveyance to take his family to church in, moderate education for a reasonable number of children, and such an income beyond that as will enable him to safely, when a young man, incur interest-bearing debt for half the value of the land which he tills, with the expectation of paying it off by the time he is fifty years old, and retiring from labor when sixty. For his own blunders or extravagance he must pay, and by as much as he expends effort in this way, by so much he should fail of earned satisfactions in other respects. I think this a just standard of life for the farmer and that the standards of other lives should be based on this.

Only the stout farmers can now attain this standard—those who would get over the heads of their fellows in any walk of life, not always, by the way, the most useful or the most amiable of men, but the most effective. It is unsafe for the farmer of ordinary abilities to incur debt, and without debt it is not usually possible for young farmers to get farms. The reason of it is that too large a portion of their income is taken by capitalists, traders, and working men, of whom the latter take by far the most, and yet not sufficient for their necessities.

It seems, then, to me, that the fundamental cause of the farmer's discontent is the absence from his own mind of a clear-cut ideal of the measure of the reward which is justly due him from society, with, in many cases, a notion more or less vague of an ideal which he could not reach without injustice to other classes; for it must not be forgotten that although the farmer is the only man necessary to our existence, and as such certainly entitled to something more than what may happen to be left after satisfying others, yet it is to other classes, after all, that we are indebted for what really makes existence worth supporting, and they are entitled to their due. To this fundamental trouble is added the fact that the farmer, through ignorance—largely willful— of facts, and his unwillingness to work with his fellows, does not in the main succeed in getting what he should and might have.

As for the "remedies" proposed or possible for this state of things I have not much to say here. Many of them are discussed in other chapters. Many of those proposed by the farmers themselves I do not think of much value. But, as I have already said, it is not the wisest farmers that we hear most from. Those who speak, who are usually the indebted, do not seem to me to always understand their own case. A great physician once said to me that he never knew a man who was smart enough to correctly count his own pulse. Perhaps most of us are not good judges of our own ailments. The articulate farmers do not say that they are weak and erring, or recognize that with the end of our new lands the era of speculative farming is over in America, and that of close business methods begun, but attribute their trouble rather vaguely to "rings," "combinations," and "monopolies." Now these agencies, so far as they exist, can inflict economic injury upon the farmer only by lowering the price of what he has to sell, or raising that of what he has to buy. As a matter of fact they have not done the latter, for all commodities, even transportation, which the farmer has to buy, are now lower than at the date when any long-standing debt was incurred. The farmer must have estimated, if he estimated at all, upon paying more for what he buys than he does pay, and in so far as the prices of what he has to sell have fallen, they have done so largely as the result of cheaper production and of overproduction in which he has himself participated, often induced by the representations of bright and unscrupulous men that the impossible would happen. I have little faith in the legislative remedies to which farmers continually turn, not even in such matters as the currency and the tariff. So only that any policy be made permanent, it seems to me that we shall all get on about alike under it. As matters stand the only remedy which I have to suggest for the next few generations is that farmers must know more and fight harder. As a farmer I desire cheap labor, cheap transportation, and cheap merchandise of all kinds that I have to buy, and high prices for everything that I have to sell. All other classes are in the same situation. It is a law of commerce that no agreement is possible between contending interests, until after an exhaustive trial of strength. It is also a law that agreement, when possible, invariably follows such contests. We are now engaged in the conflict, which I suppose will continue indefinitely, and sometime close in the usual way. None of us now here will see the close, or can even imagine the nature of a settlement. But we know that it will be made, for it is the law.

The law of nature is that the due of every man shall be all that he can get and keep. She seems to have filled the world with warring organisms, for the mere pleasure of seeing them fight, caring nothing who shall conquer. This process, which results in the untimely death of the majority, we have learned in these later years to call evolution. But evolution has at last developed a race which, having overcome all other beings, shows signs of trying conclusions with nature herself. The history of mankind is a story of constant warfare with nature, and constant victory for man. Of later years there is an increasing number of those who are determined to abolish the method of distribution of satisfactions according to strength alone, which is the cornerstone of nature's method. Who shall say that the progress of evolution shall not destroy evolution itself? Possibly it may, and then may come the end of all things. Certainly we can have no conception of life unaccompanied by struggle. The condition wherein no struggle is, we call death. This may be all wrong, since the ages which will be required for such development may be sufficient for the evolution in us of qualities which will make existence without struggle endurable. We can not tell. Such a race would, perhaps, differ no more from us than we differ from the primeval man. So far as we can now conceive, it would be a race of degenerates.

But if we fight nature, we must fight together. So long as we are divided, she is too strong for us. And yet by her own law she may be contributing to her own destruction, since, if we reach unity, it must be by the extinction of those who will not cooperate. The single man can not compete with organized man—is not now doing it successfully. The best organized are the most prosperous. Those who will not cooperate will die. The farmer, being now less organized than some others, suffers in ways that we have seen. Gradually he will cooperate more, because his environment will compel him. In part, it may be the present farmers who will do this, and, in part, the successors of those who will die because they will not cooperate. At present, his incompetence to act for the common ends with his fellows is a great cause of the farmer's discontent.

It will be seen that I do not think that the real causes of the farmer's discontent lie upon the surface, or can be removed by local treatment. If the farmer could pay all his debts with what he calls cheap money, be relieved from all taxation, and be given free transportation for himself and his produce, with a bounty on all his exports, so far as he is discontented he would be discontented still. Possibly he would be more discontented, since the seemingly more genial environment would attract to the industry still more of the weaker sort. As I have said, we do not know what qualities may yet develop in our race, but, as we are now, the rugged conditions of competition tend to eliminate discontent by the brutal process of the destruction of the discontented.

We are all children of one family governed by one law. I can conceive of no adequate treatment of the condition of the farmer except in connection with the condition of others. In the operation of the law which controls all of us, I have seemed to myself to find the real causes of the farmer's discontent, and to get some glimpse of the remedies which must cure it. The first step is to decide what is a reasonable standard of life for the industrious, honest, and thrifty, with which, when attained, we should be content, and beyond which desire is unreasonable. The next step is to learn how to control the unreasonable. Here is where we must fight nature, who favors the unreasonable, if they be strong. We do not need to concern ourselves with the lazy, the dishonest, or the unthrifty. Nature will take care of them. She will kill them.

CHARLOTTE PERKINS GILMAN

In recent years, revisionists have contrasted the historical sexism of the Midwest, particularly the states that made up the Old Northwest Territory, with the more egalitarian, less gendered traditions of the Great Plains pioneers, as celebrated short fictionist Charlotte Perkins Gilman does here. Citing "those sturdy races where the women were more like men, and the men no less manly because of it" Gilman echoes the popular *Lake Woebegone*-ism: "where all the women are strong, the men are good looking, and the children are above average."

In "The Martyr and the Pioneer" Gilman aims her feminist rhetoric at an audience of naysayers, arguing persuasively for women's fundamental equality with men. While many of the day's male conservationists and agrarians increasingly decried mechanization—a condemnation that found its most poignant collective expression in *I'll Take My Stand* in 1930—Gilman uses the advent of the machine age to leverage her argument for a woman's place in the workplace. "So the pressure of industrial conditions demands an ever-higher specialization," Gilman writes, "and tends to break up that relic of the patriarchal age, the family as an economic unit."

To agrarian traditionalists and Luddites, Gilman's triumphal sexism— "she arose, and followed her lost wheel and loom to their new place, the mill"—unfairly minimizes the pain of job losses created by the Industrial Age. Ever rhetorically able, Gilman anticipates such anxieties in "The Martyr and the Pioneer," arguing that women will always be uniquely feminine regardless of the "men's roles" they may take on. Particularly significant to the farm wife of the era, whose combination of "cook-nurse-laundress-chambermaid-housekeeper-waitress-governess ... 'jack of all trades' and mistress of none" often meant chronic overwork, Gilman's essay is informed by her own suffering, which included a nervous breakdown soon after her first marriage and child and the subsequent divorce she initiated before moving to California and leaving her daughter in the care of her ex-husband. Though most famous for her feminist short story "The Yellow Wallpaper," and her novel *Herland* (1915)—describing what happens when three men discover an extant, two thousand-year-old female-only society—Gilman's nonfiction, including *Women and Economics* (1898), established her reputation as a leading spokesperson for women's rights.

The Martyr and the Pioneer[78]

The Anglo-Saxon blood, that English mixture of which Tennyson sings, "Saxon and Norman and Dane though we be" is the most powerful expression of the latest current of fresh racial life from the north, from those sturdy races where the women were more like men, and the men no less manly because of it.[79] The strong, fresh spirit of religious revolt in the new church that protested against and broke loose from the old, woke and stirred the soul of woman as well as the soul of man, and, in the equality of martyrdom, the sexes learned to stand side by side. Then, in the daring and exposure, the strenuous labor and bitter hardship of the pioneer life of the early settlers, woman's

---

[78] Charlotte Perkins Gilman, *Women and Economics; A Study of the Economic Relation Between Men and Women* (Boston: Small, Maynard, & Co., 1899), 147–160.
[79] The title "Martyr and the Pioneer," the first named heading for a number of page-long sections occupying pages 147–160, is here deployed as an umbrella title.

very presence was at a premium, and her labor had a high economic value. Sex-dependence was almost unfelt. She who molded the bullets and loaded the guns while the men fired them was co-defender of the home and young. She who carded and dyed and wove and spun was co-provider for the family. Men and women prayed together, worked together, and fought together in comparative equality. More than all, the development of democracy has brought to us the fullest individualization that the world has ever seen. Although politically expressed by men alone, the character it has produced is inherited by their daughters. The federal democracy in its organic union, reacting upon individuals, has so strengthened, freed, emboldened, the human soul in America that we have thrown off slavery, and with the same impulse have set in motion the long struggle toward securing woman's fuller equality before the law.

This struggle has been carried on unflaggingly for fifty years, and fast nears its victorious end. It is not only in the four states where full suffrage is exercised by both sexes, nor in the twenty-four where partial suffrage is given to women, that we are to count progress but in the changes legal and social, mental and physical, which mark the advance of the mother of the world toward her full place. Have we not all observed the change even in size of the modern woman, with its accompanying strength and agility? The Gibson Girl and the Duchess of Towers, these are the new women; and they represent a noble type, indeed. The heroines of romance and drama today are of a different sort from the Evelinas and Arabellas of the last century. Not only do they look differently, they behave differently. The false sentimentality, the false delicacy, the false modesty, the utter falseness of elaborate compliment and servile gallantry which went with the other falsehoods—all these are disappearing. Women are growing honester, braver, stronger, more healthful and skilful and able and free, more human in all ways.

The change in education is in large part a cause of this, and progressively a consequence. Day by day the bars go down. More and more the field lies open for the mind of woman to glean all it can, and it has responded most eagerly. Not only our pupils, but our teachers, are mainly women. And the clearness and strength of the brain of the woman prove continually the injustice of the clamorous contempt long poured upon what was scornfully called "the female mind." There is no female mind. The brain is not an organ of sex. As well speak of a female liver.

Woman's progress in the arts and sciences, the trades and professions, is steady, but it is most unwise to claim from these relative advances the superiority of women to men, or even their equality, in these fields. What is more to the purpose and easily to be shown is the superiority of the women of today to those of earlier times, the immense new development of racial qualities in the sex. No modern proverbs, if we expressed ourselves in proverbs now, would speak with such sweeping, unbroken contumely of the women of today as did those unerring exhibitors of popular feeling in former times.

The popular thought of our day is voiced in fiction, fluent verse, and an incessant play of humor. By what is freely written by most authors and freely read by most people is shown our change in circumstances and change in feeling. In old romances the woman was nothing save beautiful, high-born, virtuous, and perhaps "accomplished." She did nothing but love and hate, obey or disobey, and be handed here and there among villain, hero, and outraged parent, screaming, fainting, or bursting into floods of tears as seemed called for by the occasion.

In the fiction of today women are continually taking larger place in the action of the story. They are given personal characteristics beyond those of physical beauty. And they are no longer content simply to be: they do. They are showing qualities of bravery, endurance, strength, foresight, and power for the swift execution of well-conceived plans. They have ideas and purposes of their own, and even when, as in so many cases described by the more reactionary novelists, the efforts of the heroine are shown to be entirely futile, and she comes back with a rush to the self-effacement of marriage with economic dependence, still the efforts were there. Disapprove as he may, use his art to oppose and contemn as he may, the true novelist is forced to chronicle the distinctive features of his time, and no feature is more distinctive of this time than the increasing individualization of women. With lighter touch, but with equally unerring truth, the wit and humor of the day show the same development. The majority of our current jokes on women turn on their "newness," their advance.

No sociological change equal in importance to this clearly marked improvement of an entire sex has ever taken place in one century. Under it all, the crux of the whole matter goes on the one great change, that of the economic relation. This follows perfectly natural lines. Just as the development of machinery constantly lowers the importance of mere brute strength of body and raises that of mental power and skill, so the pressure of industrial conditions demands an ever-higher specialization, and tends to break up that relic of the patriarchal age, the family as an economic unit.

Women have been led under pressure of necessity into a most reluctant entrance upon fields of economic activity. The sluggish and greedy disposition bred of long ages of dependence has by no means welcomed the change. Most women still work only as they "have to" until they can marry and "be supported." Men, too, liking the power that goes with money, and the poor quality of gratitude and affection bought with it, resent and oppose the change, but all this disturbs very little the course of social progress.

A truer spirit is the increasing desire of young girls to be independent, to have a career of their own, at least for a while, and the growing objection of countless wives to the pitiful asking for money, to the beggary of their position. More and more do fathers give their daughters, and husbands their wives, a definite allowance, a separate bank account, something which they can play is all their own. The spirit of personal independence in the women of today is sure proof that a change has come.

For a while the introduction of machinery which took away from the home so many industries deprived woman of any importance as an economic factor, but presently she arose and followed her lost wheel and loom to their new place, the mill. Today there is hardly an industry in the land in which some women are not found. Everywhere throughout America are women workers outside the unpaid labor of the home, the last census giving three million of them. This is so patent a fact, and makes itself felt in so many ways by so many persons, that it is frequently and widely discussed. Without here going into its immediate advantages or disadvantages from an industrial point of view, it is merely instanced as an undeniable proof of the radical change in the economic position of women that is advancing upon us. She is assuming new relations from year to year before our eyes; but we, seeing all social facts from a personal point of view, have failed to appreciate the nature of the change.

Consider, too, the altered family relation which attends this movement. Entirely aside from the strained relation in marriage, the other branches of family life feel the strange new forces, and respond to them. "When I was a girl," sighs the gray-haired mother, "we sisters all sat and sewed while mother read to us. Now every one of my daughters has a different club!" She sighs, be it observed. We invariably object to changed conditions in those departments of life where we have established ethical values! For all the daughters to sew while the mother read aloud to them was esteemed right, and, therefore, the radiating diffusion of daughters among clubs is esteemed wrong, a danger to home life. In the period of the common sewing and reading, the women so assembled were closely allied in industrial and intellectual development as well as in family relationship. They all could do the same work, and liked to do it. They all could read the same book, and liked to read it. (And reading, half a century ago, was still considered half a virtue and the other half a fine art.) Hence the ease with which this group of women entered upon their common work and common pleasure.

The growing individualization of democratic life brings inevitable change to our daughters as well as to our sons. Girls do not all like to sew, many do not know how. Now to sit sewing together, instead of being a harmonizing process, would generate different degrees of restlessness, of distaste, and of nervous irritation. And, as to the reading aloud, it is not so easy now to choose a book that a well-educated family of modern girls and their mother would all enjoy together. As the race become more specialized, more differentiated, the simple lines of relation in family life draw with less force, and the more complex lines of relation in social life draw with more force, and this is a perfectly natural and desirable process for women as well as for men.

It may be suggested, in passing, that one of the causes of "Americanitis" is this increasing nervous strain in family relation, acting especially upon woman. As she becomes more individualized, she suffers more from the primitive and undifferentiated conditions of the family life of earlier times. What "a wife" and "a mother" was supposed to find perfectly suitable, this newly specialized wife and mother, who is also a personality, finds clumsy and ill-fitting, a mitten where she wants a glove. The home cares and industries, still undeveloped, give no play for her increasing specialization. Where the embryonic combination of cook-nurse-laundress-chambermaid-housekeeper-waitress-governess was content to be "jack of all trades" and mistress of none, the woman who is able to be one of these things perfectly, and by so much less able to be all the others, suffers doubly from not being able to do what she wants to do, and from being forced to do what she does not want to do. To the delicately differentiated modern brain the jar and shock of changing from trade to trade a dozen times a day is a distinct injury, a waste of nervous force. With the larger socialization of the woman of today, the fitness for and accompanying desire for wider combinations, more general interest, more organized methods of work for larger ends, she feels more and more heavily the intensely personal limits of the more primitive home duties, interests, methods. And this pain and strain must increase with the advance of women until the new functional power makes to itself organic expression, and the belated home industries are elevated and organized, like the other necessary labors of modern life.

In the meantime, however, the very best and foremost women suffer most; and a heavy check is placed on social progress by this difficulty in enlarging old conditions to suit new powers. It should still be remembered it is not the essential relations of wife and mother which are thus injurious, but the industrial conditions born of the economic dependence of the wife and mother, and hitherto supposed to be part of her functions. The change we are making does not in any way militate against the true relations of the family, marriage, and parentage, but only against those sub-relations belonging to an earlier period and now in process of extinction. The family as an entity, an economic and social unit, does not hold as it did. The ties between brother and sister, cousins and relatives generally, are gradually lessening their hold, and giving way under pressure of new forces which tend toward better things.

The change is more perceptible among women than among men, because of the longer survival of more primitive phases of family life in them. One of its most noticeable features is the demand in women not only for their own money, but for their own work for the sake of personal expression. Those who object to women's working on the ground that they should not compete with men or be forced to struggle for existence look only at work as a means of earning money. They should remember that human labor is an exercise of faculty, without which we should cease to be human, that to do and to make not only gives deep pleasure, but is indispensable to healthy growth. Few girls today fail to manifest some signs of this desire for individual expression. It is not only in the classes who are forced to it: even among the rich we find this same stirring of normal race-energy. To carve in wood, to hammer brass, to do "art dressmaking," to raise mushrooms in the cellar, our girls are all wanting to do something individually. It is a most healthy state, and marks the development of race-distinction in women with a corresponding lowering of sex-distinction to its normal place.

In body and brain, wherever she touches life, woman is changing gloriously from the mere creature of sex, all her race-functions held in abeyance, to the fully developed human being, nonetheless true woman for being more truly human. What alarms and displeases us in seeing these things is our funny misconception that race-functions are masculine. Much effort is wasted in showing that women will become "unsexed" and "masculine" by assuming these human duties. We are told that a slight sex-distinction is characteristic of infancy and old age, and that the assumption of opposite traits by either sex shows either a decadent or an undeveloped condition. The young of any race are less marked by sex-distinction, and in old age the distinguishing traits are sometimes exchanged, as in the crowing of old hens and in the growing of the beard on old women. And we are therefore assured that the endeavor of women to perform these masculine economic functions marks a decadent civilization, and is greatly to be deprecated. There would be some reason in this objection if the common racial activities of humanity, into which women are now so eagerly entering, were masculine functions. But they are not. There is no more sublimated expression of our morbid ideas of sex-distinction than in this complacent claiming of all human life processes as sex-functions of the male. "Masculine" and "feminine" are only to be predicated of reproductive functions, processes of race-preservation. The processes of self-preservation are racial, peculiar to the species, but common to either sex.

If it could be shown that the women of today were growing beards, were changing as to pelvic bones, were developing bass voices, or that in their new activities they were manifesting the destructive energy, the brutal combative instinct, or the intense sex-vanity of the male, then there would be cause for alarm. But the one thing that has been shown in what study we have been able to make of women in industry is that they are women still, and this seems to be a surprise to many worthy souls. A female horse is no less female than a female starfish, but she has more functions. She can do more things, is a more highly specialized organism, has more intelligence, and, with it all, is even more feminine in her more elaborate and farther-reaching processes of reproduction. So the "new woman" will be no less female than the "old" woman, though she has more functions, can do more things, is a more highly specialized organism, has more intelligence. She will be, with it all, more feminine, in that she will develop far more efficient processes of caring for the young of the human race than our present wasteful and grievous method, by which we lose 50 percent of them, like a codfish. The average married pair, says the scientific dictator, in all sobriety, should have four children merely to preserve our present population, two to replace themselves and two to die, a pleasant method this, and redounding greatly to the credit of our motherhood.

The rapid extension of function in the modern woman has nothing to do with any exchange of masculine and feminine traits: it is simply an advance in human development of traits common to both sexes, and is wholly good in its results. No one who looks at the life about us can fail to see the alteration going on. It is a pity that we so fail to estimate its value. On the other hand, the growth and kindling intensity of the social consciousness among us all is as conspicuous a feature of modern life as the change in woman's position, and closely allied therewith.

THORSTEIN VEBLEN

As seen in "Conspicuous Consumption," Veblen's socio-economic theorizing is in large part based on a sometimes implicit, but more often explicit, comparison between the leisure class and the agrarian class from which, in large part, the leisure class arose, and to which it reacted. Having grown up in a Norwegian immigrant farming community in Wisconsin, Veblen's agrarian foundations were as much geographical and geocultural as ideological. His farm-bred cynicism reportedly made him a poor fit for teaching, though he held professorships at the University of Chicago, University of Wisconsin, and Stanford. In the essay that follows, Veblen uses journeyman printers, transient and urbane, to illustrate the quintessential difference between the urban worker and the yeoman firmly yoked to hearth and home.

The agrarian and conservationist, then as now, finds resonance in Veblen's analysis of "noble" and "ignoble labor" as well as the mystifying process by which the man of the leisure class recasts as recreation aspects of a rural life—hunting, for example—at which his ancestors once labored by pure necessity. What often comes of this reactionary work instinct, according to Veblen, is a series of quasi-artistic, quasi-scholarly surrogate pursuits allowing the leisured man a "make believe" world of simulated industry and workmanlike leisure. Remarkable for its examination of Jeffersonian notions of agrarian primacy viewed through an economic lens, the theory of the leisure class, by virtue of its articulation of the recreational desires of the moneyed class, provides a solid theoretical foundation for the eco- and farm tourism industries of the modern day.

Conspicuous Consumption[80]

Conspicuous consumption claims a relatively larger portion of the income of the urban than of the rural population, and the claim is also more imperative. The result is that, in order to keep up a decent appearance, the former habitually live hand-to-mouth to a greater extent than the latter. So it comes, for instance, the American farmer and his wife and daughters are notoriously less modish in their dress, as well as less urbane in their manners, than the city artisan's family with an equal income. It is not that the city population is by nature much more eager for the peculiar complacency that comes of a conspicuous consumption, nor has the rural population less regard for pecuniary decency. But the provocation to this line of evidence, as well as its transient effectiveness, are more decided in the city. This method is therefore more readily resorted to, and in the struggle to outdo one another the city population push their normal standard of conspicuous consumption to a higher point, with the result that a relatively greater expenditure in this direction is required to indicate a given degree of pecuniary decency in the city. The requirement of conformity to this higher conventional standard becomes mandatory. The standard of decency is higher, class for class, and this requirement of decent appearance must be lived up to on pain of losing caste.

Consumption becomes a larger element in the standard of living in the city than in the country. Among the country population its place is to some extent taken by savings and home comforts known through the medium of neighborhood gossip sufficiently to serve the like general purpose of pecuniary repute. These home comforts and the leisure indulged in where the indulgence is found are of course also in great

---

[80] Thorstein Veblen, *The Theory of the Leisure Class: An Economic Study in the Evolution of Institutions* (New York: Macmillan, 1899), 87–96.

part to be classed as items of conspicuous consumption; and much the same is to be said of the savings. The smaller amount of the savings laid by the artisan class is no doubt due, in some measure, to the fact that in the case of the artisan the savings are a less effective means of advertisement, relative to the environment in which he is placed, than are the savings of the people living on farms and in the small villages. Among the latter, everybody's affairs, especially everybody's pecuniary status, are known to everybody else. Considered by itself simply taken in the first degree this added provocation to which the artisan and the urban laboring classes are exposed may not very seriously decrease the amount of savings, but in its cumulative action, through raising the standard of decent expenditure, its deterrent effect on the tendency to save cannot but be very great.

   A felicitous illustration of the manner in which this canon of reputability works out its results is seen in the practice of dram-drinking, "treating," and smoking in public places, which is customary among the laborers and handicraftsmen of the towns, and among the lower middle class of the urban population generally. Journeymen printers may be named as a class among whom this form of conspicuous consumption has a great vogue, and among whom it carries with it certain well-marked consequences that are often deprecated. The peculiar habits of the class in this respect are commonly set down to some kind of an ill-defined moral deficiency with which this class is credited, or to a morally deleterious influence which their occupation is supposed to exert, in some unascertainable way, upon the men employed in it. The state of the case for the men who work in the composition and pressrooms of the common run of printing houses may be summed up as follows. Skill acquired in any printing house or any city is easily turned to account in almost any other house or city; that is to say, the inertia due to special training is slight. Also, this occupation requires more than the average of intelligence and general information, and the men employed in it are therefore ordinarily more ready than many others to take advantage of any slight variation in the demand for their labor from one place to another. The inertia due to the home feeling is consequently also slight. At the same time the wages in the trade are high enough to make movement from place to place relatively easy. The result is a great mobility of the labor employed in printing, perhaps greater than in any other equally well-defined and considerable body of workmen. These men are constantly thrown in contact with new groups of acquaintances, with whom the relations established are transient or ephemeral, but whose good opinion is valued none the less for the time being. The human proclivity to ostentation, reinforced by sentiments of good fellowship, leads them to spend freely in those directions which will best serve these needs. Here as elsewhere prescription seizes upon the custom as soon as it gains a vogue, and incorporates it in the accredited standard of decency. The next step is to make this standard of decency the point of departure for a new move in advance in the same direction, for there is no merit in simple spiritless conformity to a standard of dissipation that is lived up to as a matter of course by everyone in the trade.

The greater prevalence of dissipation among printers than among the average of workmen is accordingly attributable, at least in some measure, to the greater ease of movement and the more transient character of acquaintance and human contact in this trade. But the substantial ground of this high requirement in dissipation is, in the last analysis, no other than that same propensity for a manifestation of dominance and pecuniary decency which makes the French peasant-proprietor parsimonious and frugal, and induces the American millionaire to found colleges, hospitals, and museums. If the canon of conspicuous consumption were not offset to a considerable extent by other features of human nature, alien to it, any saving should logically be impossible for a population situated as the artisan and laboring classes of the cities are at present, however high their wages or their income might be.

But there are other standards of repute and other more or less imperative canons of conduct, besides wealth and its manifestation, and some of these come in to accentuate or to qualify the broad, fundamental canon of conspicuous waste. Under the simple test of effectiveness for advertising, we should expect to find leisure and the conspicuous consumption of goods dividing the field of pecuniary emulation pretty evenly between them at the outset. Leisure might then be expected gradually to yield ground and tend to obsolescence as the economic development goes forward, and the community increases in size while the conspicuous consumption of goods should gradually gain in importance, both absolutely and relatively, until it had absorbed all the available product, leaving nothing over beyond a bare livelihood. But the actual course of development has been somewhat different from this ideal scheme. Leisure held the first place at the start, and came to hold a rank very much above wasteful consumption of goods, both as a direct exponent of wealth and as an element in the standard of decency, during the quasi-peaceable culture. From that point onward, consumption has gained ground, until, at present, it unquestionably holds the primacy, though it is still far from absorbing the entire margin of production above the subsistence minimum.

The early ascendancy of leisure as a means of reputability is traceable to the archaic distinction between noble and ignoble employments. Leisure is honorable and becomes imperative partly because it shows exemption from ignoble labor. The archaic differentiation into noble and ignoble classes is based on an invidious distinction between employments as honorific or debasing, and this traditional distinction grows into an imperative canon of decency during the early quasi-peaceable stage. Its ascendancy is furthered by the fact that leisure is still fully as effective an evidence of wealth as consumption. Indeed, so effective is it in the relatively small and stable human environment to which the individual is exposed at that cultural stage that, with the aid of the archaic tradition which deprecates all productive labor, it gives rise to a large impecunious leisure class, and it even tends to limit the production of the community's industry to the subsistence minimum. This extreme inhibition of industry is avoided because slave labor, working under a compulsion more rigorous than that of reputability, is forced to turn out a product in excess of the subsistence minimum of the working class. The subsequent relative decline in the use of conspicuous leisure as a basis of repute is due partly to an increasing relative effectiveness of consumption as an evidence of wealth, but in part it is traceable to another force, alien, and in some degree antagonistic, to the usage of conspicuous waste.

This alien factor is the instinct of workmanship. Other circumstances permitting, that instinct disposes men to look with favor upon productive efficiency and on whatever is of human use. It disposes them to deprecate waste of substance or effort. The instinct of workmanship is present in all men, and asserts itself even under very adverse circumstances. So that however wasteful a given expenditure may be in reality, it must at least have some colorable excuse in the way of an ostensible purpose. The manner in which, under special circumstances, the instinct eventuates in a taste for exploit and an invidious discrimination between noble and ignoble classes has been indicated in an earlier chapter. Insofar as it comes into conflict with the law of conspicuous waste, the instinct of workmanship expresses itself not so much in insistence on substantial usefulness as in an abiding sense of the odiousness and aesthetic impossibility of what is obviously futile. Being of the nature of an instinctive affection, its guidance touches chiefly and immediately the obvious and apparent violations of its requirements. It is only less promptly and with less constraining force that it reaches such substantial violations of its requirements as are appreciated only upon reflection.

So long as all labor continues to be performed exclusively or usually by slaves, the baseness of all productive effort is too constantly and deterrently present in the mind of men to allow the instinct of workmanship seriously to take effect in the direction of industrial usefulness, but when the quasi-peaceable stage (with slavery and status) passes into the peaceable stage of industry (with wage labor and cash payment), the instinct comes more effectively into play. It then begins aggressively to shape men's views of what is meritorious, and asserts itself at least as an auxiliary canon of self-complacency. All extraneous considerations apart, those persons (adults) are but a vanishing minority today who harbor no inclination to the accomplishment of some end, or who are not impelled of their own motion to shape some object or fact or relation for human use. The propensity may in large measure be overborne by the more immediately constraining incentive to a reputable leisure and an avoidance of indecorous usefulness, and it may therefore work itself out in make-believe only, as, for instance, in "social duties," and in quasi-artistic or quasi-scholarly accomplishments, in the care and decoration of the house, in sewing-circle activity or dress reform, in proficiency at dress, cards, yachting, golf, and various sports. But the fact that it may, under stress of circumstances, eventuate in inanities no more disproves the presence of the instinct than the reality of the brooding instinct is disproved by inducing a hen to sit on a nestful of china eggs.

This latter-day, uneasy reaching out for some form of purposeful activity that shall at the same time not be indecorously productive of either individual or collective gain marks a difference of attitude between the modern leisure class and that of the quasi-peaceable stage. At the earlier stage, as was said above, the all-dominating institution of slavery and status acted resistlessly to discountenance exertion directed to other than naively predatory ends. It was still possible to find some habitual employment for the inclination to action in the way of forcible aggression or repression directed against hostile groups or against the subject classes within the group, and this served to relieve the pressure and draw off the energy of the leisure class without a resort to actually useful, or even ostensibly useful employments. The practice of hunting also served the same purpose in some degree. When the community developed into a peaceful industrial organization, and when fuller occupation of the land had reduced the opportunities for the hunt to an inconsiderable residue, the pressure of energy-seeking, purposeful employment was left to find an outlet in some other direction. The ignominy which attaches to useful effort also entered upon a less acute phase with the disappearance of compulsory labor, and the instinct of workmanship then came to assert itself with more persistence and consistency.

The line of least resistance has changed in some measure, and the energy which formerly found a vent in predatory activity, now in part takes the direction of some ostensibly useful end. Ostensibly purposeless leisure has come to be deprecated, especially among that large portion of the leisure class whose plebeian origin acts to set them at variance with the tradition of the *otium cum dignitate*. But that canon of reputability which discountenances all employment that is of the nature of productive effort is still at hand, and will permit nothing beyond the most transient vogue to any employment that is substantially useful or productive. The consequence is that a change has been wrought in the conspicuous leisure practiced by the leisure class, not so much in substance as in form. A reconciliation between the two conflicting requirements is effected by a resort to make-believe. Many and intricate polite observances and social duties of a ceremonial nature are developed; many organizations are founded, with some specious object of amelioration embodied in their official style and title; there is much coming and going, and a deal of talk, to the end that the talkers may not have occasion to reflect on what is the effectual economic value of their traffic. And along with the make-believe of purposeful employment, and woven inextricably into its texture, there is commonly, if not invariably, a more or less appreciable element of purposeful effort directed to some serious end.

PART II

THE GOLDEN AGE, READINGS 1900–1920

INTRODUCING THE GOLDEN AGE

The story of agrarian and conservation writing in the Progressive Era, 1900–1920, a bona fide Golden Age for both agriculture and wilderness, begins and ends with American frontier. University of Wisconsin historian Frederick Jackson Turner "closed" the frontier in his visionary 1893 speech, "The Significance of the American Frontier in History," delivered to the American Historical Association at the Chicago World's Fair, and in subsequent speeches throughout the late 1890s. In an 1896 school dedication in Turner's hometown of Portage, Wisconsin, he contemplated what had been lost with the frontier:

> Americans had a safety valve for social danger, a bank account on which they might continually draw to meet losses. This was the vast unoccupied domain that

> stretched from the borders of the settled area to the Pacific Ocean.... No grave social problem could exist while the wilderness at the edge of civilization [*sic*] opened wide its portals to all who were oppressed, to all who with strong arms and stout heart desired to hew a home and a career for themselves.[81]

As the availability of land decreased, farmers proved less likely to move on to greener pastures, and more likely to develop new methods on existing ground. In the Plains, for example, many farmers tried dryland farming and drought-resistant wheat to deal with arid conditions.[82] As the farmer was forced to reckon his future on a single plot, better soil and water conservation followed naturally. He became more specialized and, by dint of that specialization, better able to meet market demands. For preservationists, wilderness advocates, and conservationists, it paid to be out in front of these trends, aware of the land pressures a closing western frontier would create for miners, foresters, farmers, ranchers, sheep grazers, and city officials, and to prepare a strategy accordingly.

In the Progressive Era, yesterday's farmer became today's preservationist, as farmers' sons such as John Muir and Liberty Hyde Bailey became two of the most articulate spokespersons for the cause of nature in the fields and in the wilderness. Mixing familial agricultural backgrounds with cosmopolitan travels and university training produced a new breed of horticulturalist, botanist, agriculturalist, and naturalist dedicated to the average citizen and empowered by specialized training. The term *conservation*, not widely used prior to the Progressive Era, was coined by Gifford Pinchot and his team of experts to suggest the productive, almost agricultural science of resource-managing public lands—an approach also known, somewhat euphemistically, as "right

---

[81] Jim Cullen, *The American Dream: A Short History of an Idea That Shaped a Nation* (Oxford: Oxford University Press, 2003), 142.

[82] Ted Steinberg, *Down to Earth: Nature's Role in American History* (Oxford: Oxford University Press, 2002), 134.

use" or "wise use."[83] After 1905, the year in which Roosevelt appointed Pinchot to head the new U.S. Forest Service, TR consistently sided with "Pinchotism," as it came to be called, as a politically sensible middle ground. (ibid 58).

What strikes the reader of the following selections, published 1900 to 1920, are the tensions inherent in Progressivism itself. Progressivists of the William Jennings Bryan, Theodore Roosevelt, and Woodrow Wilson vintage believed wholeheartedly that technological and scientific literacy, and education more broadly, could cure a host of social, economic, and environmental ills. As a movement it borrowed liberally from Populism, as its chief advocates and administrators were often men with decidedly agrarian leanings. In his fine history of the era, George Mowry describes a cult of political leadership founded in "the image of men dedicated to the social good, an image approximating the hope of Plato for his guardians."[84] Mowry echoes Harold Frederic's characterization of these "changemakers" as "ambitious men and ruthless, but only *ruthless* in their zeal for human advancement" (ibid). However, others, notably environmental historian Samuel P. Hays, have noted the estrangement of some scientific Progressivists from the people they served, a rift exacerbated by geographical and educational difference. Hays writes of a supporting cast of hydrologists, foresters, geologists, and anthropologists for whom "loyalty to professional ideals, not close association with the grassroots public, set the tone."[85]

Indeed the Progressive Era was a time of organization, alliance, and partisanship of one stripe or another, as committed individuals looked for collective bargaining power. The Boone and Crockett Club, represented here in readings from three of its most prominent members—Theodore Roosevelt, George Bird Grinnell, and Henry

---

[83] Benjamin Kline, *First Along the River: A Brief History of the U.S. Environmental Movement*, (Lanham: Acada Books, 2000), 57.
[84] George Mowry, *The Era of Theodore Roosevelt and the Birth of Modern America* (New York: Harper, 1958), 88.
[85] Kline, *First Along the River*, 56.

Fairfield Osborn—serves as an interesting case in point. Founded in 1888 by Teddy Roosevelt in association with a founding nucleus of Elihu Root, Madison Grant, and Henry Cabot Lodge, Boone and Crockett's stated goal was the endorsement of big-game hunting. And yet, as the selections that follow make clear, its mandate was both preservationist and conservationist. Introducing the 1904 edition of the Boone and Crockett Club's *American Big Game in Its Haunts; The Book of the Boone and Crockett Club*, Grinnell celebrates the club's founder, Roosevelt, as an exceptional man—part sportsman, part conservationist, part politician, and part scientist—in short, a man for the times. Grinnell writes:

> It is not too much to say, however, that the chair of the chief magistrate has never been occupied by a sportsman whose range of interests was so wide, and so actively manifested, as in the case of Mr. Roosevelt. It is true that Mr. Harrison, Mr. Cleveland, and Mr. McKinley did much in the way of setting aside forest reservations, but chiefly from economic motives; because they believed that the forests should be preserved, both for the timber that they might yield, if wisely exploited, and for their value as storage reservoirs for the waters of our rivers. As a boy Mr. Roosevelt was fortunate in having a strong love for nature and for outdoor life, and, as in the case of so many boys, this love took the form of an interest in birds, which found its outlet in studying and collecting them.[86]

The Sierra Club, founded some five years before Boone and Crockett, also boasted a formidable leader in John Muir, whose notoriety in the first decade of the twentieth century nearly matched, in environmental circles, that of the President. The Club, originally posited by editor Robert Underwood Johnson as a Yosemite and Yellowstone defense organization, was originally founded in San Francisco in 1892

---

[86] George B. Grinnell, ed., *American Big Game in Its Haunts; The Book of the Boone and Crockett Club* (New York: Forest and Stream Publishing Company, 1904), 16.

by twenty-seven men dedicating themselves to "exploring, enjoying, and rendering accessible the mountain regions of the Pacific Coast" and to "enlist[ing] the support of the people and the government in preserving the forests and other features of the Sierra Nevada Mountains."[87] Modeled after the earlier Appalachian Mountain Club—an association of professors and mountain-lovers considered the first permanent organization of its kind in America[88]—the Sierra Club planned to do its part to resist exploitative economic interests in the far West.

Time and again the various groups and associations of the Progressive Era would lobby in the same arenas and concerning the same issues, sometimes standing together, other times not. The battle royal proved to be the proposed Hetch Hetchy Dam, a controversy which excited the passions of a nation and challenged Roosevelt's ambivalent stance on forest preservation. The Hetch Hetchy Valley had originally been preserved as part of the act creating Yosemite National Park in 1890, and the challenging of that most recent declaration demonstrated, for wildlife activists nationwide, the shiftlessness of the government. The controversy, already simmering in the late nineteenth century, came to a full boil after the devastating San Francisco fire and earthquake of 1906, which highlighted the city's urgent lack of fresh water and caused its immediate reapplication for the use of the Valley as a reservoir site.[89] The application, ultimately referred to Congress, divided Roosevelt between his friend Gifford Pinchot—of whom he once said "in all forestry matters I have put my conscience in the keeping of Gifford Pinchot" (ibid 163)—and John Muir, whose time with the President in the Yosemite in 1903 had been described by the President as "the best day of his life!" (ibid 138). Organizations such as The Sierra Club, the Appalachian Club, and the American Scenic and Historic Preservation Society, in

---

[87] Nash, *Wilderness and the American Mind*, 132.
[88] Stephen Fox, *The American Conservation Movement: John Muir and His Legacy* (Madison: University of Wisconsin Press, 1985), 61.
[89] Nash, 161.

conjunction with editorial support from publications such as *The Century* and *Outlook*, mobilized effectively against the building of the dam, stopping San Francisco's application in the Sixth Congress. Ultimately, however, the profundity of San Francisco's immediate needs, just three years after the fire and earthquake, caused even some Sierra Club and Appalachian Mountain Club members to agree with the sentiments of a *San Francisco Chronicle* article that described preservationists as "hoggish and mushy esthetes" (ibid 169). Others, including the city's engineer, Marsden Manson, struck further below the belt, decrying the selfish motives of an opposition made up of "short-haired women and long-haired men" (ibid). The debate's increasingly antagonistic tenor is shown here in Muir's cynical response to the San Francisco lobby, "The Hetch Hetchy Valley" from his 1912 book-length tribute to the park *The Yosemite*.

Ultimately, Congress would take up the issue again in the Woodrow Wilson administration in 1913, approving the Hetch Hetchy grant and authorizing the building of the dam Johnson and Muir so ardently resisted. And while the defeat crushed Muir, he consoled himself with the idea that "the conscience of the whole country had been aroused from sleep."[90] Muir felt his one-time friends and fellow wilderness advocates Theodore Roosevelt, Gifford Pinchot, and California Congressman William Kent had betrayed him. Indeed, Muir had become an increasingly polarizing figure, a fact implicit in his being denied invitation to the important Governor's Conference on the Conservation of Natural Resources, organized by Gifford Pinchot and held at the White House in 1908.

Under Roosevelt, the Progressive Era was a time for government-sponsored regulation, study, and ownership of the land. One of many high profile commissions created by Roosevelt, the Country Life Commission of 1907, consisted of agrarians Liberty Hyde Bailey and Henry Wallace alongside the father of the rural sociology discipline,

---

[90] Nash, *Wilderness and the American Mind*, 180.

Kenyon L. Butterfield, and conservation stalwart Gifford Pinchot. Earlier commissions authorized by Roosevelt suggest the President's linkage of concerns disparately related to the land and included the Public Lands Commission (1903), the Inland Waterways Commission (1907), and the Conservation Commission (1908).

Government intervention came to define the Progressive Era and the Golden Age of agriculture and conservation. Almost inexorably, Progressive priorities—social uplift, environmental conservation, socialization through association—offered relief to families struggling to make a living from the land while simultaneously provoking deep-seeded rural fears of government meddling as exemplified by the Smith-Lever and Smith-Hughes Acts of 1914 and 1917. Thus Progressivism and its champions sought to cure the less appealing aspects of turn-of-the-century country life, while, ironically, these very reformers came to be viewed by some as threats to long-standing rural folkways. "The movement," writes historian John Whiteclay Chambers, "proved both inspiring and confusing" because progressive leaders habitually "mixed new methods with old visions."[91]

Liberty Hyde Bailey, a farmer's son and horticulturist at Cornell University, embodied such tensions. The son of a Michigan apple grower, Bailey had honed his orchardist's skills growing up in a pioneering family in southern Michigan. And yet Bailey's father was one of a growing number of farmers who, to make ends meet, earned an income off the farm. Bailey Sr. and his son were, by inclination and by necessity, organizationally-minded and methodologically-curious, as their membership in the South Haven (Michigan) Pomological Society and "first premium" orchard endorsement suggests.[92] Liberty would go on to study botany at the Michigan Agricultural

---

[91] John Whiteclay Chambers, *The Tyranny of Change: America in the Progressive Era, 1900–1917* (New York: St. Martin's, 1980), 106.
[92] Cornell University Division of Rare Manuscripts. *Liberty Hyde Bailey: A Man for All Seasons*. 2004, http://rmc.library.cornell.edu/bailey/biography/index.html.

College and at Harvard where he studied under America's leading botanist, Asa Gray—experiences that solidified his view that agriculture ought to be considered an academic discipline amenable to scientific inquiry.

Because of Bailey's roots in extension work and his academic inclination, the Country Life Commission report he drafted embodied, for many rural Americans, an overeducated, somewhat paternalistic overreach. Even Bailey's description of the Commission mandate as "the working out of the desire to make rural civilization as effective and satisfying as other civilization" went beyond description to critical evaluation of rural cultures.[93] Indeed the Commission's three principle recommendations—based on dozens of public hearings and over half a million questionnaires—called for a countrywide extension service, continuing surveys, and greater support for agricultural economics and rural sociology in academe. Not surprisingly, the Commission's conclusions, excerpted in this section, closely matched chairman Bailey's personal philosophy as well as his professional bias. In fact the opinions Bailey drafted into the Commission report were critical, but decidedly pro-rural: "The city," Bailey wrote, "exploits the country; the country does not exploit the city."[94]

The irony of the Country Life Commission was that farmers overall were experiencing good times, with a period of especially dramatic prosperity from 1909 to World War I in 1914. This six-year period of profound economic gain represents historians' most narrow designation of the Golden Age. Congress had primed the prosperity pump in 1909 with an Enlarged Homestead Act expanding potential claims to 320 acres of nonirrigable land and nonmineral land in nine select states in the West, while in 1916, the cattlemen's lobby for 640-acre parcels of

---

[93] Cornell University Division of Rare Manuscripts. *Liberty Hyde Bailey: A Man for All Seasons*.

[94] James Penick Jr. "The Progressives and the Environment: Three Themes from the First Conservation Movement" in *The Progressive Era*, ed. Lewis Gould (Syracuse: Syracuse University Press, 1974), 116.

public land resulted in the Stock Raising Homestead Act.[95] The advent of more reliable gasoline-powered tractors between 1912 and 1917 cut workloads and made sodbusting easier for early adopters, though the tractor remained a rarity well into 1920s (ibid). Taken together, key agricultural improvements during the Golden Age meant farmers were, in many ways, better off than other American workers insofar as their purchasing power was concerned (ibid 221). And markets would push still higher until European recovery in 1920, a period which, despite declining prices, still valued the farmer and the food he produced, as Marion Florence Lansing's 1920 tribute essay "At a World Table" affirms.

Of course this very bonanza sharpened differences between agriculturalists—characteristically aligned with the utilitarian school of conservation—and true preservationists. For example, boom times meant that submarginal land in the Plains[96]—land only worth cultivating when prices were high—was pressed back into production, contributing to wind and soil erosion and contributing to a growing agricultural surplus. High prices also encouraged so-called "suitcase farmers"—bankers, teachers, and the like—who turned absentee farmers by planting fall wheat in the Plains and returning in the Spring to harvest their profits.[97] Much of the agricultural and horticultural writing of the time, including Edward Payson Powell's "Renovating the Deserted Homestead," Bolton Hall's "The Garden Yard," and Dallas Lore Sharp's "The Nature Movement" attempt to educate new farmers, gardeners, and nature-lovers drawn away from a professional life to a life on the land. Bolton Hall, in particular, packages a blend of utilitarian conservationism and nature study for the urban or suburban reader. While granting that Thoreau would roll over in his grave at the way "wilderness" experiences had turned to pabulum, Dallas Lore Sharp pitched his argument perfectly for city dwellers and suburbanites in his essay "The Nature Movement." Sharp opines,

---

[95] Hurt, *American Agriculture*, 241.
[96] Hurt, *American Agriculture*, 221.
[97] Steinberg, *Down to Earth*, 135.

"As a nation, we had just begun to get away from the farm and out of touch with the soil. The nature movement is sending us back in time. A new wave of physical soundness is to roll in upon us as the result, accompanied with a newness of mind and of morals."

In actuality, Sharp's valorization of farming, removed as it was from market realities, could not fully comprehend the changes sweeping the large-scale farms and ranches of the Midwest and West. Interestingly, the price correction that followed at the very end of the Golden Age turned some progressive cultivators into full-fledged conservation farmers. Walter Thomas Jack, an Iowa farmer, teacher, and agricultural writer describes the boom times before 1920 thusly in his soil conservation classic *The Furrow and Us*:

> The going was good during the lush war years and the few succeeding years put no tax on personal ingenuity, for prices were good and if yields faltered and fluctuated somewhat, the gap between actual yields and maximum yields was spanned by high prices. There seemed to be no need for worry; old methods of doing things seemed quite adequate and the complications ... of the soil seemed to be of no immediate concern. The soil had always responded to our call.[98]

Jack realized only later, as the Depression loomed, that he had "inadvertently expected too much of mother earth" and would henceforth "instead of mining her resources ... study her way of doing things, listen to her hunger signs, and appreciate the fact that she mothers us all" (ibid 25). The similarity of Jack's comments to, in particular, Liberty Hyde Bailey's, hints at a maturing conservation consciousness among the more prosperous farmers of the Middle West. Certainly Jack's wholehearted conversion was anomalous among farmers who weathered the economic downturn of the 20s. Most farmers that

---

[98] Walter Thomas Jack, *The Furrow and Us* (Philadelphia: Dorrance and Company, 1946), 23.

survived the 20s, according to scholars Randall S. Beeman and James Pritchard, did so by "adopting the methods of modern industrialized agriculture," keeping them in a "cycle of debt and dependence on expensive chemical inputs."[99] In sum, the rare economic prosperity of the Golden Age prepared the seedbed for wilderness conservation and sustainable agriculture by affording Americans the circumspection of a life lived less hand to mouth. Others found the sharp downturn of the postwar period sufficiently compelling, and sobering, to turn towards conservation practices. In any case, a permanent agriculture and forestry was urgently needed, as scientists such as Nathaniel Southgate Shaler attempted to conduct, albeit belatedly, a fair and accurate accounting of the Earth's remaining natural resources. In "Earth and Man" Shaler warns of the developed world's increasing appetite for oil and petroleum, while cautioning against the agricultural mindset involved in the extraction of such: "As soon as agriculture begins, the ancient order of the soils is subverted." Meanwhile, Great Plains farmers broke an estimated 32 million acres of sod between 1909 and 1929[100] with an army of increasingly gasoline-powered engines.

As one moved further west in what Frederick Jackson Turner claimed was a frontier-less America, conservation and agricultural practice increasingly diverged. In California, high prices for irrigated fruit and vegetables trumped those for wheat and created conditions typified by large-scale, immigrant-staffed, commercial fruit growing. The immigration of large numbers of Japanese and Mexican fieldworkers further diversified California's San Joaquin Valley, where writer Mary Hunter Austin lived and from whence she wrote the desert study "The Land of Little Rain." Austin witnessed firsthand the machinations of the Newlands Reclamation Act of 1902, which traded public land for farmers' ability to impound water for irrigation purposes over the long term. While the Reclamation Service, which had

---

[99] Randall S. Beeman and James Pritchard, *A Green and Permanent Land: Ecology and Agriculture in the Twentieth Century* (Lawrence: University of Kansas Press, 2001), 4.
[100] Hurt, *American Agriculture*, 235.

authorized twenty-four large irrigation projects in fifteen western states by 1907,[101] was considered an engineering triumph and proof positive of the role science played in fruitful farming—its obstacles: participant ignorance, participant indebtedness, and participant inexperience—were, in the very same year, identified by the Country Life Commission as national rather than regional rural problems.

Despite waving the banner of ruralism, the Country Life Commission's criticism did not sit well with many on the land. If they were heard at all, the Commission's conclusions were resisted in many parts of rural America as top-down ultimatums foreign to the real farmer's workaday existence. In fact, Liberty Hyde Bailey and his crew straddled several resurgent, often conflicting political and cultural forces—the university, the farm, and the church—unintentionally inflaming all of them. By 1900, the term "social science," used to describe a number of academic disciplines represented in the readings in this section, had entered popular usage. The ideology and methodology of these "sciences," infused not just the Commission's report, but also the Progressive political agenda. In 1914 Walter Lippman published *Drift and Mastery* in praise of science as the "only discipline which gives assurance that from the same set of facts men will come approximately to the same conclusion."[102] And yet, clearly in the case of the Country Life Commission and the National Conservation Committee, both of which began reporting their findings in 1908, opinions did differ among those living on the land and those social scientists studying it. In many ways, the farmer could not help but feel singled out by the report of the Country Life Commission, which asked of him, or hoped for him, "better farming, better living, and better business."[103] As an individual producer operating on the tightest of profit margins, the family farmer felt entitled to more empathetic consideration than that given, for example, to an environmentally-negligent timber

---

[101] Hurt, 240.

[102] George Mowry, *The Era of Theodore Roosevelt and the Birth of Modern America* (New York: Harper, 1958), 92.

[103] Danbom, *Born in the Country*, 169.

company or offensive mining outfit. No other class of worker, the farmer was quick to point out, was a victim of such careful scrutiny and eager rebuke. Questions of land use and resource allocation haunt farm and conservation writings from 1900 to 1920. In his book *The Holy Earth*, Liberty Bailey writes that the things of the earth "do not belong to man to do with them as he will. Dominion does not carry personal ownership."[104] And yet the notion of private property and the dominion guaranteed therein were sacred to the American farmer, who nevertheless, at least in his ideal incarnation, stewarded the land with the same naturalist's hand and benign green thumb Bailey possessed as a trained horticulturalist and a so-called expert on country life. If the farmer could not be trusted to use his land responsibly and to continue its benevolent ownership, would American farming ever be the same?

The daring, some would say presumptuousness, of the Country Life Commission's studying, inventorying, and mapping what had theretofore been a traditional, if not somewhat closed agrarian society encouraged a generation of social scientists towards field study. In "Country Fetes" Charles Josiah Galpin documents the country's enduring folkways. In "Local Degeneracy" Wilbert Lee Anderson pushes back against members of the urban press who decry rural poverty and small town "depletion" by validating the authority of local eyewitness. "For the man who has seen with his own eyes," Anderson writes, "is to be trusted rather than the man who has not seen." Similarly, rather than project the cause of rural deprivations onto the farmer, in "Church and Community" Warren H. Wilson, argues that country churches must do a better job serving the farmer, not vice versa. Professor of rural economics John Lee Coulter's essay "The Argument for Cooperation," on the other hand, better typifies the social scientist's response to the Commission's report, as it confirms the committee's belief that country life was indeed losing ground—and population—to city life

---

[104] Cornell University Division of Rare Manuscripts. *Liberty Hyde Bailey: A Man for All Seasons*.

on account of its own self-inflicted backwardness. Coulter argues that rural folk must become more professional, more collegial—in short, more like city businessmen. He defends his right to minister to rural Americans, writing, "It is much pleasanter (but probably less useful), to draw pictures of the beautiful flowers, to roam in the fields and listen to the twittering birds, and to enjoy the blessed sunshine, than it is to turn to the conditions surrounding us which need improvement and worry over the existing social, economic, and political problems."

In fact, Coulter and the Country Life Commission's calls for increased fellowship and cooperative exchange among rural producers was already well underway by 1902 with the formation of the Farmers' Union. The Union's platform, advocating collective action by farmers, improvements in agricultural education, and improved living conditions for farmers and their families[105] predated the Country Life Commission's report by some half dozen years. Likewise, the Non-partisan League, formed in North Dakota in 1915, went so far as to advocate a moderate form of agrarian socialism on the Plains, including the establishment of state-owned enterprises such as grain elevators and banks. Those who claimed that the farmer had no taste for, or ability in, politics were proven wrong by the NPL, who actually succeeded in electing their man, Lynn J. Frazier, as governor and passing a series of Progressivist reforms in the North Dakota legislature, including a nine-hour workday for women (ibid 262). The Farmers' Union and the NPL, while not truly national movements, nonetheless showed an organizational moxie equal to the Sierra Club, the National Conservation Association, or the American Game Protective Association—all prominent conservation players in the Progressive Era.

A number of essays in this section address the Country Life Commission's recommendations for improving family life, either amending the Country Lifers' prescriptions, or, in most cases affirming them. In particular, and in keeping with the Progressive interests in youth education as the most efficient route to societal change, the plight of

---

[105] Hurt, *American Agriculture*, 260.

rural school children receives attention. Farm women, too, another particular concern of Liberty Bailey's, receive a close and sympathetic read. Independent of Section Six of the Country Life Commission Report he edited—excerpted in the following pages as "Woman's Work on the Farm"—Bailey had, from his position as head of the Bureau of Nature Study and Farmer's Reading Course, earlier written a letter addressed directly to the farmer's wife, encouraging her to "talk back" and providing her postage to do just that. Bailey opens the letter with this stark statement: "In all the vocations of life, there are none in which success depends so much on the wife as in farming, and we never think of an unmarried farmer."[106] Bailey's domestic concern is amplified in two excerpts from William A. McKeever included here under the single title "Selections from *Farm Boys and Girls,*" readings dedicated to the well-being of the farmer's son and the farmer's wife respectively. Following suit, Martha Foote Crow's essay "The Country Girl: Where is She?" pays badly needed attention to the condition of the young woman in rural America.

While writings like these did much to publicize inequities and differences between city and country and between men and women, the real agent for change in the Golden Age, as in other epochs, proved economic. By 1905, for instance, the Post Office Department had created some 24,000 rural routes that brought catalogs, bulletins, and newspapers to many rural doors, allowing farmers to conserve the time once required for in-town shopping—a fact lamented by many village merchants dependent on rural dollars.[107] By the end of the Age of Prosperity, rural electricity and radio reached an increasing percentage of American farms, bringing, on one hand, how-to farm broadcasts and weather reports destined to improve farming practice, and, on the other, a decrease in the use of renewable energy sources traditionally used on the farm, including steam engine, windmill, and

---

[106] Cornell University Division of Rare Manuscripts. *Liberty Hyde Bailey: A Man for All Seasons.*
[107] Hurt, *American Agriculture,* 272.

water wheel (ibid 270). In any case, most scholars agree with preeminent farm historian R. Douglas Hurt's assessment that "profit, or lack of it, shaped and changed the face of the countryside" (ibid 276) more dramatically than any government report, just as economic rather than altruistic motives had powered the creation of Yellowstone and Yosemite National Parks.

In Washington, Roosevelt and his chief of the U.S. Forest Service, Gifford Pinchot, believed conservation was the duty of every citizen, though the exact meaning of conservation remained ambiguous. Pinchot's definition of conservation and forest management as "the art of producing from the forest whatever it can yield for the service of man" or, alternately, "to make the forest produce the largest amount of whatever crop or service will be most useful and keep on producing it,"[108] put him in direct conflict with, on one hand, naturalists like John Muir—who favored a less exploitative use of nature—and, on the other, William Taft and Secretary of the Interior Richard Ballinger, who leaned towards the Army Core of Engineers and the anti-conservation lobby in the West. In his essay "Moral Issue," Gifford Pinchot fashions a necessarily ambivalent definition, understandable given the political climate of the time, arguing "that it is about as important to see that the people in general get the benefit of our natural resources as to see that there shall be natural resources left."

H. W. Foght's essay "Nature Study and School Grounds" joins Progressivist concerns for the environment and efficiency with the Country Life Commission's emphasis on the schools as agents of change in rural America. Foght, in championing the belief that "we have all along relied too much on textbooks to the neglect of real, living nature" conjures an image of a revitalized rural school curriculum teaching a love of the land by firsthand experience. If children could not access nature in its grander sense, Foght suggested, school grounds themselves could be turned into makeshift botanical gardens. In this belief, Foght stood side by side with Liberty Hyde Bailey,

---

[108] Penick, "The Progressives and the Environment," 125.

perhaps the most well-known advocate of nature study, who wrote, "If it were possible for every person to own a tree and to care for it, the good results would be beyond estimation."[109]

The needs of the land consistently brought together farmers, conservationists, and preservationists during the Golden Age, 1900 to 1920. Two of the leading figures of the day, Liberty Hyde Bailey and John Muir, represented, at various stages of their evolution, all three camps. Pinchot himself, eminently influential not only in the coining of the term *conservation*, but also in the articulation of its principles, would continue to bring farmers, foresters, and business interests together after leaving government. In the National Conservation Exposition of 1913, for which Pinchot and W. J. McGee served as advisors, the stated goal was to "teach farmers and timberland owners the necessity for general cooperation."[110]

The results achieved during this dynamic period produced undeniable results. President Theodore Roosevelt, a one-time rancher and abiding country life enthusiast, would create fifty-three Federally-protected wildlife areas, beginning with Pelican Island, Florida in 1903. Under the American Antiquities Act, TR would authorize national monuments for preservation, including the first national monument, Devil's Tower, and the Petrified Forest National Monument. Many others followed, including the Grand Canyon National Monument in 1908. In 1916, the National Parks Service Act provided for the creation of the National Park Service in the Department of Interior, increasing oversight and coordination in the care and management of the nation's parks. Wildlife, too, found protection under Woodrow Wilson and Congress in the form of the Migratory Bird Act of 1913, and the Migratory Bird Treaty Act of 1918. Likewise, air and water received attention from the Progressive Era administrations via the Inland Water Commission and, in an important

---

[109] Cornell University Division of Rare Manuscripts. *Liberty Hyde Bailey: A Man for All Seasons*.
[110] Library of Congress, "Evolution of the Conservation Movement, 1850–1920," http://memory.loc.gov/ammem/amrvhtml/cnchron6.html.

first step against large-scale urban water pollution, the 1909 "Act to Prevent the Dumping of Refuse Materials in Lake Michigan at or near Chicago." Concerns for air and water quality, presciently identified by Ellen H. Richards in her essay "Protection of Water Supplies as a Conservation of Natural Resources" initiated a concern for what was called "human conservation." Human ecology's failure to capture the public interest to the extent that wildlife preservation had speaks to the tremendous disenfranchisement and voicelessness of the working poor, who bore the brunt of air and water pollution in the Golden Age. Still, a class-based explanation for the inattention to human conservation offers only a partial explanation, as Henry Wallace evinces in "The Health of Farm Folk." Wallace, trusted by rural residents sufficiently to earn the nickname Uncle Henry, explores the poor health conditions endemic to the country, citing reports by the American Medical Association and the National Educational Association attesting to poorer health among country schoolchildren than their city counterparts.

Of course, the Progressivists fell victim to their fair share of environmental oversights, among them a failure to acknowledge the good work done by conservationist farmers and the authorization in 1915, at the behest of farming and ranching interests, of the Bureau of Biological Survey to begin killing predators such as wolves and coyotes. Thus, while William Temple Hornaday's "The Former Abundance of Wildlife" chastises the American farmer and sportsman for their needless predation of America's most noble species, it does not extend the call for protection to so-called "noxious." Similarly, "Birds,"—Mary Huston Gregory's impassioned plea to farmers and gunners to spare insect-eating songbirds—conveniently excludes "birds of prey" such as hawks, eagles, and owls. Despite Hornaday's advocacy for the buffalo and Roosevelt's just-in-time creation of the National Bison Range in Montana, other animals in addition to coyotes and wolves—including mountain lions, bobcats, and others deemed injurious to livestock— were ruthlessly eliminated throughout the Progressive Age. In Wyoming alone forty thousand animals were exterminated between

1916 and 1928.[111] The effect of this wholesale slaughter would be felt throughout the West in the 1920s, as the disruption of native ecologies resulted in overpopulation of deer and, after the hunting of the hawk, plagues of mice in parts of California (ibid).

In the final analysis, neither specialization nor bureaucratization nor parochialism could prevent those who loved the land in the Golden Age of agriculture and conservation from finding one another, forming alliances, and dialoguing vigorously to achieve many firsts in the care and cultivation of national lands. And though neither the physical sciences nor the social sciences proved a panacea for farmers or for conservationists, a corps of well-respected university-trained experts, especially Wilbert Lee Anderson, Liberty Hyde Bailey, Kenyon L. Butterfield, Charles Josiah Galpin, H. W. Foght, Gifford Pinchot, Ellen H. Richards, and Dallas Lore Sharp, proved to many skeptics that academic training promised to illuminate—rather than castigate—the various ways which Americans expressed their love of the land.

FARM PAGES: THE GOLDEN AGE

The six farm and conservation pages that follow, accompanied by original, how-to prose, evidence the then-expanding audience for Golden Age farm and conservation literature. The era's improving country roads and rural mail delivery meant greater accessibility to books for rural residents while the continued settlement of the Plains and Midwest further decentralized publishing, a phenomenon suggested here by the inclusion of publications from Lincoln, Nebraska's Woodruff Collins Press and St. Paul's Webb Publishing.

The contemporary reader will note in the following images an increased emphasis on plainspoken how-to advice for the hobby farmer and nature enthusiast alike, a demand answered by major East Coast publishers such as Doubleday, McGraw-Hill, and Macmillan.

---

[111] Steinberg, *Down to Earth*, 146.

As Country Lifers—and popular nature writers such as Dallas Lore Sharp and John Burroughs—established bucolic life as in vogue, land-based writing became, paradoxically, more democratic and more elitist. As the number of country escapes and estates expanded, the audience for farm and conservation writing grew to include the small dairyman, woodlot manager, backyard horticulturalist, vintner, and apiculturalist, among other enthusiasts. The day's how-to writing recommended products (for instance, hive tools and bee brushes), often condescendingly dictated best practices (witness the "third-class man" of the blacksmithing illustration that follows), and highlighted efficiency gains (see "Two Coops from a Barrel"). The tripartite notion then embraced by the popular farm presses—prescription, evaluation, efficiency—echoed the era's voluminous USDA and extension publications produced for "real" farmers and foresters.

Marketing catchphrases such as "practical" and "everyday" abounded in Golden Age titles aimed at a general readership. Land-based writing increasingly sought to be of use, taking advantage of improving print technology to better integrate text with illustrative image. Conversely, out of this empowering do-it-yourself trend—a movement still largely with us today—grew the popular pejorative "book farmer" as uttered by those large-scale or hand-to-mouth producers who, by and by, sought to differentiate themselves from mere hobbyists and recreationists. As Thorstein Veblen's notion of conspicuous consumption and the leisure class took root in the Age of Prosperity, class and cultural differences implicit in the roles of producer and consumer drew more attention. However, the rising popularity and affordability of the automobile and the tractor at the close of the Golden Age may have mitigated diverging, class-based interests, as farmer, woodsman, and urbanite alike turned to books and diagrams (see "Farm Motors") for a firsthand understanding of the intricacies of engines and motors for farm and home.

In order of appearance, the image sources for these engravings include Hardy Campbell's *Campbell's 1907 Soil Culture Manual* (Woodruff Collins, 1907); John Holmstrom's *Standard Blacksmithing ...* (Webb Publishing, 1907); Rolfe Colbeigh's *Handy Farm Devices and How to Make Them* (Orange Judd, 1910); Ralph Clarkson's *Practical Talks on Farm Engineering* ... (Doubleday, 1915); Everett Franklin Phillips' *Beekeeping: A Discussion of the Life of the Honeybee* ... (Macmillan, 1917); and Andrey Potter's *Farm Motors, Steam and Gas Engines, Hydraulic and Electric Motors* ... (McGraw-Hill, 1917);

CAMPBELL'S SOIL CULTURE MANUAL

The stalk sent up from an imperfect seed bed is of slow growth and scant of leaves. The stalk which runs upward from a perfect seed bed spreads out and probably branches, and the leaves are abundant and strong. If there is an abundance of humus in the soil there is stooling out of the stalk so that instead of one upright stem there are two or three or mayhap a dozen stems sent up to bear flowers and grain. It therefore follows that where manure has been used in a man-

A Modern Manure Spreader.

ner to develop in the soil the greatest amount of humus, so that plant food is more than sufficient for the needs of the plant, a very much smaller quantity of seed should be sown per acre than on soil less favorable to growth. If there is too much seed per acre the grain will stool too much and make so heavy a growth that it will stand up. Heavy stooling results in the weak straw carrying down the grain, and in this condition the grain will not fiill and it often happens it cannot be harvested.

Standard Blacksmithing

When we see a mechanic with these arrangements, we know he is rated in the third class.

Correct position and composure of the body must be attained—First, for the ease it gives in performing the work second, it lends grace to the movements, which we should strive to gain, for it is very painful to look at an awkward mechanic, and such a workman will never succeed as well

Figure 2. Incorrect Position.

as the easy workman, whose movements are both correct and graceful.

In looking at Fig. 1, p. 7, you see a smith with free and easy position. Fig. 2 is the opposite to this. Here is a third-class man. Just mark his awkwardness. He presses his elbows against his rib. When you see a smith of this kind, you can rest assured he would not command a very high salary if he were to work for another

TWO COOPS FROM A BARREL

Very good coops can be made at small cost from empty barrels, as shown in this picture. First, drive shingle nails through the hoops on both sides of each stave and clinch them down on the inside. Then divide the barrel in halves, if it is big enough, by cutting through the hoops and the

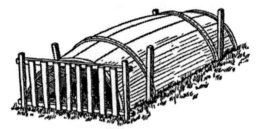

BARREL CHICKEN COOP

bottom. Drive sticks into the ground to hold the coop in place, and drive a long stick at each side of the open end just far enough from coop to allow the front door to be slipped out and in. The night door can be made of the head from the barrel or any solid board, and the slatted door, used to confine the hen, by nailing upright strips of lath to a crosslath at top and bottom.

Weak men wait for opportunities; strong men make them.—Marden.

SAND FILTER FOR RAIN WATER

In this way water is being freshly supplied to the filter at all times. A pipe from the bottom

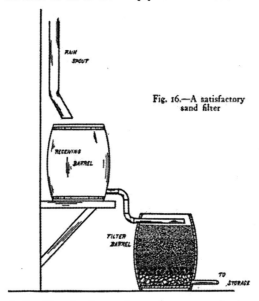

Fig. 16.—A satisfactory sand filter

of the filter leads to the main storage basin. As many receiving barrels as desired may be joined together, and more than one filter barrel may be used if it is desired to filter the water fast.

Apparatus

Black wire-cloth veils are often used and, while they are a better protection than the cloth veils, they are less convenient as they cannot so easily be thrown back.

A steel tool of some kind is needed to pry up covers and to loosen and separate frames. A screwdriver will answer but some specially devised tools (Fig. 27) may be found preferable.

Fig. 26. — Cotton netting veil with silk tulle front.

Gloves of cloth or leather are sometimes used to protect the hands. The handling of frames is less impeded if the finger ends are cut out. Gloves are hot, usually sticky or stiff, and are as a rule abandoned after the early stages of beekeeping are passed.

A brush to sweep bees from the combs is a convenience, especially in removing bees while taking frames from the hives

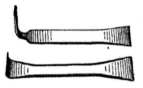

Fig. 27. — Hive tools.

at extracting time. The German brush with white bristles (Fig. 28) is perhaps the best of those manufactured, but a turkey feather, a long whisk broom or a bunch of weeds pulled as needed are as good.

Fig. 28. — German bee brush.

A tool box or portable seat (Fig. 29) and a wheelbarrow or cart for carrying supplies and honey are among the other conveniences used

FARM MOTORS

shown in Fig. 19 is used to close a pipe threaded on the inside or to close a fitting.

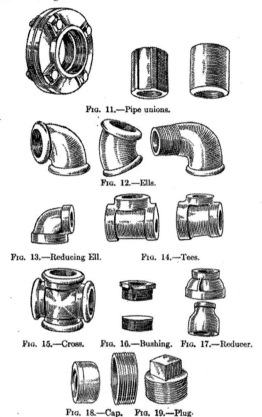

Fig. 11.—Pipe unions.

Fig. 12.—Ells.

Fig. 13.—Reducing Ell. Fig. 14.—Tees.

Fig. 15.—Cross. Fig. 16.—Bushing. Fig. 17.—Reducer.

Fig. 18.—Cap. Fig. 19.—Plug.

Valves.—The function of a valve is to control and regulate the flow of water, steam, or gas in a pipe.

Edward Payson Powell

A Presbyterian pastor, Civil War historian, and Country Life advocate, Edward Payson Powell's long life spanned the Civil War through the Progressive Era, allowing him various authorial reincarnations as evident in his earlier, historical titles *Nullification and Secession in the United States* and *Sermons Preached on Recent National Victories, and the National Sorrow.*

A New Englander, E. P. Powell's observations on the beauty of suburban and country landscaping were made possible by frequent travels throughout the region. In fact, Powell's land ethic fit very tidily into his religious views, two strands he braids in the introduction to *Hedges, Windbreaks, Shelters and Live Fences*, where he writes, "Not a spot exists on the globe that does not need exactly what God put in Eden—a man and a woman to trim and control it. A soul is needed everywhere, and a hand, but a brutish soul and a brute force hand is needed nowhere." Within this context, Powell's less-is-more, beware-the-hand-of-man stance in "Renovating the Deserted Homestead," seems more consistent with twenty-first century trends towards simple living than the interventionist Progressivism of his day. Still, Powell's agenda—a return to the natural elegance of farmer presidents Washington, Jefferson, and Madison, deference to nature, and the purposeful resettling of the country to slow migration towards the cities—is very much rooted in the Golden Age. Powell encapsulates beautifully the trend towards home improvement, a luxury made possible by the economic gains of agriculture's age of prosperity. The largesse created by expanding markets and high prices allowed many farm families to invest for the first time in indoor fineries such as drapes and carpets and outdoor improvements in landscaping. In being judicious and patient in his dealings with mother nature, the home improver and the tree cultivator becomes, in Powell's words, "a public benefactor" rather than "mere autocrat." "Horticulture," Powell writes, "consists first of all in establishing this intimate acquaintance." Powell's book *The Country Home* (1905) is his best known and, like *Hedges, Windbreaks, Shelters and Live Fences* is especially mindful of the suburban homeowner. It is worth noting that government land policy during the Gilded and Golden Age, especially the Timber Culture Act, mandated the kind of tree-planting and landscape improvement—in exchange for deed to public lands—Powell describes.

Renovating the Deserted Homestead[112]

A man's individual life is longer and wider for being lived as part of the family history. Here in this arbor sat our sainted mother; here worked in this garden corner our father. This tree was planted by a grandfather. So everything gets to have a language, if not a poetry. My own homestead was bought by my father direct from the family to whom the Indians donated the land. On a high knoll stands the group of hemlocks of which the Oneida chief, Sconondoah, said: "I am an aged hemlock. The winds of a hundred winters have whistled through my boughs." These orchard trees were planted conjointly by this same chieftain and his missionary friend, Dominie Kirkland. The soil, the brooks, the rocks, the trees, the glen, have associations that unite them together, and give them an individuality. Every man should, if possible, know the history of his own home whether he knows the history of the United States or of the Anglo-Saxon nations or not. It then falls to him to add a chapter to this history, which is inherently beautiful, and useful, and worthy of being carved into trees, hedges, stone walls and buildings.

Still you will have room for exercising the full spirit and zeal of improvement. You will doubtless find there are no driveways and hedges and shelters; or if any, that others are still needed. Windbreaks are likely to be found in abundance. Do not let an axe touch an old clump of basswood, or a thicket, or a tangled mass of hemlock and wild grape—not until you are sure they are not what you want, after they have been cleaned and ordered. A few additions, a few dead limbs cut out, and you are likely to find what nature asks for. Beware of

---

[112] Edward Payson Powell, *Hedges, Windbreaks, Shelters and Live Fences: A Treatise on the Planting, Growth and Management* (New York: Orange Judd Company, 1900), 125–137.

the professional landscape artist who comes to lay out his patented pictures on your land. He will destroy in a day what you cannot recover in a century. Above all, look out for the professional trimmer. He will, if allowed, cut your evergreens into monstrosities. He thinks it beautiful to cut out the middle branches of your spruces, or to cut up from the bottom your pines. He likes green hens on top of hedges, and if let loose, he will absolutely ruin the idea which nature has endeavored to work out. I advise everyone who is going out of the city to take up a country home to be very patient. Take time to think for yourself. Get acquainted with your land. Grow into it. If you were a boy here at one time, renew your association with the past. Plant nothing and cut nothing until you have got the whole place well-gathered into your mind. Indeed, I recommend that you do very little for the first year, except to look out for sanitation and the simplest comforts. You will then be prepared to work in shelters where they are needed; you will know where the wind strikes, and you will be able to get at a shrubbery, and gardens with hedges and appropriate drives. I am sure that by the second year you will have lost the saw and axe passion.

It will generally turn out that, by careful study, you can use a large part of what is at hand, even including some defects. A little management, and a neglected corner, with half-decayed trees and thickets of underwood, can be gently trained and taught to speak of the beautiful and the useful. If you begin with the determination of cutting away everything that, looked at in and of itself is defective, you will end by cutting down everything on the place. Remember, that that which is defective in itself may not be defective in relation to and combination with other things. Often the defective parts have so grown together as to create a unity of another sort, and while your hedges are severely overgrown by other things, you had better not interfere too sharply in your effort to restore absolute precision.

Do not mistake me when I advise you to rely largely upon yourself because you may be the very person above all others who is in need of a wise friend. I do not know you, so it may be as well to add, if you are confident that there is someone to be found who is judicious, who knows how to sympathize with nature, get him to walk with you and counsel you in forming your first impressions. Gardiner, in his *Homes and All About Them* says he would rather dig ditches for a philosopher than build palaces for a fool. There are these two classes also who wish advice about their lawns and their drives. The philosopher thinks, studies, and above all, grows. The fool knows everything at a glance. He cuts trees and he plants trees with a commodore's self-importance. It happens often that in doing this he injures his neighbors as well as himself. No man absolutely owns his acres and trees. He is under moral and sometimes legal obligation to the neighborhood. When he cuts down a grove or a windbreak, he is opening the currents that drive against other people's homes. This an honest man will consider. Let me say to anyone who is going into the country for a home: not only find the relation of the parts of your own land, but try to comprehend the relation which your property bears to that of other people about you. Consult even the prejudices of those who live adjacent. They have formed their associations, their tastes, even their characters, largely from the trees and the collocation of the natural scenery that surrounds them. Disturb them just as little as possible. Indeed, there is a certain sort of property that another man has in what you claim as your own. Emerson sings: "One harvest from your field, / Homeward brought your oxen strong, / Another crop your acres yield, / Which I gather in a song."

My plea is that you be careful of the feelings, the tastes and old associations that make up the neighborhood of which you should be a component part. Press forward even your improvements considerately. It is possible to consult those whose judgment you do not value. In the long run, if you are right, you will improve not only your own property, but all the neighborhood; if you are wrong, and the chances are you will be, you will get time to correct yourself.

Home. The final word is *home.*[113] Everything should have this in view—not a mere residence from which children can take flight,

---

[113] This sentence marks the beginning of Chapter IX, pages 131–137, "Homes," in the 1900 edition, represented here as a subheading of the same name.

but a family home made up of the best that nature gives us, and from which no one cares to go. To create such a home, everything should be made to contribute. If you purpose to grow hedges, or to plant cornfields, or to raise Holstems or Cotswolds as an end, you will prove a flat failure. If all of these things and many more are made constituent parts of homebuilding, you will succeed.

When a man feels that the time has come for him to establish himself on the earth, in other words, to create a home, the first thing he should decide to do is to develop himself into his surroundings, much as a mollusk grows a shell. Yet most people have not given a thought of what they would look like, if all their selfhood or character could be seen as you can see their faces. Understand yourself and your work, or at least set you to discussing what they are. When you have found yourself out, all you have to do is to grow. Grow out first into a house. Don't be fooled by trying to fit your soul into John Jones's shell or into David Williams's. Grow yourself into an easy-fitting, comfortable, warm, cozy jacket of a house. Have a parlor if you need it, but not for somebody else. Of course you and your wife are one, and can grow together. Anyhow, you will have to do this, and so you must let her feel easy also. But when the house is planned, or while it is growing, go on growing all over your place. Make it such that anyone coming along will say, "By George! That's Henry Owen's place! I'd know it by the cut of it!" Go slow—I mean, *grow slow*—and find out where you want a tree, or hedge, or windbreak, or even a rosebush, before you plant it. Every bush, every tree, every fence, every windbreak or hedge should be a part of yourself, and when you get through with your first season's growth it will be apparent that your place means you as much as your body means you.

Then, by and by, when you begin to cut or trim, it will be just as when you pare your nails; it will be because something has overgrown in a perfectly natural way and must be pared off. A real home, rightly planted, never needs to be revolutionized; it is always, however, undergoing evolution. Having started right, you will see something to be added and something to be improved upon each year. A commonsense planter always works with a memorandum—a pocket memory.

Whenever he is about his property he jots down what he sees is needed—every little trifle and every suggested improvement. Every night he looks over his memoranda and marks what is to be done the next day. In this way nothing is overlooked, and fully five times as much progress will be worked in. Nor will breakages and little leakages be overlooked. He will know that a board is loose, that a graft is to be waxed, that the *aphis* have made lodgment on one of his trees, that a new disease is to be fought with Bordeaux, that the time has come for battling the currant worms, or that a brook is washing into his garden, or that his strawberries are in need of water. In this way the mind is everywhere, without too much friction and without too severe a tax of the brain. The owner knows, every minute, everything about his place, and is never compelled to say of anything that is damaged that he had not knowledge of it in due time. I shall place as much emphasis as possible on this point, because I am convinced that no one will succeed with a beautiful rural home in any other way.

Nature takes care to put us into types, but she takes equal care to give us all individuality in features. She says look at your faces, and just take notice how vast the number of copies I can make, and, in all the dissimilarity, I shall not destroy the similarity. Work after the same manner. Do you see that you do not simply try to make what someone else has made; and yet I wish you to follow the general type so as not to create monstrous things—like stone dogs and hedge roosters. John Burroughs says, "One of the greatest pleasures of life is to build a house for one's self," but it is a greater pleasure to build a home. The house of a wise horticulturist is only one of his windbreaks and shelters. It is not here that he should exhaust his cash, but he should expend with equal liberality outdoors and indoors.

I object to outdoor parlors, but I believe in outdoor and indoor sitting rooms. About a beautiful home there is never any occasion for putting up "Keep off the grass." Every lawn should be free to the children and to visitors—at least to the children. But for all that there should be order and system about your home. The best plan is to prepare for games and sports from the very outset—lawn tennis, or croquet, or quoits, or all together. These will naturally draw the young gamesters away from the shrubbery and flower gardens when they wish to romp and play. A croquet ground should be absolutely level, and kept level by nice stone wall, which should rise high enough to stop the balls from rolling into the grass. It should be graded with fine shale and not a weed allowed to grow. Then plant a windbreak, or plant it behind the windbreak. Much of the fun of such a game is spoiled if we cannot play it on cool or windy days. Beside my own ground is a great living arbor in which are chairs, where those who need shade can get it. You will lose nothing by thus making your whole property homeful. You will have kept your boy and girls with you, and no possible influence can attract them away. In other words, they find you yourself everywhere, with your love and your smile.

What we wish to have the common folks see is that the end of home-getting is not to buy someone else's house, not even to have a house that you have built yourself, but that a man or woman who would have a home must begin to live himself or herself outdoors until the grounds are a part of the habitation. Whoever proposes to build a house must rather say, not what will the house cost, but what will the homestead cost and estimate altogether the cost of the planting of live trees as well as the sawing and hammering together of dead ones. If you spend less on dust-holding carpets and curtains, on bric-a-brac furnishings, and more on beautiful grounds, you will live longer and more happily. If a real home grows rather than happens, there will always be present a sense of rest and repose. Hedges, windbreaks, coverts, shelters, suggest protection and comfort, if not they should never exist. The difficulty with many so-called homes is that everything is on edge all the while. You feel the constant presence of shears, and you hear the everlasting and detestable lawn mower—the one implement that never points to rest and to peace, but to clatter and toil. I smell sweat whenever I see one.

A man who builds a house without a room in it except for work and sleep has made exactly the same blunder as he who plants his acres for nothing but work and food. It is an old law that man cannot live by bread alone. A right sort of home should, from its inception, include as an object the beautiful as well as the useful, expecting the two, in combination, to create the good. It is hardly necessary to add that with this idea of home operative, there is no room for mere display. Home wraps one around as clothes wrap a sensible person. They are put on for comfort and good taste, not to exploit wealth. Gardens, trees, hedges, orchards, buildings, say plainly, not I am rich, but I am *at home*. Nothing of this sort can be accomplished in the way of making a true home without sympathy with nature.

Perhaps I have said enough to make it unnecessary to say here that nothing of this sort can be accomplished in the way of making a true home without sympathy with nature. A person who understands a bush is in love with it, and knows what to do with it, and it must be understood that every bush has a character of its own. You may almost say that every tree has a moral character of its own. It is good in one place, and it is bad in another. Horticulture consists first of all in establishing this intimate acquaintance. If it is not established, you can do nothing in the way of wise planting. A city girl visiting my place enjoyed it immensely, but, after running about, picking flowers, and eating fruit for some hours, she sat down on the steps of the house, and taking a survey of the whole, said, "Well, it's immensely pretty, but it must be awful lonely here."

"To be sure," I said, "to you. But don't you see, you don't know anybody here. But to us all these trees and plants have souls. We are all acquainted, and we all understand each other out here. The bushes, and the hedges, and the trees make good company. Your friends all put on golf suits, but mine grow golf suits." The poor girl could not have possibly enjoyed the most beautiful country life for over one day. Her character had never grown a bush; her soul had never developed a rosebud.

Mary Hunter Austin

A gifted writer, naturalist, and folklorist among many—Georgia O'Keefe and D. H. Lawrence, to name two—who would spend a substantial part of their artistic lives in New Mexico, Mary Hunter Austin was one of a number of midwesterners whose family tried its hand at homesteading in the West. A graduate of Blackburn College in her native

Illinois, Hunter displayed a love of the land, and of writing, early on, taking a Chautauqua course in geology when she was twelve and simultaneously declaring her intention to be a writer. Not wholly a farm girl nor a product of the wilderness, Hunter's early adulthood, after homesteading with her family in California's San Joaquin Valley, was filled with a healthy dose of both ranch life and natural study outdoors. After marrying and separating from her husband, a frustrated vineyardist and homesteader, Austin made ends meet by teaching before settling in Carmel, California in 1900, where she helped found an artists' colony. There she rubbed shoulders with, among others, Jack London, whose *Call of the Wild* would be published in 1903 and become a bestseller in an era hungry for tales of primitivism. Hunter's own writing career would likewise be launched in 1903 by the publication of *The Land of Little Rain*, the same year that the newly created National Reclamation Bureau moved to reclaim, for thirsty Los Angeles, the Owens Valley that Austin called home.

While Austin confessed *The Land of Little Rain* took her only a month to write, she spent twelve years, she claimed, "peeking and prying" while it simmered. Like John Muir, Austin came to the attention of Theodore Roosevelt, who solicited her opinion on forest lands and grazing. *The Land of Little Rain* beautifully and impressionistically documents the desert regions of California, particularly Death Valley and the Mojave. In the passage that follows, Austin proves herself a transcendent chronicler of the desert's underappreciated diversity of species and its essential power to romance as well as to seduce. Sympathetic to the region's Indians as well as expert in them, she offers a portrait of the "loneliest land that ever came out of God's hands," a landscape foreign to many easterners and midwesterners because of its ability to dictate terms to would-be settlers. "There are hints to be had here," Austin writes, "of the way in which a land forces new habits on its dwellers." In so saying, the author attends closely not only to the plant and animal inhabitants of this harsh, illusory landscape, but also to the Indian tribes, Anglo prospectors, borax miners, and descendants of Hispanic settlers who attempted to plumb its mysteries. In the few years after the publication of

The Land of Little Rain, Congress would create the Reclamation Service, thereby beginning a decades-long process by which the desert West would be made arable through hydroelectric power and large-scale irrigation.

The Land of Little Rain[114]

If you have any doubt about it, know that the desert begins with the creosote. This immortal shrub spreads down into Death Valley and up to the lower timberline, odorous and medicinal as you might guess from the name, wand-like, with shining fretted foliage. Its vivid green is grateful to the eye in a wilderness of gray and greenish white shrubs. In the spring it exudes a resinous gum which the Indians of those parts know how to use with pulverized rock for cementing arrow points to shafts. Trust Indians not to miss any virtues of the plant world!

Nothing the desert produces expresses it better than the unhappy growth of the tree yuccas. Tormented, thin forests of it stalk drearily in the high mesas, particularly in that triangular slip that fans out eastward from the meeting of the Sierras and coastwise hills where the first swings across the southern end of the San Joaquin Valley. The yucca bristles with bayonet-pointed leaves, dull green, growing shaggy with age, tipped with panicles of fetid, greenish bloom. After death, which is slow, the ghostly hollow network of its woody skeleton, with hardly power to rot, makes the moonlight fearful. Before the yucca has come to flower, while yet its bloom is a creamy cone-shaped bud of the size of a small cabbage, full of sugary sap, the Indians twist it deftly out of its fence of daggers and roast it for their own delectation. So it is that in those parts where man inhabits one sees young plants of *Yucca arborensis* infrequently. Other yuccas, cacti, low herbs, a thousand sorts, one finds journeying east from the coastwise hills. There is neither poverty of soil nor species to account for the sparseness of desert growth, but simply that each plant requires

---

[114] Mary Austin, *The Land of Little Rain* (Boston: Houghton Mifflin, 1903), 10–21.

more room. So much earth must be preempted to extract so much moisture. The real struggle for existence, the real brain of the plant, is underground; above there is room for a rounded perfect growth. In Death Valley, reputed the very core of desolation, are nearly two hundred identified species.

Above the lower tree line, which is also the snow line, mapped out abruptly by the sun, one finds spreading growth of pinion, juniper, branched nearly to the ground, lilac and sage, and scattering white pines.

There is no special preponderance of self-fertilized or wind-fertilized plants, but everywhere the demand for and evidence of insect life. Now where there are seeds and insects there will be birds and small mammals, and where these are, will come the slinking, sharp-toothed kind that prey on them. Go as far as you dare in the heart of a lonely land, you cannot go so far that life and death are not before you. Painted lizards slip in and out of rock crevices and pant on the white-hot sands. Birds, hummingbirds even, nest in the cactus scrub; woodpeckers befriend the demoniac yuccas; out of the stark, treeless waste, rings the music of the night-singing mockingbird. If it be summer and the sun well down, there will be a burrowing owl to call. Strange, furry, tricksy things dart across the open places, or sit motionless in the conning towers of the creosote.

The poet may have "named all the birds without a gun," but not the fairy-footed, ground-inhabiting, furtive, small folk of the rainless regions. They are too many and too swift; how many you would not believe without seeing the footprint tracings in the sand. They are nearly all night workers, finding the days too hot and white. In mid-desert where there are no cattle, there are no birds of carrion, but if you go far in that direction the chances are that you will find yourself shadowed by their tilted wings. Nothing so large as a man can move unspied upon in that country, and they know well how the land deals with strangers. There are hints to be had here of the way in which a land forces new habits on its dwellers. The quick increase of suns at the end of spring sometimes overtakes birds in their nesting and effects a reversal of the ordinary manner of incubation. It becomes necessary to keep eggs cool rather than warm. One hot, stifling spring in the Little Antelope I had occasion to pass and repass frequently the nest of a pair of meadowlarks, located unhappily in the shelter of a very slender weed. I never caught them sitting except near night, but at midday they stood, or drooped above it, half fainting with pitifully parted bills, between their treasure and the sun. Sometimes both of them together with wings spread and half-lifted continued a spot of shade in a temperature that constrained me at last in a fellow feeling to spare them a bit of canvas for permanent shelter. There was a fence in that country shutting in a cattle range, and along its fifteen miles of posts one could be sure of finding a bird or two in every strip of shadow; sometimes the sparrow and the hawk, with wings trailed and beaks parted, drooping in the white truce of noon.

If one is inclined to wonder at first how so many dwellers came to be in the loneliest land that ever came out of God's hands, what they do there and why stay, one does not wonder so much after having lived there. None other than this long, brown land lays such a hold on the affections. The rainbow hills, the tender bluish mists, the luminous radiance of the spring, have the lotus charm. They trick the sense of time, so that once inhabiting there you always mean to go away without quite realizing that you have not done it. Men who have lived there, miners and cattlemen, will tell you this, not so fluently, but emphatically, cursing the land and going back to it. For one thing there is the divinest, cleanest air to be breathed anywhere in God's world. Some day the world will understand that, and the little oases on the windy tops of hills will harbor for healing its ailing, house-weary broods. There is promise there of great wealth in ores and earths, which is no wealth by reason of being so far removed from water and workable conditions, but men are bewitched by it and tempted to try the impossible.

You should hear Salty Williams tell how he used to drive eighteen and twenty mule teams from the borax marsh to Mojave, ninety miles, with the trail wagon full of water barrels. Hot days the mules would go so mad for drink that the clank of the water bucket set them into an uproar of hideous, maimed noises, and a tangle of harness chains while Salty would sit on the high seat with the sun glare heavy in his eyes, dealing out curses of pacification in a level, uninterested voice until the clamor fell off from sheer exhaustion. There was a line of shallow graves along that road; they used to count on dropping a man or two of every new gang of coolies brought out in the hot season. But when he lost his swamper, smitten without warning at the noon halt, Salty quit his job; he said it was "too durn hot." The swamper he buried by the way with stones upon him to keep the coyotes from digging him up, and seven years later I read the penciled lines on the pine headboard, still bright and unweathered.

But before that, driving up on the Mojave stage, I met Salty again crossing Indian Wells, his face from the high seat, tanned and ruddy as a harvest moon, looming through the golden dust above his eighteen mules. The land had called him.

The palpable sense of mystery in the desert air breeds fables, chiefly of lost treasure. Somewhere within its stark borders, if one believes report, is a hill strewn with nuggets; one seamed with virgin silver; an old clayey water bed where Indians scooped up earth to make cooking pots and shaped them reeking with grains of pure gold. Old miners drifting about the desert edges, weathered into the semblance of the tawny hills, will tell you tales like these convincingly. After a little sojourn in that land you will believe them on their own account. It is a question whether it is not better to be bitten by the little horned snake of the desert that goes sidewise and strikes without coiling than by the tradition of a lost mine.

And yet—and yet—is it not perhaps to satisfy expectation that one falls into the tragic key in writing of desertness? The more you wish of it the more you get, and in the mean time lose much of pleasantness. In that country which begins at the foot of the east slope of the Sierras and spreads out by less and less lofty hill ranges toward the Great Basin, it is possible to live with great zest, to have red blood and delicate joys, to pass and repass about one's daily performance an area that would make an Atlantic seaboard state, and that with no peril, and, according to our way of thought, no particular difficulty. At any rate, it was not people who went into the desert merely to write it up who invented the fabled Hassaympa, of whose waters, if any drink, they can no more see fact as naked fact, but all radiant with the color of romance.

I, who must have drunk of it in my twice seven years' wanderings, am assured that it is worthwhile.

For all the toll the desert takes of a man, it gives compensations, deep breaths, deep sleep, and the communion of the stars. It comes upon one with new force in the pauses of the night that the Chaldeans were a desert-bred people. It is hard to escape the sense of mastery as the stars move in the wide clear heavens to risings and settings unobscured. They look large and near and palpitant, as if they moved on some stately service not needful to declare. Wheeling to their stations in the sky, they make the poor world fret of no account. Of no account you who lie out there watching, nor the lean coyote that stands off in the scrub from you and howls and howls.

INTRODUCING THE BOONE AND CROCKETT CLUB—GEORGE BIRD GRINNELL, HENRY FAIRFIELD OSBORN, AND THEODORE ROOSEVELT

The following three, short essays, introduced under the heading of The Boone and Crockett Club, belong to a triumvirate of the club's most prominent members, including George Bird Grinnell, the editor of this, one of several volumes published by the Club before 1920. The 1904 collection highlights the club's conservationism via a variety of articles on hunting, wildlife, and wilderness preservation written by celebrated members, including club founder, Theodore Roosevelt.

Boone and Crockett was one of a number of sportsmen's clubs and associated publications to merge conservationist and sporting concerns, including George Oliver Shield's League of American Sportsmen and its related magazines, *Recreation* and *Shield's Magazine.* Roosevelt, in particular, was sensitive to the class critique of such clubs, usually subscribed to exclusively by well-to-do white men not dependent on wildlife for survival. Roosevelt's vulnerability to such criticism is evident in his comment: "The work of preservation must be carried on in such a way as to make it evident that we are working in the interest of the people as a whole." A short listing of the Boone and Crockett Club's constitutional objectives typifies the eclectic mission of the sportsmen's club prior to the fracturing of its concerns into separate wildlife, conservation, and hunting-fishing associations:

1. To promote manly sport with the rifle.
2. To promote travel and exploration in the wild and unknown, or but partially known, portions of the country.
3. To work for the preservation of the large game of this country, and, so far as possible, to further legislation for that purpose, and to assist in enforcing the existing laws.
4. To promote inquiry into, and to record observations on, the habits and natural history of the various wild animals.
5. To bring out among the members the interchange of opinions and ideas on hunting, travel and exploration; on the various kinds of hunting rifles; on the haunts of game animals, etc.

Also included was an unusual membership requirement stating, "No one shall be eligible for regular membership who shall not have killed with the rifle, in fair chase, by still-hunting or otherwise, at least one individual of each of three of the various kinds of American large game."

The first reading, excerpted from George Bird Grinnell's contribution to the volume he edited, is entitled "The Mountain Sheep and Its Range." Unlike William Temple Hornaday, likewise an advocate for wilderness preservation, writer-editor-publisher George Bird Grinnell embraces the appellation "sportsman," finding in it room for both the naturalist and the scientist. In fact, it was this coalescence of responsible big game hunting and scientific inquiry that distinguished the members of the Boone and Crockett Club from their antithesis: the market gunner. "The Mountain Sheep and Its Range," while displaying Grinnell's preoccupation with the habits and haunts of big game for hunting purposes, is, on closer read, a compassionate, sometimes tender account of what Grinnell insists is the ongoing, tragic extermination of a species. In sum, Grinnell documents the alarming biological and physiological changes wrought by settling of the Great Plains, using the once gentle mountain sheep as a case study and litmus test. Significantly, Grinnell draws his accounts mostly from the farmers and ranchers then learning to share the range with, at least, the more docile big game species.

The second reading, "Life of the Sequoia and the History of Thought," was written by paleontologist, and then-director of the American Museum of Natural History, Henry Fairfield Osborn. Osborn was a controversial figure in his day, though not primarily for the pro-Sequoia stance he articulates here, but for his rejection of the idea of primate ancestry in favor of Non-Darwinian theories centering on will and genetic improvement—theories that would eventually result in his role in the Scopes Monkey Trial. Osborn's keen interest in species origins and his knowledge of ancient and classical history is evident in "Life of the Sequoia and the History of Thought." In this short homage, Osborn departs the animal kingdom to argue lovingly for the preservation of the ancient Sequoia, whose cause John Muir had by this time made public. Osborn belies a modicum of distrust of the American public, as he calls for the passage of urgent legislation protecting the tree while praising the average American for his commendable "veneration of age." Osborn cites the number of Sequoias alive in 1900 at five hundred and the number of logging companies engaged in cutting them down at forty—clearly alarming odds for Sequoia defenders in Congress, in sportsmen's clubs, and among the general public.

Despite his sometimes troubling and unpopular political and racial beliefs, beliefs informed by his religious fundamentalism, Osborn's success as director of the American Museum of Natural History resulted in his being named to the American Association of Museum's Centennial Honor Roll in 2005. The Honor Roll pays tribute to one hundred of America's museum champions who have worked during the past one hundred years to "innovate, improve and expand how museums in the United States serve the public." Importantly, Henry Fairfield Osborn would also pass his love of natural history on to his son, Henry Fairfield Osborn Jr., who later served as president of New York Zoological Society, founder of the Conservation Foundation—which would later become the World Wildlife Fund—and a member of the Save the Redwoods League.

The third and final reading from the Boone and Crockett Club comes from the hand of its founder, Theodore Roosevelt. TR's wildlife encounter narrative cannot help but charm the reader who remembers Roosevelt the bully bloviator. The piece begins in typical Rooseveltian fashion, setting forth an almost apologetic conservationism intended to placate right-wingers. But this personable accounting of his 1903 trip to Yellowstone with the great naturalist John Burroughs suggests a kind of conversion, whereby Roosevelt the hunter gives way to Roosevelt the nature observer. And yet for all its intimacy, Roosevelt's disdain for the mountain lion relative to his obvious appreciation for the park-tamed pronghorn and mountain sheep, bespeaks a president invested in the alteration of nature as well as in the admiration of it. Still, George Bird Grinnell's endorsement of Roosevelt expresses the delight many conservationists, sportsmen, and outdoorsmen experienced at Roosevelt's sudden elevation to the highest office in the land:

> Of the Presidents of the United States not a few have been sportsmen, and sportsmen of the best type. The love of Washington for gun and dog, his interest in fisheries, and especially his fondness for horse and hound, in the chase of the red fox, have furnished the theme for many a writer; and recently Mr. Cleveland and Mr. Harrison have been more or less celebrated in the newspapers, Mr. Harrison as a gunner, and Mr. Cleveland for his angling, as well a his duck-shooting proclivities. It is not too much to say, however, that the chair of the chief magistrate has never been occupied by a sportsman whose range of interests was so wide, and so actively manifested, as in the case of Mr. Roosevelt

The Mountain Sheep and its Range[115] *by George Bird Grinnell*

As in Asia, so in America, the wild sheep is an inhabitant of the high grass land plateaus. It delights in the elevated prairies, but near these prairies it must have rough or broken country to which it may retreat when pursued by its enemies. Before the days of the railroad and the settlements in the West, the sheep was often found on the prairie. It was then abundant in many localities where today farmers have their wheatfields, and to some extent shared the feeding ground of the antelope and the buffalo. Many and many a time while riding over the prairie, I have seen among the antelope that loped carelessly out of the way of the wagon before which I was riding, a few sheep, which would finally separate themselves from the antelope and run up to rising ground, there to stand and call until we had come too near them, when they would lope off and finally be seen climbing some steep butte or bluff, and there pausing for a last look, would disappear.

Those were the days when if a man had a deer, a sheep, an antelope, or the bosset ribs of a buffalo cow on his pack or in his wagon, it did not occur to him to shoot at the game among which he rode. I have seen sheep feeding on the prairies with antelope, and in little groups by themselves in North Dakota, Montana, and Wyoming, and men whose experience extends much further back than mine—men, too, whose life was largely devoted to observing the wild animals among which they lived—unite in telling me that they were commonly found in such situations. Personally, I never saw sheep among buffalo, but knowing as I do the situations that both inhabited and the ways of life of each, I am confident that sheep were often found with the buffalo, just as were antelope.

The country of northwestern Montana, where high prairie is broken now and then by steep buttes rising to a height of several hundred feet,

---

[115] George B. Grinnell, ed. *American Big Game in Its Haunts; The Book of the Boone and Crockett Club* (New York: Forest and Stream Publishing Company, 1904), 273–278.

and by little ranges of volcanic uplifts like the Sweet Grass Hills, the Bear Paw Mountains, the Little Rockies, the Judith, and many others, was a favorite locality for sheep, and so, no doubt, was the butte country of western North Dakota, South Dakota and Nebraska, this being roughly the eastern limit of the species. In general it may be said that the plains sheep preferred plateaus much like those inhabited by the mule deer, a prairie country where there were rough broken hills or buttes, to which they could retreat when disturbed. That this habit was taken advantage of to destroy them will be shown further on.

Today, if one can climb above timberline in summer to the beautiful, green alpine meadows just below the frowning snow-clad peaks in regions where sheep may still be found, his eye may yet be gladdened by the sight of a little group resting on the soft grass far from any cover that might shelter an enemy. If disturbed, the sheep get up deliberately, take a long careful look, and walking slowly toward the rocks, clamber out of harm's way. It will be labor wasted to follow them.

Such sights may be witnessed still in portions of Montana and British Columbia, Idaho, Wyoming and Colorado, where bald, rolling mountains, showing little or no rock, are frequented by the sheep, which graze over the uplands, descending at midday to the valleys to drink, and then slowly working their way up the hills again to their illimitable pastures.

Of Dall's sheep, the white Alaskan form, we are told that its favorite feeding grounds are bald hills and elevated plateaus, and although when pursued and wounded it takes to precipitous cliffs, and perhaps even to tall mountain peaks, the land of its choice appears to be not rough rocks, but rather the level or rolling upland. The sheep formerly was a gentle, unsuspicious animal, curious and confiding rather than shy; now it is noted in many regions for its alertness, wariness, and ability to take care of itself.

Richardson, in his *Fauni-Boreali Americana* says:

> Mr. Drummond informs me that in the retired part of the mountains, where hunters had seldom penetrated,

> he found no difficulty in approaching the Rocky Mountain sheep, which there exhibited the simplicity of character so remarkable in the domestic species; but that where they had been often fired at they were exceedingly wild, alarmed their companions on the approach of danger by a hissing noise, and scaled the rocks with a speed and agility that baffled pursuit.

The mountain men of early days tell precisely the same thing of the sheep. Fifty or sixty years ago they were regarded as the gentlest and most unsuspicious animal of all the prairie, excepting, of course, the buffalo. They did not understand that the sound of a gun meant danger, and, when shot at, often merely jumped about and stared, acting much as in latter times the elk and the mule deer acted.

We may take it for granted that, before the coming of the white man, the mountain sheep ranged over a very large portion of western America, from the Arctic Ocean down into Mexico. Wherever the country was adapted to them, there they were found. Absence of suitable food, and sometimes the presence of animals not agreeable to them, may have left certain areas without the sheep, but, for the most part, these animals no doubt existed from the eastern limit of their range clear to the Pacific. There were sheep on the plains and in the mountains; those inhabiting the plains when alarmed sought shelter in the rough badlands that border so many rivers, or on the tall buttes that rise from the prairies, or in the small volcanic uplifts which, in the north, stretch far out eastward from the Rocky Mountains.

While some hunters believe that the wild sheep were driven from their former habitat on the plains and in the foothills by the advent of civilized man, the opinion of the best naturalists is the reverse of this. They believe that over the whole plains country, except in a few localities where they still remain, the sheep have been exterminated, and this is probably what has happened. Thus Dr. C. Hart Merriam writes me:

> I do not believe that the plains sheep have been driven to the mountains at all, but that they have been

exterminated over the greater part of their former range. In other words, that the form or subspecies inhabiting the plains (*auduboni*) is now extinct over the greater part of its range, occurring only in the localities mentioned by you. The sheep of the mountains always lived there, and, in my opinion, has received no accession from the plains. In other words, to my mind it is not a case of changed habit, but a case of extermination over large areas. The same I believe to be true in the case of elk and many other animals.

That this is true of the elk—and within my own recollection—is certainly the fact. In the early days of my western travel, elk were reasonably abundant over the whole plains as far east as within 120 miles of the city of Omaha on the Missouri River, north to the Canadian boundary line—and far beyond—and south at least to the Indian territory. From all this great area as far west as the Rocky Mountains they have disappeared, not by any emigration to other localities, but by absolute extermination.

Life of the Sequoia and the History of Thought[116]
by Henry Fairfield Osborn

The National and Congressional movement for the preservation of the Sequoia in California represents a growth of intelligent sentiment.[117] It is the same kind of sentiment which must be aroused, and aroused in time, to bring about government legislation if we are to preserve our native animals. That which principally appeals to us in the Sequoia is its antiquity as a race, and the fact that California is its last refuge.

---

[116] George B. Grinnell, ed. *American Big Game in Its Haunts; The Book of the Boone and Crockett Club* (New York: Forest and Stream Publishing Company, 1904), 349–353.

[117] The passage excerpted here as "Life of the Sequoia and the History of Thought" includes three closely related but distinctly titled sections: "Sentiment and Science," "Life of the Sequoia and the History of Thought," and "Veneration of Age."

As a special and perhaps somewhat novel argument for preservation, I wish to remind you of the great antiquity of our game animals, and the enormous period of time which it has taken nature to produce them. We must have legislation, and we must have it in time. I recall the story of the judge and jury who arrived in town and inquired about the security of the prisoner, who was known to be a desperate character; they were assured by the crowd that the prisoner was perfectly secure because he was safely hanging to a neighboring tree. If our preservative measures are not prompt, there will be no animals to legislate for.

The sentiment which promises to save the Sequoia is due to the spread of knowledge regarding this wonderful tree, largely through the efforts of the Division of Forestry. In the official chronology of the United States Geological Survey—which is no more nor less reliable than that of other geological surveys, because all are alike mere approximations to the truth—the Sequoia was a well-developed race ten million years ago. It became one of a large family, including fourteen genera. The master genus—the Sequoia—alone includes thirty extinct species. It was distributed in past times through Canada, Alaska, Greenland, British Columbia, across Siberia, and down into southern Europe. The Ice Age, and perhaps competition with other trees more successful in seeding down, are responsible for the fact that there are now only two living species—the "redwood," or *Sequoia sempervirens*, and the giant, or *Sequoia gigantea*. The last refuge of the *gigantea* is in ten isolated groves, in some of which the tree is reproducing itself, while in others it has ceased to reproduce.

In the year 1900 forty mills and logging companies were engaged in destroying these trees.

All of us regard the destruction of the Parthenon by the Turks as a great calamity; yet it would be possible, thanks to the laborious studies which have chiefly emanated from Germany, for modern architects to completely restore the Parthenon in its former grandeur, but it is far beyond the power of all the naturalists of the world to restore one of these Sequoias, which were large trees, over one hundred feet in height, spreading their leaves to the sun, before the Parthenon was even conceived by the architects and sculptors of Greece.

In 1900 five hundred of the very large trees still remained, the highest reaching from 320 to 325 feet. Their height, however, appeals to us less than their extraordinary age, estimated by Hutchins at three thousand, six hundred years or by John Muir, who probably loves them more than any man living, at from four thousand to five thousand years. According to the actual count of Muir of four thousand rings, by a method which he has described to me, one of these trees was one thousand years old when Homer wrote *The Iliad*; one thousand five hundred years of age when Aristotle was foreshadowing his evolution theory and writing his history of animals; two thousand years of age when Christ walked upon the earth; nearly four thousand years of age when the *Origin of Species* was written. Thus the life of one of these trees spanned the whole period before the birth of Aristotle (384 B.C.) and after the death of Darwin (A.D. 1882), the two greatest natural philosophers who have lived.

These trees are the noblest living things upon earth. I can imagine that the American people are approaching a stage of general intelligence and enlightened love of nature in which they will look back upon the destruction of the Sequoia as a blot on the national escutcheon.

The veneration of age sentiment which should, and I believe actually does, appeal to the American people when clearly presented to them even more strongly than the commercial sentiment, is roused in equal strength by an intelligent appreciation of the race longevity of the larger animals which our ancestors found here in profusion, and of which but a comparatively small number still survive. To the unthinking man, a bison, a wapiti, a deer, a pronghorn antelope, is a matter of hide and meat; to the real nature lover, the true sportsman, the scientific student, each of these types is a subject of intense admiration. From the mechanical standpoint they represent an architecture more elaborate than that of Westminster Abbey, and a history beside which human history is as of yesterday.

Wilderness Reserves[118] by Theodore Roosevelt

The practical common sense of the American people has been in no way made more evident during the last few years than by the creation and use of a series of large land reserves—situated for the most part on the Great Plains and among the mountains of the West—intended to keep the forests from destruction, and therefore to conserve the water supply. These reserves are created purely for economic purposes. The semiarid regions can only support a reasonable population under conditions of the strictest economy and wisdom in the use of the water supply, and, in addition to their other economic uses, the forests are indispensably necessary for the preservation of the water supply and for rendering possible its useful distribution throughout the proper seasons. In addition, however, to the economic use of the wilderness by preserving it for such purposes where it is unsuited for agricultural uses, it is wise here and there to keep selected portions of it—of course only those portions unfit for settlement in a state of nature—not merely for the sake of preserving the forests and the water, but for the sake of preserving all its beauties and wonders unspoiled by greedy and short-sighted vandalism. These beauties and wonders include animate as well as inanimate objects.

The wild creatures of the wilderness add to it by their presence a charm which it can acquire in no other way. On every ground it is well for our nation to preserve, not only for the sake of this generation, but above all for the sake of those who come after us, representatives of the stately and beautiful haunters of the wilds which were once found throughout our great forests, over the vast lonely plains, and on the high mountain ranges, but which are now on the point of vanishing save where they are protected in natural breeding grounds and nurseries. The work of preservation must be carried on in such

---

[118] George Bird Grinnell, ed. *American Big Game in Its Haunts; The Book of the Boone and Crockett Club* (New York: Forest and Stream Publishing Company, 1904), 25–29.

a way as to make it evident that we are working in the interest of the people as a whole, not in the interest of any particular class, and that the people benefited beyond all others are those who dwell nearest to the regions in which the reserves are placed. The movement for the preservation by the nation of sections of the wilderness as national playgrounds is essentially a democratic movement in the interest of all our people.

On April 8, 1903, John Burroughs and I reached the Yellowstone Park and were met by Major John Pitcher of the regular army, the superintendent of the park. The Major and I forthwith took horses, he telling me that he could show me a good deal of game while riding up to his house at the Mammoth Hot Springs. Hardly had we left the little town of Gardiner and gotten within the limits of the Park before we saw prong-buck. There was a band of at least a hundred feeding some distance from the road. We rode leisurely toward them. They were tame compared to their kindred in unprotected places—that is, it was easy to ride within fair rifle range of them—but they were not familiar in the sense that we afterwards found the bighorn and the deer to be familiar. During the two hours following my entry into the Park, we rode around the plains and lower slopes of the foothills in the neighborhood of the mouth of the Gardiner and we saw several hundred—probably a thousand all told—of these antelope. Major Pitcher informed me that all the pronghorns in the Park wintered in this neighborhood. Toward the end of April or the first of May, they migrate back to their summering homes in the open valleys along the Yellowstone and in the plains south of the Golden Gate. While migrating they go over the mountains and through forests if occasion demands. Although there are plenty of coyotes in the Park, there are no big wolves, and save for very infrequent poachers, the only enemy of the antelope, as indeed the only enemy of all the game, is the cougar.

Cougars, known in the Park as elsewhere through the West as "mountain lions," are plentiful, having increased in numbers of recent years. Except in the neighborhood of the Gardiner River, that is within a few miles of Mammoth Hot Springs, I found them feeding on elk, which in the Park far outnumber all other game put together, being so numerous that the ravages of the cougars are of no real damage to the herds. But in the neighborhood of the Mammoth Hot Springs the cougars are noxious because of the antelope, mountain sheep, and deer which they kill, and the superintendent has imported some hounds with which to hunt them. These hounds are managed by Buffalo Jones, a famous old plainsman, who is now in the Park taking care of the buffalo. On this first day of my visit to the Park, I came across the carcasses of a deer and of an antelope which the cougars had killed. On the great plains cougars rarely get antelope, but here the country is broken so that the big cats can make their stalks under favorable circumstances. To deer and mountain sheep, the cougar is a most dangerous enemy—much more so than the wolf.

The antelope we saw were usually in bands of from twenty to one hundred and fifty, and the traveled strung out almost in single file, those in the rear would sometimes bunch up. I did not try to stalk them, but got as near them as I could on horseback. The closest approach I was able to make was to within about eighty yards of two which were by themselves—I think a doe and a last year's fawn. As I was riding up to them, although they looked suspiciously at me, one actually lay down. When I was them at about eighty yards distance, the one became nervous, gave a sudden jump, and away the two went at full speed.

Why the prong bucks were so comparatively shy I do not know, for right on the ground with them we came upon deer and, in the immediate neighborhood, mountain sheep, which were absurdly tame. The mountain sheep were nineteen in number, for the most part does and yearlings with a couple of three-year-old rams, but not a single big fellow—for the big fellows at this season are off by themselves, singly or in little bunches, high up in the mountains. The band I saw was tame to a degree matched by but few domestic animals. They were feeding on the brink of a steep washout at the upper edge of one of the benches on the mountainside just below where the abrupt slope began. They were alongside a little gully with sheer walls. I rode my horse to within forty yards of them, one of them occasionally looking up and at once continuing to feed. Then they moved slowly off and leisurely crossed the gully to the other side. I dismounted, walked around the head of the gully, and, moving cautiously but in plain sight, came closer and closer until I was within twenty yards, where I sat down on a stone and spent certainly twenty minutes looking at them. They paid hardly any attention whatever to my presence— certainly no more than well-treated domestic creatures would pay. One of the rams rose on his hind legs, leaning his forehoofs against a little pine tree, and browsed the ends of the budding branches. The others grazed on the short grass and herbage or lay down and rested—two of the yearlings several times playfully butting at one another. Now and then one would glance in my direction without the slightest sign of fear—barely even of curiosity. I have no question whatever but that with a little patience this particular band could be made to feed out of a man's hand. Major Pitcher intends during the coming winter to feed them alfalfa—for game animals of several kinds have become so plentiful in the neighborhood of the Hot Springs, and the Major has grown so interested in them, that he wishes to do something toward feeding them during the severe winter. After I had looked at the sheep to my heart's content, I walked back to my horse, my departure arousing as little interest as my advent.

Nathaniel Southgate Shaler

Published a year before his death in 1906, *Man and Earth* exhibits Shaler's ideologically-conflicted experience as a Harvard geologist dedicated, on one hand, to the careful survey of natural resources as agents of national supremacy and, on the other, as aspects of pure science. Shaler's statistical worrying over depleted coal, iron, and petroleum reserves echoes, sometimes uncannily, the obsessions of twenty-first century national planners. Still, one of Shaler's longest studies was also one of his most local—an extended geological survey of his home state of Kentucky, where he had been stationed as an artillery officer for the federal militia during the Civil War. As a professor of paleontology, geology, and later dean of sciences at Harvard, Shaler was mentored by Louis Agassiz, the legendary Swiss geologist credited with demonstrating that glaciers once covered the earth. Agassiz, who established a zoological laboratory off the coast of Massachusetts to study animals in their natural environment, left a conflicted scientific inheritance for his disciple, who begrudgingly accepted the basic tenets of Darwinism that Agassiz had rejected outright.

Purposefully "forelooking," Shaler the conservationist imagines in "Earth and Man" a future generation, that, having survived his own generation's profligate waste, will "date the end of barbarism from the time when the generations began to feel that they rightfully had no more than a life estate in this sphere, with no right to squander the inheritance." Though primarily recognized as a geological and conservationist tract, Shaler attends closely to agriculture in the passage below, particularly intensive cropping, as both the hope of a crowded world and a danger to its future productivity. Identifying erosion and mineral-depleted soils as twin problems facing the American tiller, Shaler conjures a sobering image of early twentieth-century America where even the fertile soils of the Mississippi Valley show signs of dramatic fertility loss. Particularly global and empirical in its take, Shaler's work is remarkable not only for its portrayal of man "the spoiler" but also man the redeemer, where redemption is enabled and informed by science. A sampling of Shaler's most popular publications places him firmly in the camp of scholars who arrived at their conservationist views via a vigorous interdisciplinarity: *Kentucky: A Pioneer Commonwealth* (1884), *Nature and Man in America* (1891), *The United States of America; A Study of the American Commonwealth, Its Natural Resources* ... (1894), *American Highways: A Popular Account of Their Conditions, and of the Means by which They May Be Bettered* (1896) and *The Neighbor; The Natural History of Human Contacts* (1904). Shaler's most widely held book, *The Autobiography of Nathaniel Southgate Shaler* (1909), demonstrates the American public's interest in this eclectic, sometimes controversial, figure.

Earth and Man[119]

The situation of man with reference to the material resources of the earth deserves more attention than has been given to it. Here and there students of the mineral deposits of certain countries, especially those of Great Britain, have computed the amounts of coal and iron within limited fields and estimated the probable time when those stores would be exhausted, but a general account of the tax that civilization makes on the fields it occupies and a forecast as to their endurance of the present and prospective demand on them is lacking. It is evident that such a forelooking should be one of the first results of high culture. We may be sure those who look back upon us and our deeds from the centuries to come will remark upon the manner in which we use our heritage, and theirs, as we are now doing, in the spendthrift's way, with no care for those to come. They will date the end of barbarism from the time when the generations began to feel that they rightfully had no more than a life estate in this sphere, with no right to squander the inheritance of their kind.

To see our position with reference to the resources of the earth, it is well to begin by noting the fact that the lower animals, and primitive men as well, make no drain on its stores. They do not lessen the amount of soil or take from the minerals of the under-earth: in a small way they enrich it by their simple lives, for their forms are contributed to that store of chemically organized matter which serves the needs of those that come after them. With the first step upward, however, and ever in increasing measure as he mounts toward civilization, man becomes a spoiler. As soon as he attains the grade of a hunter, he begins to disturb the balance of the life about him, and, in time, he attains such success in the art that he exterminates the larger, and therefore the rarer, beasts. Thus when our genus homo comes into view, elephants of various species existed in considerable numbers in all the continents except Australia. Its first large accomplishment appears to have

---

[119] National Southgate Shaler, *Man and the Earth* (New York, Fox, Duffield & Company, 1905), 1–12.

consisted in the extermination of these noble beasts in the Americas, in Europe, and in northern Asia. There is no historic record of this work, but the disappearance of the elephants can be well explained only by the supposition that they went down before the assault of vigorous men, as has been the case with many other species of large land animals.

So long as men remained in the estate of the hunter, the damage they could do was limited to the destruction of the larger beasts and the birds, such as the moa, that could not fly. Prolific species, even of considerable size, such as the bison, if they were nimble and combative, seem to have been able to hold the field against the attacks of primitive hunters. While in this station the tribes of men are never very numerous, for their wars, famines, and sorceries prevent their increase, which, under the most favorable conditions, is never rapid among savages. As soon, however, as stone implements begin to be replaced by those of metal, man begins to draw upon the limited stores of the under-earth, and with each advance in his arts the demand becomes the greater. In the first centuries of the Iron Age the requisition was much less than a pound each year for each person. Four centuries ago it probably did not exceed, even in the most civilized countries, ten pounds per capita each year. It appears to have been at something like that rate when the English colonies were founded in North America. At the present time in the United States, it is at the average rate of about five hundred pounds per annum for every man, woman, and child in the land, and the demand is increasing with startling rapidity. It seems eminently probable that before the end of the present century, unless checked by a great advancement of cost, it will require a ton of iron each year to meet the progressive desires of this insatiable man.

Of the other, long-used metals and other earth resources, the increase in consumption is, with slight exceptions, as notable as in the case of iron; within a generation, mainly because of the use of the metal in electrical work, the need of copper has augmented even more rapidly than that of iron and the gain in the requirements is going on with exceeding speed. So, too, the demand for the other base metals long in use, zinc and tin, has been in nowise lessened by the more extended use of iron and copper; they are ever finding new places in the arts and a larger demand in the markets. As regards the so-called noble metals, silver and gold, the demand from the beginning has not been distinctly related to use, but to unlimited desire. Men have always wrested all they could of them from the earth or from each other, with little reference to the profit they won in the process. There has been of late something like a halt in the production of silver, except when it comes as a byproduct, because it has generally been abandoned as a standard of value, but, taken together, the production of these precious metals has in modern times increased about as rapidly as that of iron. It is likely, however, that they will in time become of no economic importance.

As regards the earth's resources in the way of fuel—coal, oil, wood, petroleum, and peat—the history of the modern increase in demand is as evident and menacing as in the case of the metals. When the American English colonies were founded, coal had hardly begun to become into use in any country. It is doubtful if the output of the world amounted at that time to one hundred thousand tons, possibly to not more per capita of the folk in Europe than a pound, or about the same as iron at that late period in the so-called Iron Age. At the present time the total production of Europe and North America amounts to an average of at least two tons per each unit of the population, and the increase goes on at a high ratio. Petroleum, practically unknown to the occidental peoples until about half a century ago, has, with wonderful rapidity, become a necessity to all civilized and many barbaric peoples; the increase in the rate of consumption is swifter than that of any other earth product. Timber and peat, the primitive resources for light and heat, are the only earth products for which the demand has not greatly extended in modern times; it appears, indeed, to have shrunk in most civilized countries with the cheapening and diffusion of coal, due to the lessened cost of mining and of transportation.

The increase in the tax of the earth's resources is seen also in the very great number of substances which were unknown to the ancients, or disregarded by them, but which now find a large place in our arts. A comparison of the demands of three centuries ago with those of our day is interesting. In, say, 1600, when men were very much alive to the question of what they could gain, there were only about twenty substances, other than precious stones, for which they looked the underground realm. Clays for the potter and bricklayer, whetstones and millstones, iron, copper, tin, gold, silver, lead, sand for glass, mica, coal, peat, salt, and mercury make up all the important elements of this list. At the present time, we more or less seriously depend on what is below the ground for several hundred substances or their immediate derivatives which find a place in our arts.

Petroleum alone has afforded the basis of far more earth products than were in use at the time of the discovery of America. It gives us a large number of dyes and a host of medicines. It is indeed likely that the products immediately derived from the mineral oils exceed all those obtained from the earth at the time of Columbus—and each year brings additions to the demand.

The advance in needs of dynamic power, in modern times, has been even greater than in ponderable things. Even two centuries ago the energy available for man's work was mainly limited to that obtained from domesticated animals. The wind served in a small measure through the sails of ships and of windmills, and there were water-wheels, but the average amount of energy at his service was certainly less than one horsepower per capita. At the present time it may safely be reckoned that in the United States and in European countries on a similar economic basis, the average amount is at least ten times as great, and the present rate of increase quite as high as in the case of mineral resources. It is true, that, so far as water is concerned, this increase in the demand for energy in the arts does not come as a tax on the store of the under-earth, as it is obtained through solar energy which would otherwise be dissipated in space. But the use of falling water as a source of power, though rapidly increasing, does not keep pace with that of coal, which is obtained from a store which is in process of rapid exhaustion, one that cannot be relied on for more than a few hundred years to come: if the world keeps the rate of consumption with which it enters the twentieth century, it will be exhausted before the twenty-third.

The problem of the underground store of wealth, though as we shall see on more detailed examination it is very serious, is not so immediate or menacing as that afforded by the question of food supply. As far as man is concerned, the supply has to come from two sources—the tilled soil and the waters, especially the sea. While it is possible by a widely extended system of fish culture greatly to increase the amount of food derived from the waters, experience does not warrant the supposition that the supply from this source can be manifold. The life of the oceans, as of the primeval lands, is already packed to the utmost point. We cannot hope to double the number of edible fishes without reducing the number of their enemies or of the other creatures which compete with them for subsistence. Neither of these things can we at present see the way to do. It is to the soil, to the tilled soil alone, that we are to look for the body of the food that is to feed man for all the time he abides on this sphere.

In the life below man, the relation of the creatures to the soil had been beautifully adjusted. The plants, by associated action, formed on all the land surfaces, except in very arid regions, a mat of roots and stems which served to defend the slowly decaying rock against the attack of the rainwater. This adjustment is so perfect that, in a country bearing its primeval vegetation, the eroding of the soil is essentially limited to what is brought about by the dissolving action of the water which creeps through the earth and there takes the substances of the rocks into solution; very little goes away, in suspension, in the form of mud. In these conditions the slowly decaying rock passes very gradually to the sea; for a long time it bides in the soil layer where, with the advance in its decomposition, it affords the mineral substances needed by the plants that protect it. Thus until man disturbs the conditions of forest and prairie, the soils tend to become deep and rich, affording the best possible sustenance to the plants which feed in them. In their normal state they represent the preserved waste of hundreds, or it may be, thousands of feet of rocks which have gradually worn down by being dissolved in the rainwater that creeps through them.

As soon as agriculture begins, the ancient order of the soils is subverted. In order to give his domesticated plants a chance to grow, the soil tiller has to break up the ancient protective mantle of plants, which through ages of natural selection became adjusted to their task, and to expose the ground to the destructive action of the rain. How great this is may be judged by inspecting any newly plowed field after a heavy rain. If the surface has been smoothed by the roller, we may note that where a potsherd or a flat pebble has protected the soil it rests on top of a little column of earth, the surrounding material having been washed away to the streams where it flows onward to the sea. A single heavy rainstorm may lower the surface of a tilled field to the amount of an inch, a greater waste than would, on the average, be brought about in natural conditions in four or five centuries. The result is that in any valley in which the soils are subjected to an ordinary destructive tillage the deportation of the material goes on far more rapidly than their restoration by the decay of the underlying rocks. Except for the alluvial plains, whereupon the flood waters lay down the waste of fields of the upper country, nearly all parts of the arable lands which have been long subjected to the plow are thinned so that they retain only a part of their original, food-yielding capacity. Moreover, the process of cropping takes away the soluble minerals more rapidly than they are prepared, so that there is a double waste in body and in the chemical materials needed by the food-giving plants.

There is no question that the wasting of soils under usual tillage conditions constitutes a very menacing evil. Whoever will go, with his eyes open to the matter, about the lands bordering on the Mediterranean, will see almost everywhere the result of this process. Besides the general pauperizing of the soils, he will find great areas where the fields have prevailingly steep slopes from which the rains have stripped away the coating down to the bedrock. In Italy, Greece, and Spain, this damage has gone so far that the food-producing capacity of those countries has been greatly reduced since they were first subjected to general tillage. There is no basis for an accurate reckoning, but it seems likely from several local estimates that the average loss of tillage value of the region about the Mediterranean exceeds one-third of what it was originally. In sundry parts of the United States, especially in the hilly country of Virginia and Kentucky, the depth and fertility of the soil has, in about one hundred and fifty years, been shorn away in like great measure. Except in a few regions, as in England and Belgium, where the declivities are prevailingly gentle, it may be said that the tilled land of the world exhibits a steadfast reduction in those features which give it value to man. Even when the substance of the soil remains in unimpaired thickness, as in the so-called prairie lands of the Mississippi Valley, the progressive decrease on the average returns to cropping shows that the impoverishment is steadfastly going on.

In considering the struggle which men have to make in the time to come in order to maintain the food-giving value of the earth, it is well to keep in mind the fact that the battle is with one of the inevitables—with gravitation, which urges everything ponderable down into the sea. What we know as soil is rock material on its way to the deep, but considerably restrained in its going by the action of the plants which form a mat upon it. All the materials which go into solution naturally pass in that state on the same way; thus whatever we do, we cannot expect to effect anything more than a retardation of the process to that point where the decay of the bedrocks will effectively restrain the wasting process, so that the loss may be made good. It is indeed not desirable to arrest this passage of earth material to the sea. So far as that passage is here and there effected by natural processes we find that, in time, the soil loses its fertility because the necessary mineral constituents are exhausted. Thus in the case of the coal beds, the swamp-bottoms in which the plants grew did not have their materials renewed by the decay of the underlying rock and so were, in time, exhausted by the drain upon them and became too unfertile to maintain vegetation. The preservation of the food-giving value of the soil as used by civilized man depends on the efficiency of the means by which he keeps the passage of the soil to the sea at a rate no greater than that at which it is restored by the decay of the materials on which it rests.

Some of those who have essayed a forecast of the future of man have felt that the prospect was shadowed by a doubt as to his permanence.

Seeing, as we do, that the life of this earth is characteristically temporary, the species of any geological period rarely enduring to the next, it is a natural conclusion that our own kind will share the fate of others and, in a geological sense of the word, soon pass away. Closer attention to the matter leads us to believe that the genus homo is one of those exceptional groups, of which there are many, which have a peculiar capacity for withstanding those influences which bring about the death of species. There are a number of such forms in most of the classes of animals, creatures which have existed, it may be, from Paleozoic times, perhaps for fifty or more million years, so little changed that the earliest of them seem as nearly akin to the latest as are the diverse species of mankind. Man has been upon the earth certainly for two geological periods. He withstood the colossal accident of the last glacial epoch. He is by his intellectual quality exempted from most of the agents that destroy organic groups. So we may fairly reckon that he is not to pass from the earth in all foreseeable time, but is to master it and himself for ages of far-reaching endeavor. The limits set to him are not those set by the death of his species, but by the endurance of the earth to the demands his progressive desire make upon it.

WILBERT LEE ANDERSON

Wilbert Lee Anderson's *The Country Town: A Study of Rural Evolution* is one of the most curious, and most underestimated, sociological explorations of the early Progressive period. A New Englander living in Exeter, New Hampshire, Anderson's interest in small, rural towns and their spirited brand of self-government seems logical. And yet "Local Degeneracy" belies Anderson's agonizing ideological, spiritual, and economical uncertainties about the health and welfare of small town America. A religious man, Anderson nevertheless takes issue with the sometimes grave analysis of Reverend Josiah Strong's introduction to his volume. Favoring demographic and economic data over pseudo-sociological narrative, Anderson endorses local "eyewitness"— the man or woman on the ground—even as he demonstrates how such eyewitness misleads. An optimist who goes to great lengths to show rural and small town America as equal to, if not superior to, urban America, a close read of "Local Degeneracy" bespeaks growing threats to agrarian towns Anderson describes as positioned precariously on an upward incline, forever in danger of falling into a cultural, intellectual abyss. Indeed, New England farm centers suffered especially in the early twentieth century, as competition from the Midwest forced many farmers in the region to abandon large cropping and livestock operations altogether or to consolidate them into specialized dairying.

Despite his distrust of literary men, journalists, and humorists turned amateur sociological observers, Anderson's writing in this passage displays an unusual lyricism and emotive power that reminds of sociologically-informed contemporary agrarian studies such as Osha Gray Davidson's *Broken Heartland: The Rise of the Rural Ghetto* and Victor Davis Hanson's *Fields Without Dreams: Defending the Agrarian Ideal*. Closer to his own time, Anderson's provocative, almost tortured ambivalence concerning the small towns of his native region reminds of James Agee's *Let Us Now Praise Famous Men*. Anderson's commitment to the Progressivist spirit of social uplift is evident in his preface:

> Even scientific diagnosis avails nothing unless remedies are applied, and certainly to refute the pessimist when the hour demands the rescue of a civilization would be no better than fiddling while Rome burned. If this book had the gift of prophecy and knew all mysteries and all knowledge, if it had all faith so as to remove mountains, and did not prompt the deeds of love, it would be nothing.

Local Degeneracy[120]

When we ask how the country people are acquitting themselves in their moral warfare, the question concerns the fortune of half mankind

---

[120] Wilbert Lee Anderson, *The Country Town: A Study of Rural Evolution* (New York: Baker and Taylor, 1906), 96–108.

in our part of the world.[121] By actual count of heads, the people of farm and village deserve equal consideration with the more conspicuous masses of the city. In our own country three in five of the people, be it remembered, still live in towns of less than two thousand, five hundred inhabitants,[122] and until there are great changes in the form of civilization, it will be necessary for a man to live somewhere in the country as the partner of the man who lives in the city. Insignificant as a single country town may be as compared with great cities, the aggregate rural population is vast and imposing.

The significance of the rural condition is enhanced, and for many this is the chief matter by the moral contribution of the country to the city. The city is better equipped for independence than when Emerson wrote: "The city would have died out, rotted, and exploded, long ago, but that it was reinforced from the fields,"[123] for in modern conditions the natural increase of its own people would keep it full and its own social forces would check decadence. Thus far, nevertheless, country-bred men have dominated our entire civilization, and therefore any suggestion that the country has become impoverished, and that the rural sources of intelligence and character are approaching exhaustion excites grave apprehension.

Our inquiry has to do primarily with the depleted town. Were all country towns growing, it would occur to no one to ask whether they, in distinction from the rest of the world, are making moral progress. The pessimist and the optimist, doubtless, would wax earnest in the familiar debate concerning the character of modern civilization, and each would be able to find material for his argument in every part of the land. It is the thinning out of the rural population that has startled the observer and aroused the theorist. It is, of course, possible for a depleted town to make intellectual and moral gains. If farms were

---

[121] For the sake of concision and readability, the passage begins with the second full sentence of the second paragraph of Chapter 5 as it appears in the 1906 edition.
[122] Anderson's original footnote here read "Cf. p. 42."
[123] Anderson's original footnote here read "'Essays,' Second Series, Manners, Riverside Ed., Vol. III, p. 126."

consolidated by the sale of unprofitable lands to the richest farmers, if the new owners introduced improved methods of cultivation, if social intercourse were skillfully adapted to the changed conditions, if the educational and ecclesiastical structure were remodeled, the population might be depleted to almost any extent without social or moral injury. At the best, however, such adjustment will be slowly wrought and many communities will fail of it altogether.

On the other hand, the growing town possesses no moral insurance. The increase of population may be due to some industry that supports an alien and inferior labor colony. In that case the decay of the native stock, the weakening of social institutions, and all the consequences of depletion may be the same as in towns that have not received this increase of population. The census is by no means an infallible guide to the area within which the changes under consideration are found.

When matters of fact are to be determined, the authority is the eyewitness, for the man who has seen with his own eyes is to be trusted rather than the man who has not seen, whatever may be the seeming wisdom of his speculations. If it were possible to summon competent witnesses from a sufficient number and variety of localities, their testimony would be final, and social theorizing would be restricted by these determining observations. It is fair to say, however, that the most honest eyewitness cannot free himself wholly from personal bias, for some facts appeal to him and others escape his notice as certainly as he possesses a distinctive inclination and interest. The picture that he paints represents his selection, grouping, and interpretation of the items actually before his eyes. If he is of a pessimistic temper, or if he is an artist delighting in somber effects, his handling of light and shade will not present the moral landscape exactly as it is. In point of fact almost all of those who feel the artistic interest in an impression do deepen the shadows by emphasizing the remnant of a larger population, the decay of schools where once were crowds of children, the abandonment of farms, the forsaking of churches, changes incident to modern progress. The literary artist has full right to these shadows; all that is pointed out is that he is likely to avail himself of them in such a manner that his canvas will have a look unlike a more scientific presentation of the same conditions. And to this it must be added that in serious discussion there is no equivalent for the literary appeal to the imagination and the feeling, as the tragedy of some broken life is set forth, or the quaint backwardness of a decadent region is portrayed. The artist has the great human interests on his side, and summaries and generalizations, however carefully and wisely framed, are ineffective in comparison.

An example may make the matter plain. Some years ago a correspondent of the *New York Evening Post* wrote at length and with much telling detail concerning the New England churches. The article appeared with the following headlines: "Decay of Religion in Rural Communities—A Mournful Review—Changes of Twenty Years—Pitiful State of a Christian Minister—The Root of the Trouble." Here we have the deepening of the shadow by the artful use of social change, and the vivid portrayal of local conditions, and the appeal to sympathy by a pathetic personal example. The account of "church buildings crumbling and old church societies dead," of "family after family and hamlet after hamlet that have not seen a church service in years," of "communities that might as well have been in the middle of the Dark Continent so far as Christianizing influences are concerned," of the "social chaos" about these "deserted houses of worship" was so graphic and so startling that it remained in my memory, although a masterly reply filed with it was forgotten. So alarming was this representation that a citizen of New York, who knew New England well, was stirred up to make an investigation. He sent out two hundred letters to the most intelligent men whose names were obtainable in each of the New England states, asking for an opinion based upon their observation of actual conditions. About one hundred and fifty replies were received, and their comments on the situation were a complete exposure of the false impression produced by an adroit use of selected facts. In the main it was found that conditions had not been misrepresented in their actual local character, although the personal touch was gained by repeating the story of an insane minister as told by himself. The items in this debate need not concern us; reference is made to it only to show how much less impressive is general prosperity than local decadence. It would not be possible to tell the story of religious life as it goes forward in hundreds of churches in a way to arrest attention or charge the memory. There is no news value perhaps, no literary or artistic value in, the wholesome averages of life.

With such precautions as have been suggested, there is no reason to discredit the tales of rural decadence told by eyewitnesses. Certainly the existence of an abandoned meeting house is a fact to be accepted, if a truthful observer has seen it. The same may be said of "twenty-five wretched families of the degenerate type in a single town," and of a "family in which grandparents, adult children, and grandchildren are all foolish." Degenerate families by the score, and families with three generations of idiots are easily observed, labeled, and certified. It is idle to ask whether degenerate families or three generations of idiots can be matched in the city; it is sufficiently shocking that they are found in the country. Reasoning from such facts is precarious, but facts, once correctly observed, must take their place in social estimates. What then are the facts observed in country towns, exceptional perhaps and less significant than might be supposed, but unquestionable facts that must be taken into account?

The *New York Evening Post*, commenting on the article to which reference was made above, said, "Rural life is monotonous and hard, with nothing in it to stimulate the imagination or refine the taste. The closest observers will soonest admit that by consequence the grosser forms of vice and crime in rural communities abound more than in great cities with all the slums counted in." Although the comparison with cities is not sustained by statistics,[124] this arraignment of rural communities for grosser forms of vice and crime has much local justification. In its issue of January 2, 1904, the *St. Albans* (VT) *Messenger* quoted from the *Montpelier Argus*:

> The two recent murder trials in Bennington County
> and the one now in progress in Windsor County reveal

---

[124] Anderson's original footnote here read "Cf. p. 113."

> a surprising state of depravity, a condition of affairs that might be expected to lead to murder in any community. Lack of education, lack of moral training, and lack of uplifting environment, as much as lack of individual character and stamina, account for the crimes. Such conditions are not usual in Vermont, but they appear here as elsewhere, and bring the usual and expected results.

The following editorial comment was added, with more to the same effect: "And the *Messenger* contends that it is not vilifying the state, it is not proclaiming her degenerate, for public-spirited newspapers to point out this fact to their readers and to emphasize it so often and so vigorously that the popular mind will be induced to pay that heed to it that it deserves."

An editorial in the *Boston Herald* of March 9,1903, places two prosperous New England states side by side in this deplorable local degeneracy. It says in part:

> A few years ago a writer in the *Atlantic Monthly* gave a picture of the deterioration of the quality of character and life in the remote hill towns of Massachusetts. His statements were received with incredulity, and were sharply contradicted by others who thought they knew better than he did the real conditions in such communities. One would suppose that what was said less than two weeks ago about the degeneracy of rural Connecticut in an address in New Haven before the Connecticut Bible Society, by the Reverend H. L. Hutchins, was new and strange, judging by the attention it has received. This attention is possibly more keen than it would have been if the teacher had not since died suddenly, making the deliverance seem like a prophet's last word of instruction and warning. It was a very solemn and important message, the substantial truth of which is accepted by the Connecticut newspapers with a unanimity that would hardly have been thought probable. He tells a story of degeneracy and immorality which is rather worse than the story

told by the writer of the articles in the *Atlantic Monthly*, to which we have referred. Yet no one is assailing it as a misstatement, and Massachusetts newspapers, in commenting on it, take pains to say that we have no cause to congratulate ourselves on the existence of better conditions here. We observe no such swift disposition to contradict the truth of the dreadful message as was manifest formerly. Does this mean that in the interval a conviction has spread that the wickedness of some of our rural neighborhoods is not exaggerated by those who have probed them to find the truth? The address of Mr. Hutchins was supplemented by an article which he wrote on the day of his unexpected death. The two constitute a body of testimony which neither the social nor the religious world can safely ignore.

The articles in the *Atlantic Monthly*[125] were written with high color, being in the best style of the literary impressionist. The decadence described was easily recognized as characteristic of other places, if not of "Sweet Auburn." Nothing criminal was alleged, but rather incompetence, dullness, decay lapsing into idiocy. The narrowness of the horizon, the pettiness of the endless gossip, the absence of ambition, the feebleness of faculties as if the string had broken and the bundle fallen apart, were graphically presented and evidently drawn from life.

The testimony of Mr. Hutchins has every title to respect: it was based upon the observation of many years during which it had been his business to go from house to house upon an errand that made him watchful of moral conditions; more important still, in its utterance there is a note of sharp pain, as if a constantly confronted horror had burned his soul. His story dwells upon the prevalent ignorance, the inroads of vice, the open contempt of marriage, the increase of

---

[125] The footnote here originally read "Rollin Lynde Hartt, 'A New England Hill Town,' *Atlantic Monthly*, Vol. LXXXIIL."

idiocy, the feebleness and backwardness of schools, the neglect of the church, the swift lapse into virtual heathenism in whole sections once occupied by the best type of Christian manhood.[126] The last phrase is not stronger than one used by President William D. Hyde, who, writing of religious conditions in Maine, launched a magazine article under the title, "Impending Paganism in New England."[127]

The chapter in *The New Era* in which Dr. Strong treats the problem of the country is deeply shadowed.[128] The melancholy results of depletion, with the emphasis upon the neglect of religion, are vividly portrayed. Then comes the summary under five heads: road deteriorates; property depreciates; churches and schools are impaired; the foreign crowds out the native stock; isolation increases with a tendency towards degeneration and demoralization. The conclusion is stated in the terms of momentum and velocity that distinguish the writings of this modern prophet, to whom nothing is stationary but all things in swift rush towards destiny. "If this migration continues," he says, "and no new preventive measures are devised, I see no reason why isolation, irreligion, ignorance, vice, and degradation should not increase in the country until we have a rural American peasantry, illiterate and immoral, possessing the rights of citizenship, but utterly incapable of performing or comprehending its duties." Doubtless, Dr. Strong has hope that "preventive measures" will be devised, but he leaves us with the thrill of that frightful slide towards bottomless misery upon us. That should stir us to seek "preventive measures," for it is impossible to deny this moral gravitation. There is in full action such a tendency as this towards degradation; in the main it may be counteracted it is possible that providence has set in operation "preventive measures" of high efficiency. What we are here to recognize is the inevitable movement towards decadence, a movement everywhere felt, and

---

[126] The footnote here originally read "Such local reports as these do not mean that the country churches of Connecticut as a whole are not vigorous and successful in their work."
[127] The footnote here originally read "*The Forum*, June, 1892."
[128] The footnote here originally read "See particularly pp. 170–174."

where not resisted, producing that visible degeneracy described by so many witnesses. Rural life is everywhere on an inclined plane; if it climbs instead of slides, great should be our thankfulness. Beyond question, in some places it slides.

We have no call to deny such alleged facts as have been presented in the preceding paragraphs. It is not our purpose, on the other hand to establish them, since so much of this sort has been written that it is quite unnecessary to cover the ground again. Enough has been presented to enable the reader to conceive this aspect of the rural situation. We shall not credit every gruesome tale; we shall allow something for panic, and something for art, and something for the intense realization of the reformer. The residuum of truth finds its place in that general conception of a worldwide revolution in rural life which is a vital part of modern progress. It is a thing to be expected that where the change has been made with least success, where the strain has been greatest, serious decadence should appear. Indeed it is no exaggeration to say that tendencies are at work which, unchecked, would ruin all rural communities. The inclined plane on which every country town must secure its footing is no fiction; towns that slip may slide far.

In his essay, "The Mission of Humor," the best of whose delightful surprises are odd glimpses of truth, Dr. Crothers points out a most interesting characteristic of the "prosaic theorist."[129] "He tries to get all the facts under one formula, which is a very 'ticklish business' being indeed 'like the game of pigs in clover.'" "He gets all the facts but one into the inner circle. By a dexterous thrust he gets that one in, and the rest are out." By dint of great persistence and cunning, the theorist sometimes gets all rural communities into clover, but that ends the game. Usually there is no such success, for as the recalcitrant specimen is thrust in, a scampering company breaks out with much

---

[129] This sentence marks the beginning of Chapter 6 "The Main Trend" in the original 1906 edition. Chapter 5 and the first five pages of Chapter 6 are, in subject matter and tone, best considered of a piece.

noise and raising of dust. He who would maintain that all country towns are prosperous, or that all are decadent, will find the crowd of irrepressible facts too much for his dexterity and vigilance. Nevertheless the seduction of a general formula is not easily resisted, like a modern navy, once possessed it must be used. Charles Lamb is held up by Dr. Crothers as a wholesome model of a philosopher who saves himself no end of trouble by leaving the last pig out. "He does not try to fit all the facts to one theory. That seems to him too economical, when theories are so cheap. With large-hearted generosity he provides a theory for every fact. He clothes the ragged exception with all the decent habiliments of a universal law. He picks up a little ragamuffin of a fact, and warms its heart and points out its great relations."[130]

We are willing to be as generous as Charles Lamb; we will provide a special theory for the decadent country town. Have we not sought to "clothe this ragged exception with all the decent habiliments of a universal law." What more can be asked than that we recognize local degeneracy, equip it with an explanation, and give it the standing of an inevitable fact under modern conditions? But when we have done this, it is enough. We refuse to introduce all facts into the same society. We are willing to take this little vagabond of a fact, make the most of him, warm his heart, tell the tale of great relations, but, when all is done, we brand him a ragamuffin, although a very interesting one, and shut the door of good society in his face. Why should all pigs be in clover, except in a game? And why should every uncouth fact be permitted to cherish the dream of universal equality? The humorist finds in exceptions and odd remainders and things left over in the cosmic workshop abundant materials for his art. The decadent town is such an intractable remainder in the evolution of civilization, more pathetic than humorous, but above all things calling for the sympathetic treatment that recognizes the unique exception.

---

[130] The footnote here originally read "Samuel McChord Crothers, 'The Gentle Reader,' pp. 84–85."

We proceed, therefore, to arrange rural facts in two systems, the system of decadence and the system of progress. Some communities are in the main trend of prosperity; others are off the track of the modern movement in various stages of deterioration or mere backwardness. It is obvious that descriptions of one social condition will not apply to the other. It has been customary, nevertheless, to treat all rural facts in one class, especially in statistical summaries. If it is remembered that communities are either adjusted or not adjusted to modern conditions, that they are in or out of the trend of progress, that they are decadent or prosperous, it will appear of how little use the ordinary general formula is. The average is an abstraction, useful for comparison but rarely available as an account of fact as it is in itself. The average elevation of the earth's surface would be interesting to know, but land, being solid and not flowing to a level like the sea, towers in mountain and sinks in valley, so that scarcely a square mile is spread out at the average altitude. Thus uneven is human society, with elevation matching depression, and few communities poised midway on the scale.

Dallas Lore Sharp

An emphasis on what he called "back pasture" nature meant unusual popular success for Dallas Lore Sharp, a professor of English at Boston University and one of the foremost traveler-naturalists of his era. Though Sharp did much of his nature observing on his own eight-acre woodlot, the New Jersey native wrote memorably of several trips to the West in such books as *Where Rolls the Oregon* (1914)—his tribute to the wonders of the Oregon mountains and the Columbia River Valley—and *The Better Country* (1928). Sharp's most popular book of natural history, *The Face of the Fields* (1911), consists of nature vignettes, including the chapter "Turtle Eggs for Agassiz," in which the author relates how his biology teacher, J. Jenks of Middleboro, had once been given three hours to supply the legendary Louis Agassiz with fresh specimens for study.

Praising Thoreau, Emerson, and Burroughs for showing Americans that a homely but no less miraculous nature lay no further than a short walk or train ride away from their homes, Sharp's words particularly encouraged nature-loving Americans of modest means while trying to protect them from sensationalized accounts proffered by fraudulent "naturalists." Sharp was preceded in his counterattacks against "nature-fakers" by John Burroughs, whose 1903 *Atlantic Monthly* essay "Real and Sham Natural History" had previously implicated such well-loved nature writers as Ernest Thompson Seton and William J. Long. The controversy, which demonstrated that nature and conservation writing had become sufficiently popular to produce competing factions and membership anxieties, would cause conservationists nationwide to take sides; Roosevelt and Sharp sided with Burroughs. The hullabaloo would resurface five years later when part-time illustrator turned nature experimenter Joseph Knowles returned to Boston claiming he had survived sixty days living *au naturel*, "as Adam lived," in the wilds of Maine, though many discounted his story as an apocryphal publicity stunt.

Like Liberty Hyde Bailey, Sharp viewed nature appreciation as an antidote for increasingly hectic times, a kind of tonic that, once imbibed, would restore in his countrymen such virtues as thoughtfulness, sympathy, and simplicity. In "The Nature Movement" Sharp presents such nature-inspired traits as quintessentially American, unprecedented in the history of the world. Pointing to settlers who arrived to a natural Eden sufficiently miraculous as to make their affectionate nature study inevitable, Sharp characterizes America's love of the outdoors as both fated and inborn. A keen observer of the farm life and the farm press, especially naturalist farmers of the sort he describes in the following pages, Sharp, like Thoreau, bridges the gap between farmer and woodsman, village dweller and rural resident, drawing epiphanies from all sides.

The Nature Movement[131]

I was hurrying across Boston Common. Two or three hundred others were hurrying with me. But ahead, at the union of several

---

[131] Dallas Lore Sharp, *The Lay of the Land* (Boston: Houghton Mifflin, 1908), 114–126.

paths, was a crowd, standing still. I kept hurrying on, not to join the crowd, but simply to keep up the hurry. The crowd was not standing still, it was hurrying, too, scattering as fast as it gathered, and as it scattered I noticed that it wore a smile. I hastened up, pushed in, as I had done a score of times on the Common, and got my glimpse of the show. It was not a Mormon preaching, not a single-taxer, not a dogfight. It was Billy, a gray squirrel, taking peanuts out of a bootblack's pocket. And every age, sex, sort, and condition of Bostonian came around to watch the little beast shuck the nuts and bury them singly in the grass of the Common.

"Ain't he a cute little cuss, mister?" said the boy of the brush, feeling the bottom of his empty pocket, and looking up into the prosperous face of Calumet and Hecla at his side. C. and H. smiled, slipped something into the boy's hand with which to buy another pocketful of peanuts for Billy, and hurried down to State Street.

This crowd on the Common is nothing exceptional. It happens every day, and everywhere, the wide country over. We are all stopping to watch, to feed, and—to smile. The longest, most far-reaching pause in our hurrying American life today is this halt to look at the outdoors, this attempt to share its life, and nothing more significant is being added to our American character than the resulting thoughtfulness, sympathy, and simplicity—the smile on the faces of the crowd hurrying over the Common.

Whether one will or not, he is caught up by this nature movement and set adrift in the fields. It may, indeed, be "adrift" for him until he gets thankfully back to the city. "It was a raw November day," wrote one of these new nature students, who happened also to be a college student, "and we went for our usual Saturday's birding into the woods. The chestnuts were ripe, and we gathered a peck between us. On our way home, we discovered a small bird perched upon a cedar tree with a worm in its beak. It was a hummingbird, and after a little searching we found its tiny nest close up against the trunk of the cedar, full of tiny nestlings just ready to fly."

This is what they find, many of these who are caught up by the movement toward the fields, but not all of them. A little five-year-old from the village came out to see me recently, and while playing in the orchard she brought me five flowers, called them by their right names, and told me how they grew. Down in the loneliest marshes of Delaware Bay, I know a lighthouse keeper and his solitary neighbor, a farmer: Both have been touched by this nature spirit. Both are interested, informed, and observant. The farmer there on the old Zane's place, is no man of books, like the rector of Selborne, but he is a man of birds and beasts, of limitless marsh and bay and sky, of everlasting silence and wideness and largeness and eternal solitude. He could write a natural history of the Maurice River marshes.

These are not rare cases. The nature books, the nature magazines, the nature teachers, are directing us all outdoors. I subscribe to a farm journal (club rates, twenty-five cents a year!) in which an entire page is devoted to "nature studies," while the whole paper is remarkably fresh and odorous of the real fields. In the city, on my way to and from the station, I pass three large bookstores, and from March until July, each of these shops has a big window given over almost continuously to "nature books." I have before me from one of these shops a little catalogue of nature books—"a select list" for 1907 containing 233 titles, varying in kind all the way from *The Tramp's Handbook* to one (to a dozen) on the very stable subject of "The Farmstead." These are all distinctively "nature books," books with an appeal to sentiment as well as to sense, and very unlike the earlier desiccated, unimaginative treatises.

There are a multitude of other signs that show as clearly as the nature books how full and strong is this tide that sets toward the open fields and woods. There are as many and as good evidences, too, of the genuineness of this interest in the outdoors. It may be a fad just now to adopt abandoned farms, to attend parlor lectures on birds, and to possess a how-to-know library. It is pathetic to see "nature study" taught by schoolmarms who never did and who never will climb a rail fence; it is sad, to speak softly, to have the makers of certain animal books preface the stories with a declaration of their absolute truth; it is passing sad that the unnatural natural history, the impossible outdoors, of some of the recent nature books, should have been created. But fibs and failures and impossibilities aside, there still remains the thing itself—the widespread turning to nature, and the deep, vital need to turn.

The note of sincerity is clear, however, in most of our nature writers; the faith is real in most of our nature teachers; and the love—who can doubt the love of the tens of thousands of those whose feet feel the earth nowadays, whose lives share in the existence of some pond or wood or field? And who can doubt the rest, the health, the sanity, and the satisfaction that these get from the companionship of their field or wood or pond?

There is no way of accounting for the movement that reflects in the least upon its reality and genuineness. It may be only the appropriation by the common people of the world that the scientists have discovered to us; it may be a popular reaction against the conventionality of the eighteenth century; or the result of our growing wealth and leisure; or a fashion set by Thoreau and Burroughs—one or all of these may account for its origin—but nothing can explain the movement away, or hinder us from being borne by it out, at least a little way, under the open of heaven, to the great good of body and soul.

Among the cultural influences of our times that have developed the proportions of a movement, this so-called nature movement is peculiarly American. No such general, widespread turning to the outdoors is seen anywhere else; no other such body of nature literature as ours; no other people so close to nature in sympathy and understanding, because there is no other people of the same degree of culture living so close to the real, wild outdoors.

The extraordinary interest in the outdoors is not altogether a recent acquirement. We inherited it. Nature study is an American habit. What else had the pioneers and colonists to study but the outdoors? And what else was half as wonderful? They came from an old urban world into this new country world, where all was strange, unnamed, and unexplored. Their chief business was observing nature, not as dull savages, nor as children born to a dead familiarity with their surroundings, but as interested men and women, with a need and a desire to know. Their coming was the real beginning of our nature movement; their observing has developed into our nature study habit.

Our nature literature also began with them. There is scarcely a journal, a diary, or a set of letters of this early time in which we do not find that careful seeing, and often that imaginative interpretation, so characteristic of the present day. Even the modern animal romancer is represented among these early writers in John Josselyn and his delicious book *New England's Rarities Discovered.*

It was not until the time of Emerson and Bryan and Thoreau, however, that our interest in nature became general and grew into something deeper than mere curiosity. There had been naturalists such as Audubon (he was a poet, also), but they went off into the deep woods alone. They were after new facts, new species. Emerson and Bryan and Thoreau went into the woods, too, but not for facts, nor did they go far, and they invited us to go along. We went, because they got no farther than the back-pasture fence. It was not to the woods they took us, but to nature; not a-hunting after new species in the name of science, but for new inspirations, new estimates of life, new health for mind and spirit.

But we were slow to get as far even as their back-pasture fence, slow to find nature in the fields and woods. It was fifty years ago that Emerson tried to take us to nature, but fifty years ago how few there were who could make sense out of his invitation, to say nothing of accepting it! And of Thoreau's first nature book, *A Week on the Concord and Merrimack Rivers*, there were sold, in four years after publication, two hundred and twenty copies. But two hundred and twenty of such books at work in the mind of the country could leaven, in time, a big lump of it. And they did. The outdoors, our attitude toward it, and our literature about it have never been the same since.

Even yet, however, it is the few only who respond to Thoreau, Emerson, and Burroughs, who can find nature, as well as birds and trees, who can think and feel as well as wonder and look. Before we can think and feel, we must get over our wondering, and we must get entirely used to looking. This we are slowly doing—slowly, I say, for it is the monstrous, the marvelous, the unreal that most of us still go out into the wilderness for to see—bears and wolves, foxes, eagles, orioles, salmon, mustangs, porcupines of extraordinary parts and powers.

There came to my desk, tied up with the same string, not long since, three nature books of a sort to make Thoreau turn over in his grave—accounts of beasts and birds such as old Thetbaldus gave us in his *Physiologus*, that pious and marvelous bestiary of the Dark Ages. These three volumes that I refer to are modern and about American animals, but they, too, might have been written during the Dark Ages. All three have the same solemn preface, declaring the absolute truth of the observations that follow (as if we might doubt), and piously pointing out their high moral purpose; all three likewise start out with the same wonderful story, an animal biography: one of a slum cat born in a cracker box. Among the kittens of the cracker box was an extraordinary kitten of "pronounced color" who survives and comes to glory. The next book tells the biography of a fox, born in a hole among the Canadian hills. Among the pups born in this hole was one extraordinary pup "more finely colored" than the others, who survives and comes to glory. The third book tells the biography of a wolf, born in a cave among the rocks, still farther north. Among the cubs born in this cave was one extraordinary cub, "larger than the others," who survives and, as is to be expected of a wolf, comes to more glory than the cracker-box kitten or the fox pup of the hills.

Such are the stories that are made into texts and readers for our public schools; such are the animals that go roaming through the woods of the American child's imagination. But no such kittens or cubs or pups lurk in my eight-acre woodlot. I have seen several (six, to be exact) fox pups, but never did I see this overworked, extraordinary, cum laude pup of the recent nature books.

So long as we continue to read and believe such accounts, just so long shall we find it impossible to go with Audubon and Thoreau and Burroughs, for they have no place to take us, nothing to show us when we arrive. Their real world does not exist.

But we know that a real, ordinary, yet a marvelous world does exist, and right at hand. The present great nature movement is an outgoing to discover it—its trees, birds, flowers, its myriad forms. This is the meaning of the countless manuals, the how-to-know books, and the nature study of the public schools. And this desire to know Nature is the reasonable, natural preparation for the deeper insight that leads to communion with her—a desire to be traced more directly to Agassiz, and the hosts of teachers he inspired, perhaps, than to the poet-essayists like Emerson and Thoreau and Burroughs.

Let us learn to see and name first. The inexperienced, the unknowing, the unthinking, cannot love. One must live until tired, and think until baffled, before he can know his need of Nature, and then he will not know how to approach her unless already acquainted. To expect anything more than curiosity and animal delight in a child is foolish, and the attempt to teach him anything more at first than to know the outdoors is equally foolish. Poets are born, but not until they are old.

But if one got no farther than his how-to-know book would lead him, he still would get into the fields—the best place for him this side of heaven—he would get ozone for his lungs, red blood, sound sleep, and health. As a nation, we had just begun to get away from the farm and out of touch with the soil. The nature movement is sending us back in time. A new wave of physical soundness is to roll in upon us as the result, accompanied with a newness of mind and of morals.

For, next to bodily health, the influence of the fields makes for the health of the spirit. It is easier to be good in a good body and an environment of largeness, beauty, and peace—easier here than anywhere else to be sane, sincere, and "in little thyng have suffisaunce." If it means anything to think upon whatsoever things are good and lovely, then it means much to own a how-to-know book and to make use of it.

This is hardly more than a beginning, however, merely satisfying an instinct of the mind. It is good if done afield, even though such classifying of the outdoors is only scraping an acquaintance with nature. The best good, the deep healing, comes when one, no longer a stranger, breaks away from his getting and spending, from his thinking with men, and camps under the open sky, where he knows without thinking, and worships without priest or chant or prayer.

The world's work must be done, and only a small part of it can be done in the woods and fields. The merchants may not all turn plowmen and woodchoppers. Nor is it necessary. What we need to do, and are learning to do, is to go to nature for our rest and health and recreation.

BOLTON HALL

Like Edward Faulkner's *Plowman's Folly* in the early 1940s, Bolton Hall's *Three Acres and Liberty* reached not only the farmer but the average backyard gardener as well. A Malthusian, Hall believed that population growth would eventually outpace food supplies and give rise to apocalyptic shortages. In *Three Acres and Liberty*, he argues that the urban escapist could, especially in cooperation with his neighbors, meet his own food needs on a mere few acres, claims sufficient to spur a number of communal experiments around the country. Hall's focus on the garden as a societal balm was not original, although it was persuasive. George Maxwell, a California lawyer, likewise lead a national movement to use the garden to relieve social tensions.

Here in "The Garden Yard," Hall's waste-not-want-not mantra focuses on the yard itself as a hidden source of profit. Where writers like Edward Payson Powell saw aesthetic and recreational pleasure, Hall saw a potential means of survival. "I want to give the plain man or woman, who has a backyard or a back lot, out of which he might make part of a living or more of a living, a book that will show how to do it," Hall writes in his preface. Like Powell's, Hall's writing was in the vanguard, anticipating the Victory Gardens of World War I, and the increasing number of farmers who, in the Great Depression and after, would increasingly bankroll their hobby farming with off-the-farm income. In "The Garden Yard" Hall is, as usual, provocative, brooking no criticism from purists who believed that a true producer must work a minimum number of acres before he could rightly call himself a farmer. Hall writes, "If anybody sneers at your gardening as being 'book farming,' let them sneer; a fool never understands what a wise man is doing."

The Garden Yard[132]

No man knows, nor can know, the capacity of a garden yard, for it may be unlimited, just as the speed of the engine is unlimited. Just as with the engine, the only question is whether it would pay to make it do anymore—it may cost too much. Where land is cheap, labor is high; there intelligent cultivation will pay, but intensive cultivation will not. That is the place where the field crops should be raised.

But the garden crops should be raised right round the towns and cities and it is foolish to get to a distance from them. Stay right where you are and get the piece of land that is best for your purpose; buy it, if you can without paying too much for it; if not, rent it for as long a term as you can; or get permission to use the bit of land, the vacant lots—there are plenty even in the most crowded cities—and raise your truck and your income on those lots. Without separating yourself from your acquaintances or exiling your wife and children, learn to get your living out of the earth.

Suppose that a man owns his house, even if it be but a bit of a bungalow, and suppose he has a little bit of land on which he can raise the most of what the family eats; he may have to work hard, especially if his family cannot help in the work, but at least he is independent; at least panics, lock-outs, change of circumstances or even loss of health will not reduce him to starvation.

If you have a farm, intensive cultivation should interest you all the more. Every farm is full of opportunities to make good money; but you must not make the usual mistake of half working a big piece of land; that means that you will always be overworked, always have a lot of things that you know ought to be done, but cannot find time to do; always have common grade crops that bring common prices. Everyone that is overworked is underpaid, for he cannot do his best work.

---

[132] Bolton Hall, *The Garden Yard: A Handbook of Intensive Farming* (Philadelphia: D. McKay, 1909), 21–25.

Use the big fields for pasture, or for raising fine horses, or for pigs or Angora goats or even for sheep; you had better let the fields run wild rather than half cultivate them.

Keep accounts and watch your chance to sell all the land that does not pay well. It may be that you are missing a fortune in the old neglected orchard, or in the chestnut or hickory grove. The black walnuts or butternuts that are usually left for the neighbors' boys may be the most profitable part of the farm.

The woodlot may have possibilities for barrel hoops, which may be sold to the improvement of the timber. It may need only thinning to bring you a steady income while it increases in value.

Fine apples grafted on the old trees that now bear only cider apples, if properly sprayed and thinned so as to give first-class fruit, may sell for more than all the corn you can raise.

The "pesky briers" that the farmer struggles with year by year, may be the raspberries and blackberries that will sell readily for good prices, when they are cultivated, to the summer residents or boarding houses. Your exposure and soil may be just the place for the fine strawberries with which, when nicely separated from the second and third grades, no market is ever overstocked.

But if you are always behind with work and always short of cash or worried to pull through, you have no time to think of these things and no means to hire labor nor to develop them.

That pond may be needed, if it were cleared out, for a profitable ice supply, furnishing paying work in the winter. The stream may be a valuable water power or at least bring a high-priced crop of watercress; or it may be the very water needed, when properly distributed, to make yours the most fertile land in the country. The bit of swamp land, that raises nothing but mosquitoes, may need only a few dollars' worth of cranberry sets to be the best paying acre in the countryside.

There may be a veritable gold mine in a neglected quarry, or brick-clay pit, or kaolin clay deposit, or in a sand bank, or a vein of marl.

Possibly you could rent the farmhouse or let camping sites for the summer to people who would pay city prices for much of your stuff, so that you could afford to keep help enough to leave only the easy work of superintendence for you. Brains save more work than machines.

If you are raising the same crops that your neighbors do, harvesting at the same time, and getting the same prices that everyone else does, you may be sure that you are neglecting your chances.

The money is in finding things to raise that will sell, and that do not have to compete with all the others.

Says the *Farm Journal*:

> Farmers need more time to plan their work and look after the business and economic end of their calling. The employer who makes a full hand in barn and field from 5 AM till 8 AM, has no other time to devote to the real business of the farm than the hours in which nature imperatively calls upon him to rest, and a man with aching muscles and tired limbs is not in condition to think clearly or plan intelligently. It is poor economy for a farmer to take the place of a dollar-a-day man in the field, when in so doing he has left no leisure in which to work out the details of his operations.

Think—think—it is true that we ourselves must work with the men if we are to get the best work out of them; there is a big difference between saying "Go, do that," and "Come, and let us do this." But it is not enough to work; any jackass can do that.

You know the old fable: "A farmer got his wheels stuck fast in a miry road. The man knelt down in the mud crying to Hercules to come and help him. Said Hercules, "Get up and put your shoulder to the wheel, I help only those who help themselves!"

INTRODUCING THE COUNTRY LIFE COMMISSION

The following two selections, consisting of Theodore Roosevelt's prefatory special message to the Senate and House of Representatives, dated February 9, 1909, and one section, numbered six, of the Commissioners' Report dealing with rural "deficiencies," are excerpted from the Spokane Washington, Chamber of Commerce reprinting of the *Report of the Country Life Commission* authorized by William Taft. In his message Roosevelt largely summarizes the Commission's findings for consumption by Congress, though he does ask for an appropriation of some $25,000 to allow the Commission to continue its work and to digest and disseminate its "harvest of suggestions." Of particular interest in the President's message is "Appendix A" which relays the unintentionally humorous, utterly homespun responses of one Missouri farmer to the Commission's questionnaire. This lengthy appendix shows not only Roosevelt's brand of populism and earthiness, but also his readerly interests in agrarianism, as he recommends Jennie Buell's portrait of a socially-minded farm wife, *One Woman's Work for Farm Women*.

Despite its sometimes judgmental tone, the *Report of the Country Life Commission* that follows was intended as benevolent rather punitive, as its chair, Liberty Hyde Bailey of Cornell University's horticulture department, believed that reinvigorating rural institutions could slow migration to the cities. Much to Liberty Hyde Bailey's chagrin, zealous Country Life reformers used the Commission's report as further evidence—much of it self-reported by farmers and their families—of a litany of rural derelictions, choosing to emphasize the document's prescriptive rather than descriptive potential. Though many rural residents pushed back against the Commission's findings regarding school consolidation and curriculum, the supposed shortcomings in rural education would, five years after the report's publication, result in the Smith-Lever Act of 1914 and, later, the Smith-Hughes Act of 1917. Both acts provided federal funds for far-reaching, educative outreach in the country—federal aid to schools offering instruction in vocational agriculture and home economics in the case of Smith-Lever, and farm extension agents in the case of Smith-Hughes. "Woman's Work on the Farm," included here after President Roosevelt's introductory remarks from the White House, demonstrates a rededication to the welfare of farm women and families, populations whose betterment, the Commission hoped, would spur greater cooperative efficiency among farm men. Though drafted by Liberty Hyde Bailey, the report reflects the sentiments of other prominent commissioners, most especially Kenyon L. Butterfield, Gifford Pinchot, and Henry Wallace, and, to a lesser extent, Charles S. Barrett, William A. Beard, and Walter H. Page.

To the Senate and House of Representatives[133], *A Special Message from Theodore Roosevelt*

I transmit herewith the report of the Commission on Country Life. At the outset I desire to point out that not a dollar of the public money has been paid to any commissioner for his work on the Commission.

The report shows the general condition of farming life in the open country, and points out its larger problems; it indicates ways in which the government, national and state, may show the people how to solve some of these problems, and it suggests a continuance of the work which the Commission began.

Judging by thirty public hearings, to which farmers and farmers' wives from forty states and territories came, and from one hundred and twenty thousand answers to printed questions sent out by the Department of Agriculture, the Commission finds that the general level of country life is high compared with any preceding time or with any other land. If it has in recent years slipped down in some places, it has risen in more places. Its progress has been general, if not uniform.

Yet farming does not yield either the profit or the satisfaction that it ought to yield and may be made to yield. There is discontent in the country and, in places, discouragement. Farmers as a class do not magnify their calling, and the movement to the towns, though, I am happy to say, less than formerly, is still strong.

Under our system, it is helpful to promote discussion of ways in which the people can help themselves. There are three main directions in which the farmers can help themselves: namely, better

---

[133] *Report of the Country Life Commission and Special Message from the President of the United States* (Spokane, WA: Chamber of Commerce, 1909), 44–49.

farming, better business, and better living on the farm. The National Department of Agriculture, which has rendered services equaled by no other similar department in any other time or place; the state departments of agriculture; the state colleges of agriculture and the mechanic arts, especially through their extension work; the state agricultural experiment stations; the Farmers' Union; the Orange; the agricultural press; and other similar agencies have all combined to place within the reach of the American farmer an amount and quality of agricultural information which, if applied, would enable him, over large areas, to double the production of the farm.

The object of the Commission on Country Life therefore is not to help the farmer raise better crops, but to call his attention to the opportunities for better business and better living on the farm. If country life is to become what it should be, and what I believe it ultimately will be—one of the most dignified, desirable, and sought-after ways of earning a living—the farmer must take advantage not only of the agricultural knowledge which is at his disposal, but of the methods which have raised and continue to raise the standards of living and of intelligence in other callings.

Those engaged in all other industrial and commercial callings have found it necessary, under modern economic conditions, to organize themselves for mutual advantage and for the protection of their own particular interests in relation to other interests. The farmers of every progressive European country have realized this essential fact and have found in the cooperative system exactly the form of business combination they need.

Now whatever the state may do toward improving the practice of agriculture, it is not within the sphere of any government to reorganize the farmers' business or reconstruct the social life of farming communities. It is, however, quite within its power to use its influence and the machinery of publicity which it can control for calling public attention to the needs and the facts. For example, it is the obvious duty of the government to call the attention of farmers to the growing monopolization of water power. The farmers above all should have that power, on reasonable terms, for cheap transportation, for lighting their homes, and for innumerable uses in the daily tasks on the farm.

It would be idle to assert that life on the farm occupies as good a position in dignity, desirability, and business results as the farmers might easily give it if they chose. One of the, chief difficulties is the failure of country life, as it exists at present, to satisfy the higher social and intellectual aspirations of country people. Whether the constant draining away of so much of the best elements in the rural population into the towns is due chiefly to this cause or to the superior business opportunities of city life may be open to question. But no one at all familiar with farm life throughout the United States can fail to recognize the necessity for building up the life of the farm upon its social as well as upon its productive side.

It is true that country life has improved greatly in attractiveness, health, and comfort, and that the farmers earnings are higher than they were. But city life is advancing even more rapidly because of the greater attention which is being given by the citizens of the towns to their own betterment. For just this reason the introduction of effective agricultural cooperation throughout the United States is of the first importance.

Where farmers are organized cooperatively they not only avail themselves much more readily of business opportunities and improved methods, but it is found that the organizations which bring them together in the work of their lives are used also for social and intellectual advancement.

The cooperative plan is the best plan of organization wherever men have the right spirit to carry it out. Under this plan any business undertaking is managed by a committee, every man has one vote and only one vote, and everyone gets profits according to what he sells or buys or supplies. It develops individual responsibility and has a moral as well as a financial value over any other plan.

I desire only to take counsel with the farmers as fellow citizens. It is not the problem of the farmers alone that I am discussing with them, but a problem which affects every city as well as every farm in the country. It is a problem which the working farmers will have to solve for themselves, but it is a problem which also affects, in only less degree, all the rest of us, and therefore if we can render any help toward its solution, it is not only our duty but our interest to do so.

The foregoing will, I hope, make it clear why I appointed a commission to consider problems of farm life which have hitherto had far too little attention, and the neglect of which has not only held back life in the country, but also lowered the efficiency of the whole nation. The welfare of the farmer is of vital consequence to the welfare of the whole community. The strengthening of country life, therefore, is the strengthening of the whole nation.

The Commission has tried to help the farmers to see clearly their own problem and to see it as a whole; to distinguish clearly between what the government can do and what the farmers must do for themselves; and it wishes to bring not only the farmers but the nation as a whole to realize that the growing of crops, though an essential part, is only a part of country life. Crop growing is the essential foundation, but it is no less essential that the farmer shall get an adequate return for what he grows, and it is no less essential, indeed it is literally vital, that he and his wife and children shall lead the right kind of life.

For this reason, it is of the first importance that the United States Department of Agriculture, through which as prime agent the ideas the Commission stands for must reach the people, should become without delay in fact a Department of Country Life, fitted to deal not only with crops, but also with all the larger aspects of life in the open country.

From all that has been done and learned three great general and immediate needs of country life stand out:

First, effective cooperation among farmers, to put them on a level with the organized interests with which they do business.

Second, a new kind of schools in the country, which shall teach the children as much outdoors as indoors and perhaps more, so that they will prepare for country life, and not as at present, mainly for life in town.

Third, better means of communication, including good roads and a parcels post, which the country people are everywhere, and rightly, unanimous in demanding.

To these may well be added better sanitation, for easy preventable diseases hold several million country people in the slavery of continuous ill health.

The Commission points out, and I concur in the conclusion, that the most important help that the government, whether national or state, can give is to show the people how to go about these tasks of organization, education, and communication with the best and quickest results. This can be done by the collection and spread of information. One community can thus be informed of what other communities have done, and one country of what other countries have done. Such help by the people's government would lead to a comprehensive plan of organization, education, and communication, and make the farming country better to live in, for intellectual and social reasons as well as for purely agricultural reasons.

The government, through the Department of Agriculture, does not cultivate any man's farm for him. But it does put at his service useful knowledge that he would not otherwise get. In the same way the national and state governments might put into the people's hands the new and right knowledge of school work. The task of maintaining and developing the schools would remain, as now, with the people themselves.

The only recommendation I submit is that an appropriation of $25,000 be provided to enable the Commission to digest the material it has collected, and to collect and to digest much more that is within its reach, and thus complete its work. This would enable the Commission to gather in the harvest of suggestion which is resulting from the discussion it has stirred up. The Commissioners have served without compensation, and I do not recommend any appropriation for their services, but only for the expenses that will be required to finish the task that they have begun.

To improve our system of agriculture seems to me the most urgent of the tasks which lie before us. But it can not, in my judgment, be effected by measures which touch only the material and technical side of the subject; the whole business and life of the farmer must also be taken into account. Such considerations led me to appoint the Commission on Country Life. Our object should be to help develop in the country community the great ideals of community life as well as of personal character. One of the most important adjuncts to this end must be the country church, and I invite your attention to what the Commission says of the country church and of the need of an extension of such work as that of the Young Men's Christian Association in country communities. Let me lay special emphasis upon what the Commission says at the very end of its report on personal ideals and local leadership. Everything resolves itself in the end into the question of personality. Neither society nor government can do much for country life unless there is voluntary response in the personal ideals of the men and women who live in the country. In the development of character, the home should be more important than the school, or than society at large. When once the basic material needs have been met, high ideals may be quite independent of income, but they can not be realized without sufficient income to provide adequate foundation, and where the community at large is not financially prosperous it is impossible to develop a high average personal and community ideal. In short, the fundamental facts of human nature apply to men and women who live in the country just as they apply to men and women who live in the towns. Given a sufficient foundation of material well-being, the influence of the farmers and farmers' wives on their children becomes the factor of first importance in determining the attitude of the next generation toward farm life. The farmer should realize that the person who most needs consideration on the farm is his wife. I do not in the least mean that she should purchase ease at the expense of duty. Neither man nor woman is really happy or really useful save on conditions of doing his or her duty. If the woman shirks her duty as housewife, as homekeeper, as the mother whose prime function it is to bear and rear a sufficient number of healthy children, then she is not entitled to our regard. But, if she does her duty, she is more entitled to our regard even than the man who does his duty, and the man should show special consideration for her needs.

I warn my countrymen that the great recent progress made in city life is not a full measure of our civilization, for our civilization rests at bottom on the wholesomeness, the attractiveness, and the completeness, as well as the prosperity, of life in the country. The men and women on the farms stand for what is fundamentally best and most needed in our American life. Upon the development of country life rests ultimately our ability, by methods of farming requiring the highest intelligence, to continue to feed and clothe the hungry nations; to supply the city with fresh blood, clean bodies, and clear brains that can endure the terrific strain of modern life; we need the development of men in the open country, who will be in the future, as in the past, the stay and strength of the nation in time of war, and its guiding and controlling spirit in time of peace.

Appendix A

One of the most illuminating and incidentally one of the most interesting and amusing series of answers sent to the Commission was from a farmer in Missouri. He stated that he had a wife and eleven living children, he and his wife being each fifty-two-years-old; and that they owned 520 acres of land without any mortgage hanging over their heads. He had himself done well, and his views as to why many of his neighbors had done less well are entitled to consideration. These views are expressed in terse and vigorous English; they can not always be quoted in full. He states that the farm homes in his neighborhood are not as good as they should be because too many of them are encumbered by mortgages; that the schools do not train boys and girls satisfactorily for life on the farm, because they allow them to get an idea in their heads that city life is better, and that to remedy this practical farming should be taught. To the question whether the farmers and their wives in his neighborhood are satisfactorily organized, he answers: "Oh, there is a little one-horse Grange gang in our locality, and every darned one thinks they ought to be a king." To the question, *Are the renters of farms in your neighborhood making a satisfactory living?* he answers: "No; because they move about so much hunting a better job." To the question, *Is the supply of farm labor in your neighborhood satisfactory?* the answer is: "No; because the people have gone out of the baby business," and when asked as to the remedy he answers, "Give a pension to every mother who gives birth to seven living boys on American soil." To the question, *Are the conditions surrounding hired labor on the farm in your neighborhood satisfactory to the hired men?* he answers: "Yes, unless he is a drunken cuss," adding that he would like to blow up the stillhouses and root out whisky and beer. To the question, *Are the sanitary conditions on the farms in your neighborhood satisfactory?* he answers: "No; too careless about chicken yards, and the like, and poorly covered wells, in one well on neighbor's farm I counted seven snakes in the wall of the well, and they used the water daily, his wife dead now and he is looking for another." He ends by stating that the most important single thing to be done for the betterment of country life is good roads, but in his answers he shows very clearly that most important of all is the individual equation of the man or woman. The humor of this set of responses must not blind us to the shrewd common sense and good judgment they display. The man is a good citizen, his wife is a good citizen, and their views are fundamentally sound. Very much information of the most valuable kind can be gathered if the Commission is given the money necessary to enable it to arrange and classify the information obtained from the great mass of similar answers which they have received. But there is one point where the testimony is as a whole in flat contradiction to that contained above. The general feeling is that the organizations of farmers, the Granges and the like, have been of the very highest service not only to the farmers, but to the farmers' wives, and that they have conferred great social as well as great industrial advantages. An excellent little book has recently been published by Miss Jennie Buell, called *One Woman's Work for Farm Women*. It is dedicated "To farm women everywhere," and is the story of Mary A. Mayo's part in rural social movements. It is worthwhile to read this little volume to see how much the hard-working woman who lives on the farm can do for herself when once she is given sympathy, encouragement, and occasional leadership.

Woman's Work on the Farm[134] *by Liberty Hyde Bailey and the Country Life Commission*

Realizing that the success of country life depends in very large degree on the woman's part, the Commission has made special effort to ascertain the condition of women on the farm. Often this condition is all that can be desired, with home duties so organized that the labor is not excessive, with kindly cooperation on the part of husbands and sons, and with household machines and conveniences well provided. Very many farm homes in all parts of the country are provided with books and periodicals, musical instruments, and all the necessary amenities. There are good gardens and attractive premises and a sympathetic love of nature and of farm life on the part of the entire family.

On the other hand, the reverse of these conditions often obtains, sometimes because of pioneer conditions and more frequently because of lack of prosperity and of ideals. Conveniences for outdoor work are likely to have precedence over those for household work.

The routine work of woman on the farm is to prepare three meals a day. This regularity of duty recurs regardless of season, weather, planting, harvesting, social demands, or any other factor. The only differences in different seasons are those of degree rather than of kind. It follows, therefore, that whatever general hardships, such as poverty, isolation, and lack of labor-saving devices, may exist on any given farm, the burden of these hardships falls more heavily on the farmer's wife than on the farmer himself. In general, her life is more monotonous and the more isolated, no matter what the wealth or the poverty of the family may be.

The relief to farm women must come through a general elevation of country living. The women must have more help. In particular, these matters may be mentioned: development of a cooperative spirit in the home, simplification of the diet in many cases, the building of convenient and sanitary houses, providing running water in the house

---

[134] "Woman's Work on the Farm," was originally a subheading for section six of the Commission Report.

and also more mechanical help, good and convenient gardens, a less exclusive ideal of money-getting on the part of the farmer, providing better means of communication, as telephones, roads, and reading circles, and developing of women's organizations. These and other agencies should relieve the woman of many of her manual burdens on the one hand and interest her in outside activities on the other. The farm woman should have sufficient free time and strength so that she may serve the community by participating in its vital affairs.

We have found good women's organizations in some country districts, but as a rule such organizations are few or even none, or where they exist they merely radiate from towns. Some of the stronger central organizations are now pushing the country phase of their work with vigor. Mothers' clubs, reading clubs, church societies, home economics organizations, farmers' institutes, and other associations can accomplish much for farm women. Some of the regular farmers organizations are now giving much attention to domestic subjects, and women participate freely in the meetings. There is much need among country women themselves of a stronger organizing sense for real cooperative betterment. It is important also that all rural organizations that are attended chiefly by men should discuss the homemaking subjects, for the whole difficulty often lies with the attitude of the men.

There is the most imperative need that domestic, household, and heath questions be taught in all schools. The home may well be made the center of rural school teaching. The school is capable of changing the whole attitude of the home life and the part that women should play in the development of the best country living.

The General Corrective Forces that Should Be Set in Motion

The ultimate need of the open country is the development of community effort and of social resources.[135] Here and there the

---

[135] This sentence marks the beginning of the section originally assigned roman numeral III and likewise titled "The General Corrective Forces that Should be Set in Motion."

Commission has found a rural neighborhood in which the farmers and their wives come together frequently and effectively for social intercourse, but these instances seem to be infrequent exceptions. There is a general lack of wholesome societies that are organized on a social basis. In the region in which the Grange is strong this need is best supplied.

There is need of the greatest diversity in country life affairs, but there is equal need of a social cohesion operating among all these affairs and tying them all together. This life must be developed, as we have said, directly from native or resident forces. It is neither necessary nor desirable that an exclusive hamlet system be brought about in order to secure these ends. The problem before the Commission is to suggest means whereby this development may be directed and hastened directly from the land.

The social disorder is usually unrecognized. If only the farms are financially profitable, the rural condition is commonly pronounced good. Country life must be made thoroughly attractive and satisfying, as well as remunerative, and able to hold the center of interest throughout one's lifetime. With most persons this can come only with the development of a strong community sense of feeling. The first condition of a good country life, of course, is good and profitable farming. The farmer must be enabled to live comfortably.

Much attention has been given to better farming, and the progress of a generation has been marked. Small manufacture and better handicrafts need now to receive attention, for the open country needs new industries and new interests. The schools must help to bring these things about.

The economic and industrial questions are, of course, of prime importance, and we have dealt with them, but they must all be studied in their relations to the kind of life that should ultimately be established in rural communities. The Commission will fail of its purpose if it confines itself merely to providing remedies or correctives for the present and apparent troubles of the farmer, however urgent and important these troubles may be. All these matters must be conceived of as incidents or parts in a large constructive program. We must begin a campaign for rural progress.

To this end local government must be developed to its highest point of efficiency, and all agencies that are capable of furthering a better country life must be federated. It will be necessary to set the resident forces in motion by means of outside agencies, or at least to direct them, if we are to secure the best results. It is specially necessary to develop the cooperative spirit, whereby all people participate and all become partakers.

The cohesion that is so marked among the different classes of farm folk in older countries can not be reasonably expected at this period in American development, nor is it desirable that a stratified society should be developed in this country. We have here no remnants of a feudal system, fortunately no system of entail, and no clearly drawn distinction between agricultural and other classes. We are as yet a new country with undeveloped resources, many faraway pastures which, as is well known, are always green and inviting. Our farmers have been moving, and numbers of them have not yet become so well settled as to speak habitually of their farm as home. We have farmers from every European nation and with every phase of religious belief often grouped in large communities, naturally drawn together by a common language and a common faith, and yielding but slowly to the dominating and controlling forces of American farm life. Even where there was once social organization, as in the New England town (or township), the competition of the newly settled West and the wonderful development of urban civilization have disintegrated it. The middle-aged farmer of the central states sells the old homestead without much hesitation or regret and moves westward to find a greater acreage for his sons and daughters. The farmer of the Middle West sells the old home and moves to the Mountain states, to the Pacific coast, to the South, to Mexico, or to Canada.

Even when permanently settled, the farmer does not easily combine with others for financial or social betterment. The training of generations has made him a strong individualist, and he has been obliged to rely mainly on himself. Self-reliance being the essence of his nature, he does not at once feel the need of cooperation for business purposes or of close association for social objects. In the main, he has been prosperous, and has not felt the need of cooperation. If he is a strong man, he prefers to depend on his own ability. If he is ambitious for social recognition, he usually prefers the society of the town to that of the country. If he wishes to educate his children, he avails himself of the schools of the city. He does not, as a rule, dream of a rural organization that can supply as completely as the city the four great requirements of health, education, occupation, and society. While his brother in the city is striving by moving out of the business section into the suburbs to get as much as possible of the country in the city, he does not dream that it is possible to have most that is best of the city in the country.

The time has come when we must give as much attention to the constructive development of the open country as we have given to other affairs. This is necessary not only in the interest of the open country itself, but for the safety and progress of the nation.

It is impossible, of course, to suggest remedies for all the shortcomings of country life. The mere statement of the conditions, as we find them, ought of itself to challenge attention to the needs. We hope that this report of the Commission will accelerate all the movements that are now in operation for the betterment of country life. Many of these movements are beyond the reach of legislation. The most important thing for the Commission to do is to apprehend the problem and to state the conditions.

The philosophy of the situation requires that the disadvantages and handicaps that are not a natural part of the farmer's business shall be removed, and that such forces shall be encouraged and set in motion as will stimulate and direct local initiative and leadership.

The situation calls for concerted action. It must be aroused and energized. The remedies are of many kinds, and they must come slowly. We need a redirection of thought to bring about a new atmosphere, and a new social and intellectual contact with life. This means that the habits of the people must change. The change will come gradually, of course, as a result of new leadership, and the situation must develop its own leaders.

Care must be taken in all the reconstructive work to see that local initiative is relied on to the fullest extent, and that federal and even state agencies do not perform what might be done by the people in the communities. The centralized agencies should be simulative and directive, rather than mandatory and formal. Every effort must be made to develop native resources, not only of material things, but also of people.

It is necessary to be careful, also, not to copy too closely the reconstructive methods that have been so successful in Europe. Our conditions and problems differ widely from theirs. We have no historical, social peasantry, a much less centralized form of government, unlike systems of land occupancy, wholly different farming schemes, and different economic and social systems. Our country necessities are peculiarly American.

The correctives for the social sterility of the open country are already in existence or underway, but these agencies all need to be strengthened and especially to be coordinated and federated, and the problem needs to be recognized by all the people. The regular agricultural departments and institutions are aiding in making farming profitable and attractive, and they are also giving attention to the social and community questions. There is a widespread awakening as a result of this work. This awakening is greatly aided by the rural free delivery of mails, telephones, the gradual improvement of highways, farmers institutes, cooperative creameries and similar organizations, and other agencies.

The good institutions of cities may often be applied or extended to the open country. It appears that the social evils are in many cases no greater in cities in proportion to the number of people than in country districts, and the very concentration of numbers draws attention to the evils in cities and leads to earlier application of remedies. Recently much attention has been directed, for example, to the subject of juvenile crime, and the probation system in place of jail sentences for young offenders is being put into operation in many places. Petty crime and immorality are certainly not lacking in rural districts, and it would seem that there is a place for the extension of the probation system to towns and villages.

Aside from the regular churches, schools, and agricultural societies, there are special organizations that are now extending their work to the open country, and others that could readily be adapted to country work. One of the most promising of these newer agencies is the rural library that is interested in its community. The libraries are increasing, and they are developing a greater sense of responsibility to the community, not only stimulating the reading habit and directing it, but becoming social centers for the neighborhood. A library, if provided with suitable rooms, can afford a convenient meeting place for many kinds of activities and thereby serve as a coordinating influence. Study clubs and traveling libraries may become parts of it. This may mean that the library will need itself to be redirected so that it will become an active rather than a passive agency; it must be much more than a collection of books.

Another new agency is the county work of the Young Men's Christian Association, which, by placing in each county a field secretary, is seeking to promote the solidarity and effectiveness of rural social life, and to extend the larger influence of the country church. The Commission has met the representatives of this county work at the hearings and is impressed with the purpose of the movement to act as a coordinating agency in rural life.

The organizations in cities and towns that are now beginning to agitate the development of better play, recreation, and entertainment offer a suggestion for country districts. It is important that recreation be made a feature of country life, but we consider it to be important that this recreation, games, and entertainment be developed as far as possible from native sources rather than to be transplanted as a kind of theatricals from exotic sources.

Other organizations that are helping the country social life, or that might be made to help it, are women's clubs, musical clubs, reading clubs, athletic and playground associations, historical and literary societies, local business men's organizations and chambers of commerce, all genuinely cooperative business societies, civic and village improvement societies, local political organizations, Granges and other fraternal organizations, and all groups that associate with the church and school.

There is every indication, therefore, that the social life of the open country is in process of improvement, although the progress at the present moment has not been great. The leaders need to be encouraged by an awakened public sentiment, and all the forces should be so related to each other as to increase their total effectiveness while not interfering with the autonomy of any of them.

The proper correctives of the underlying structural deficiencies of the open country are knowledge, education, cooperative organizations, and personal leadership.

H. W. FOGHT

Harold Waldstein Foght, a professor of education at the Normal School in Kirksville, Missouri and elsewhere in the Plains, was a teacher's teacher and, as such, a representation of the early century's increased emphasis on the teaching professional schooled in pedagogy and classroom management. It was this very kind of teacher, a specialist rather than a generalist, that the Country Lifers hoped to introduce to allegedly backward rural schools across the country. Most closely associated with the state of Missouri, Foght taught "Rural Problems" to the next generation of teachers at the practice school in Kirksville. Not long after the publication of *The American Rural School*, which remains his most influential title, Foght worked for the Education Bureau in Washington DC, where he would author *The Rural School System of Minnesota: A Study of School Efficiency* in 1915, among others.

In "Nature Study and School Grounds," Foght proves exceptional in his allegiance to the rural practitioner, a sentiment he inscribes in dedicating his book to the "thousands of hard-working and conscientious teachers who are consecrating their lives to labor in rural communities." In his advocacy for nature studies, Foght reflects Country Life Commission Chair Liberty Hyde Bailey's promotion of nature study as the most apt route by which rural youth would come to appreciate the naturalness of farm life. While Country Lifers shared Bailey's and Foght's objective—namely that rural children ought to be taught ways to better esteem their culture—most in the movement favored agricultural courses and home economics as the means to that end. In the essay below, Foght argues that country life is, in fact, "the normal life—the best life attainable in this greatest of agricultural nations" and argues for an aesthetic training wherein the teacher, acting as a leader for numerous field trips, serves as tour guide and chief experimenter. In this regard, Foght practiced what he preached, authoring in 1906 *The Trail of the Loup; Being a History of the Loup River Region*, a far-ranging, earth-centered cultural and natural history of Nebraska's Loup River region, as well as a 1928 travelogue co-authored with his wife, Alice Mabel Robbins Foght, *Unfathomed Japan; A Travel Tale in the Highways and Byways of Japan and Formosa*. A world traveler and a publisher of more than a dozen books, Foght's prolific writings have been largely overlooked by contemporary historians, though a brief biographical sketch was completed by the Kansas Federal Writers Project.

Nature Study and School Grounds[136]

All our school work has been too formal and bookish.[137] We have all along relied too much on textbooks to the neglect of real living nature. Happily we are beginning to realize the importance of the love and study of nature, and are coming to see that from it have sprung love of art, science, and religion. Paradoxical as it may sound, the farm child has lived in the very heart of nature and yet remained a stranger there. In the struggle to subdue forest and plain, his father and grandfather before him had scant time for anything but to wring a living from the soil. Naturally enough, he inherits certain "practical" traits which make him prone to judge nature by the commercial standard rather than to love it for its own sake. To change these misconceptions, the new teacher must be able to take the child in its own little world and

---

[136] H. W. Foght, *The American Rural School: Its Characteristics, Its Future, and Its Problems* (New York: Macmillan, 1910), 69–74, 154–161.
[137] This sentence marks the beginning of the second full paragraph of Chapter IX, "Nature Study; School Grounds" as it appears in the 1910 edition.

lead it along the pathway of life, directing its native adaptabilities, sentiments, and powers, and there develop, in the child breast, a sympathy with its environment and, in the child mind, an understanding of nature and nature's ways—then, once awakened to the surpassing beauties of rural environments, the American boy and girl will no longer be in danger of deserting the farm for the man-made glitter of the city.

We may find a solution for many of our present difficulties in schoolwork in what is generally called nature study. This is not so much an attempt to add another subject to an already overcrowded curriculum. It is rather a new direction given to old subjects—a leaven infused into old forms—than anything else. It applies in great measure to the entire course of study, since it is possible to encourage the child to close and careful observation of nature through a properly directed lesson in English composition as readily almost as through lessons in geography and elementary science. Most satisfactory, perhaps, is the definition of Dr. Clifton F. Hodge in his well-known book, *Nature Study and Life*. He formulates it, as "learning those things in nature that are best worth knowing, to the end of doing those things that make life most worth living." In rural communities those things are manifestly best worth knowing which tend to make people there content with their lot, which help them to realize that rural life is for Americans the normal life—the best life attainable in this greatest of agricultural nations.

The values of nature study to the rural child are many and far-reaching. Writers offer various methods of classification. While some make use of two divisions only—aesthetic and scientific—others go farther and give as many as five or more. For convenience we may classify these values as (1) economic, (2) aesthetic, (3) social and ethical, (4) religious, and (5) educational.

The economic is counted the first, though certainly not the highest, nature-study value. With increase in population, farming must become intensive and scientific. In the past we have been wasteful and prodigal of our great resources, but we are learning new lessons in economy every day. Increasing cost of farmlands demands greater returns from the soil. To accomplish this we must study nature and learn from it how to provide against needless waste and insure increased productiveness.

We must begin at the beginning and study from the bottom up. As a nation, Americans are not intimate with nature. Our school children have been kept busy at tasks little calculated to make them familiar with the common goods in nature or with its evil things. Children should know the value of pure air and pure water, the influence of sheltering forests and shade trees, the importance to life on the farm of beneficent birds, insects, and animals. They should, on the other hand, be familiar with the pests constantly menacing life everywhere, such as destructive insects, birds, and other animals, noxious weeds and multiform vegetable disease.

It appeals strongly to a farming community to have their children accomplish real, tangible results. The effect is to draw ever closer the ties which bind the schoolhouse and farm home through kindred interests. Out of such beginnings, higher motives will eventually develop. At all events, the study of real nature opens possibilities for the farm child hitherto unknown. It is a grand thing to learn in school, and on excursions with the teacher into the woods and over the hills, the thousand and one things which make life worth living. The farmer will take a renewed interest in the school that can teach his children things of practical value for the farm. There are things for him to learn, too. It is doubtful whether the average farmer realizes the harm done by the unsightly weeds, fungus growths, and the like, to be seen about the premises, or the value of birds and toads and beneficent insects in saving the orchard and field from ravage and devastation. Someday his children will come home from the new rural school—the modern complement of farm life—and teach him.

All mankind love the beautiful. It appeals to their sense of the perfect. The human mind receives an uplift in the harmony and symmetry coming through the unification of diverse elements, manifesting itself in outward contentment and happiness. As soon as a people has subdued primitive nature with which it has had to contend and has wrung from it a sustenance, it seeks to surround itself with the beautiful in nature, thereby satisfying an instinctive craving to get above the sordid in life. The pages of history furnish us untold illustration. The rock-ribbed tombs of Egypt bear silent witness to this love of the beautiful in a nation living five thousand years ago. Late excavations at Nippur tell the story of marvelous gardens and parks which six thousand years ago gladdened the hearts of the Euphrates dwellers. Nebuchadnezzar constructed the marvelous Hanging Gardens of Babylon to console his queen pining for the wild beauties of her native Median hills. Aesthetic culture, with us, will teach the country folk to love their native woods and prairies; it will make them content to dwell there and long for them when away. To attain this end it is not enough to talk about the wonders of nature or its sublime influence; we must study and dig and plant. At home and at school the small, still voice of nature should be permitted to commune with us through beautiful flowers and waving grasses, sheltering shrubs and spreading trees. A forlorn, windswept school ground is more than we can realize the first cause to weary the boy and girl of country schools and country life. Teach them the surpassing beauty of rural environment; on the school grounds and in the school gardens teach them to dig and plant; as teacher, dig and plant side by side with them. Then the very earth shall preach her sermons in their ears and make them strong in their love to dwell close to nature's heart.

Properly directed, nature study may do much to teach children to respect the rights of others. The sooner a child learns that there are social and moral obligations which he is in duty bound to respect, the better it will be for that child. Every boy and girl is full of energy. The surplus will find a vent somehow, and be put to use, good or evil, as directed. If they are early led to love nature, they will learn to protect it. Such children will never vandalize nature by destroying planted trees or other useful flora. Birds and insects will be safe from their molestation; the insectivorous toads will no longer fear their clods and sticks. When grown up, they will wage relentless war against the many disease-breeding pests found in the fence corners, along the public highway, or in the barnyard, at the present time so little known and less heeded. Growth in respect for social and ethical law is sadly needed in our country, but communion with nature and nature's God may do much to ameliorate existing conditions.

To love nature is to love nature's God. No human being can continue in adoration of living, teeming nature without feeling a growing adoration and love for that which created all the wonders of earth, giving them to man to keep and hold dominion over. The race in its infancy sought the creator through worship of natural phenomena. Even yet "To him who in the love of nature holds Communion with her visible forms, she speaks a various language." The teacher's manifest opportunity is to take advantage of the "still voice" of nature to reach the inner recesses of the child soul to instill there a love for well-doing in looking after the happiness of God's created things, thereby attaining the child's happiness.

Finally, nature study has educational value of utmost importance. The naturalistic tendency in education has seen slow growth. Rousseau, as its first advocate, held "that the educational material should be the facts and phenomena of nature, that it should consist chiefly in an inquiry into nature's laws, and should be through an intimate, fearless, and constant association with nature rather than man." Pestalozzi saw clearly that "nature develops all the forces of humanity by exercising them." "The exercise of man's faculties and talents, to be profitable, must follow the course laid down by nature for the education of humanity." The first fruits of the new century have been to realize much that was advocated by the early educational seers. The disproportion between the formal and the practical in teaching is still very great and presumably will remain so for a long time to come. But beginnings are made in many schools which will eventually end in a satisfactory equilibrium being struck.

Just what topics should be included in the nature-study course in rural schools and what left out will be determined by the essential and fundamental things in rural life. They will center largely about the useful and practical in the local environment—in a study of the trees and flowers on the school grounds or out by the roadside, of the robin and the wren building on the grounds in trees and bird houses—these and similar topics may be studied with profit. Nature study will find concrete expression in planning, platting, and keeping school grounds, and in school garden culture, and will eventually lead to studies in elementary agriculture.

The Rural School Teacher, His Training. The problem of all the problems which await our solution in these same schools is the teaching problem.[138] It would avail but little were all other conditions satisfactory, if the teacher, on whom, after all, the great responsibility of education rests, does not measure up to the required standard. The old saying that "as the teacher, so is the school" is as true today as it was a hundred years ago. If we would have our rural schools measure up with the city schools, we must provide as good teachers for the rural districts as may now be found in the cities. And this can only be accomplished after surmounting many vexing difficulties. Do not misunderstand this statement. All rural teachers are not poor teachers, nor are all rural schools bad. Far from it! Our country districts have thousands of conscientious, hard-working teachers who have fought their way, through many difficulties, to the professional plane. These are generally progressive and take advantage of every means for self-improvement placed at their disposal. All honor is due them for much really good work already accomplished. For their protection and the welfare of rural schools, the difficulties in the way of satisfactory work must be removed. Of these we have already considered: poor unit organization and indifferent administration, insufficient school support, and insufficient supervision. To these we now add indifferent professional preparation of teacher, low salary, unsatisfactory tenure of office, short terms and irregular attendance, low educational ideals, and lack of appreciation of importance of teachers' work.[139]

---

[138] This sentence marks the second sentence of Chapter V, "The Rural School Teacher, His Training" in the 1910 edition and is preceded here by a subheading of the same name. These two short chapters have been combined as a result of their closely related subjects.

[139] A single-sentence paragraph reading "These questions will be discussed in turn, beginning with the teacher's professional preparation" has been removed for readability.

Some people will never get tired of telling us that "teachers are born, not made," and not altogether without reason, for some innate qualities are essential for the making of, at any rate, the best teachers. That *all* teachers are not "born" is obvious. The main trouble is that the "born" teachers are not born fast enough to supply the ever increasing demand. This leaves us the alternative either to "make" teachers or to get along with "makeshift" teachers. We do both. Hundreds of permanent training schools throughout the country are at work to "make" teachers and aid "born" teachers.

Unfortunately, however, many so-called teachers of the present day can neither be said to have been "born" or to have been "made." They are neither natural teachers nor professionally trained teachers—they are mere makeshifts, who neither pursue their work for the love of it nor because they are especially equipped, but simply because they must do *something*. These hangers-on, using teaching as a stepping stone to something better, are the individuals forever throwing obstacles in the way of teaching's becoming a real profession.

No one should enter lightly upon the work of teaching, as this is assuredly the most glorious of callings, and also one of the most exacting. Let everyone consider well the great opportunities and responsibilities involved in teaching children, in molding their lives, in preparing them for their great heritage. The Pestalozzis and Froebels of history have invariably entered upon the work with prayerful hearts, in full realization of their own unworthiness. Let none of us do less. No young person should venture to teach who is not satisfied of his own fitness for the calling. Certain natural qualifications are essential in the make-up of every successful teacher. Here, too, must be added a reasonable degree of academic and professional training. Thus equipped, young men and young women may face the future with fair reason to believe that success will crown their efforts.

Our theme is the professional training of rural teachers, but, first, let us enumerate some of the qualifications that every teacher must have to be a worthy teacher. For natural qualifications, he must have a sound body and good health, good common sense, natural aptitude and insight into things educational, a social and agreeable nature, patience, sympathy, and love for children. He must be tactful and logical, genuine, whole-souled, and manly, frank and unsuspicious, firm and self-reliant, altruistic.

The mere possession of these natural qualities, while very essential, is not in itself sufficient to make the teacher. There must be added an acquired training both *academic* and *professional.*

In general, no person should be permitted to teach school who has not completed a high school course or its equivalent. The high school graduate may have pursued many subjects which he will never be called upon to teach in rural schools, but such subjects are certain to furnish him with a valuable reserve store of energy to draw upon as occasion may direct. A teacher so equipped is reasonably safe from the pitfalls and ruts ever threatening his coworker whose educational horizon is narrower and less distinct.

No teacher can get too thorough an academic training. "Thorough mastery of the academic knowledge of subjects," says Dr. Levi Seeley, "is absolutely essential, and no methods or schoolroom devices or superficial tactics can take its place. More teachers fail from ignorance of the subject matter than from any other cause."

But if teaching is to be established on a professional basis, a specific knowledge of the science and art of teaching is indispensable. What person would for a moment think of becoming a surgeon and try his skill upon the human anatomy without first pursuing a course of study in some reputable school of medicine, albeit a college-bred man? We answer: no one. No more, it seems to me, should a teacher, untrained professionally, be permitted to learn his art in the schoolroom, through experimentation on human minds and souls. Every teacher, indeed, from the ungraded rural school to the college, should know something about the professional subjects—psychology and child study, philosophy of education, history of education, methods of teaching, school management, school economics, and school law.

The degree of proficiency required in these subjects will naturally depend upon the kind of school for which the teacher is preparing. It is evident that if all this agitation concerning redirecting and revitalizing the rural schools shall ever produce concrete results, we must have teachers equipped to make the rural school a natural expression of life in the average rural community. Such teachers are not yet very plentiful. As a matter of fact, we are not suffering so much from a dearth of teachers with a good academic preparation, as we are from a lack of teachers professionally trained to take hold of the new trend of education in rural communities. A majority of rural teachers have a fair knowledge of subjects, gained usually in city schools and in city environments. This is an unfortunate circumstance. For it is difficult for young teachers whose very lives are centered, or have been centered, on the city, to enter into the spirit of the new rural life. The few teachers who are reared on the farm are no better situated, for they are usually defective both in academic and professional training. Many of the normal schools, while beginning to grasp the significance of the farm movement, have not, up to the present time, made any provisions worth mention for training rural teachers, or they are already taxed to their full capacity to supply the demands for better-paid city teachers. Evidently, it will be necessary to make a radical extension in the normal school to meet the needs of the rural teacher or to establish altogether new training schools for this purpose. It may become necessary to do both.

GIFFORD PINCHOT

At the time President Roosevelt created the Country Life Commission and asked Gifford Pinchot to serve as a commissioner, Pinchot was the chief forester of the United States and an expert on economic efficiency. Like Roosevelt, at whose pleasure he served, Pinchot was a true public servant who believed that the government's job was to responsibly steward public lands for the benefit of the individual citizen rather than the monopolist or private interest. Pinchot early converted himself to the cause of renewable energy, having seen in his lifetime the exhaustion of coal mines in Iowa, Missouri, and elsewhere in the Mississippi Valley, large-scale deforestation in the South, and the precious, middle American topsoil washed yearly into the Gulf of

Mexico in amounts equal to twice the volume of the Panama Canal. "When the natural resources of any nation become exhausted," Pinchot warned, "disaster and decay in every department of national life follow as a matter of course." A product of a wealthy eastern family, Pinchot's belief in high-minded conservation sometimes put him at odds with the "marvelous hopefulness" of the American people, which he writes could be as "short-sighted as it [was intense]."

"The Moral Issue" nearly collapses under the weight of its own ideological tensions. At times Pinchot sounds notes of enlightened environmental protectionism while, at others, he resorts to capitalistic verbiage promising the average American "his fair share of the benefits" and "what he needs and all he needs." In walking the conservation tightrope between those, such as John Muir, on his left, and those, such as Taft, on his right, Pinchot could not help but become a controversial figure in Washington DC, making an enemy of Taft, among others. Though he would leave the Taft administration in protest in 1910, Pinchot is remembered by environmental scholar James Penick Jr. as "one of the most creative and innovative leaders of his generation." "The Moral Issue" serves as a succinct manifesto for the conservation movement, a movement that was, Pinchot recognized, political and moral as well as environmental. Pinchot's articulation of the public's rightful claim to water power and water rights predicts his stand in favor of the Hetch Hetchy Dam. Finally, his statement "conservation is the most democratic movement" boldly anticipates the future mainstreaming of environmental awareness.

The Moral Issue[140]

The central thing for which conservation stands is to make this country the best possible place to live in, both for us and for our descendants. It stands against the waste of the natural resources which cannot be renewed, such as coal and iron; it stands for the perpetuation

---

[140] Gifford Pinchot, *The Fight for Conservation* (New York: Doubleday, Page & Company, 1910), 79–88.

of the resources which can be renewed, such as the food-producing soils and the forests; and most of all it stands for an equal opportunity for every American citizen to get his fair share of benefit from these resources, both now and hereafter.

Conservation stands for the same kind of practical commonsense management of this country by the people that every businessman stands for in the handling of his own business. It believes in prudence and foresight instead of reckless blindness; it holds that resources now public property should not become the basis for oppressive private monopoly; and it demands the complete and orderly development of all our resources for the benefit of all the people, instead of the partial exploitation of them for the benefit of a few. It recognizes fully the right of the present generation to use what it needs and all it needs of the natural resources now available, but it recognizes equally our obligation so to use what we need that our descendants shall not be deprived of what they need.

Conservation has much to do with the welfare of the average man of today. It proposes to secure a continuous and abundant supply of the necessaries of life, which means a reasonable cost of living and business stability. It advocates fairness in the distribution of the benefits which flow from the natural resources. It will matter very little to the average citizen, when scarcity comes and prices rise, whether he can not get what he needs because there is none left or because he can not afford to pay for it. In both cases the essential fact is that he can not get what he needs. Conservation holds that it is about as important to see that the people in general get the benefit of our natural resources as to see that there shall be natural resources left.

Conservation is the most democratic movement this country has known for a generation. It holds that the people have not only the right, but the duty to control the use of the natural resources, which are the great sources of prosperity. And it regards the absorption of these resources by the special interests, unless their operations are under effective public control, as a moral wrong. Conservation is the application of common sense to the common problems for the common good, and I believe it stands nearer to the desires, aspirations, and purposes of the average man than any other policy now before the American people.

The danger to the conservation policies is that the privileges of the few may continue to obstruct the rights of the many, especially in the matter of water power and coal. Congress must decide immediately whether the great coalfields still in public ownership shall remain so, in order that their use may be controlled with due regard to the interest of the consumer, or whether they shall pass into private ownership and be controlled in the monopolistic interest of a few.

Congress must decide also whether immensely valuable rights to the use of water power shall be given away to special interests in perpetuity and without compensation instead of being held and controlled by the public. In most cases actual development of water power can best be done by private interests acting under public control, but it is neither good sense nor good morals to let these valuable privileges pass from the public ownership for nothing and forever. Other conservation matters doubtless require action, but these two, the conservation of water power and of coal, the chief sources of power of the present and the future, are clearly the most pressing.

It is of the first importance to prevent our water powers from passing into private ownership as they have been doing, because the greatest source of power we know is falling water. Furthermore, it is the only great unfailing source of power. Our coal, the experts say, is likely to be exhausted during the next century, our natural gas and oil in this. Our rivers, if the forests on the watersheds are properly handled, will never cease to deliver power. Under our form of civilization, if a few men ever succeed in controlling the sources of power, they will eventually control all industry as well. If they succeed in controlling all industry, they will necessarily control the country. This country has achieved political freedom; what our people are fighting for now is industrial freedom. And unless we win our industrial liberty, we can not keep our political liberty. I see no reason why we should deliberately keep on helping to fasten the handcuffs of corporate control upon ourselves for all time merely because the few men who would profit by it most have heretofore had the power to compel it.

The essential things that must be done to protect the water powers for the people are few and simple. First, the granting of water powers forever, either on non-navigable or navigable streams, must absolutely stop. It is perfectly clear that one hundred, fifty, or even twenty-five years ago our present industrial conditions and industrial needs were completely beyond the imagination of the wisest of our predecessors. It is just as true that we can not imagine or foresee the industrial conditions and needs of the future. But we do know that our descendants should be left free to meet their own necessities as they arise. It can not be right, therefore, for us to grant perpetual rights to the one great permanent source of power. It is just as wrong as it is foolish, and just as needless as it is wrong, to mortgage the welfare of our children in such a way as this. Water powers must and should be developed mainly by private capital and they must be developed under conditions which make investment in them profitable and safe. But neither profit nor safety requires perpetual rights, as many of the best water power men now freely acknowledge.

Second, the men to whom the people grant the right to use water power should pay for what they get. The water power sites now in the public hands are enormously valuable. There is no reason whatever why special interests should be allowed to use them for profit without making some direct payment to the people for the valuable rights derived from the people. This is important not only for the revenue the nation will get. It is at least equally important as a recognition that the public controls its own property and has a right to share in the benefits arising from its development. There are other ways in which public control of water power must be exercised, but these two are the most important.

Water power on non-navigable streams usually results from dropping a little water a long way. In the mountains, water is dropped many hundreds of feet upon the turbines which move the dynamos that produce the electric current. Water power on navigable streams is usually produced by dropping immense volumes of water a short distance, as twenty feet, fifteen feet, or even less. Every stream is a unit from its source to its mouth, and the people have the same stake in the control of water power in one part of it as in another. Under the Constitution, the United States exercises direct control over navigable streams. It exercises control over non-navigable and source streams only through its ownership of the lands through which they pass, as the public domain and national forests. It is just as essential for the public welfare that the people should retain and exercise control of water power monopoly on navigable as on non-navigable streams. If the difficulties are greater, then the danger that the water powers may pass out of the people's hands on the lower navigable parts of the streams is greater than on the upper, non-navigable parts, and it may be harder, but in no way less necessary, to prevent it.

It must be clear to any man who has followed the development of the conservation idea that no other policy now before the American people is so thoroughly democratic in its essence and in its tendencies as the conservation policy. It asserts that the people have the right and the duty, and that it is their duty no less than their right, to protect themselves against the uncontrolled monopoly of the natural resources which yield the necessaries of life. We are beginning to realize that the conservation question is a question of right and wrong, as any question must be which may involve the differences between prosperity and poverty, health and sickness, ignorance and education, well-being and misery, to hundreds of thousands of families. Seen from the point of view of human welfare and human progress, questions which begin as purely economic often end as moral issues. Conservation is a moral issue because it involves the rights and the duties of our people—their rights to prosperity and happiness, and their duties to themselves, to their descendants, and to the whole future progress and welfare of this nation.

JOHN LEE COULTER

Professor of rural economics at the University of Minnesota during the writing of *Co-operation Among Farmers* and a special agent in charge of the Agricultural Census in Washington DC shortly thereafter, Dr. John Lee Coulter believed business savvy, more than newfangled agricultural technique, would win the day for the American farmer. Coulter's dissertation, *Industrial History of the Red River Valley of the North*, established its author as one of a generation of emerging, virtuoso scholars able to think across agriculturally- and environmentally-related disciplines. *Industrial History* takes a holistic approach to the study of land use in the wheat country of Minnesota and North Dakota, featuring subsections on climate, agriculture, and population. The monograph drew the attention of the Census Bureau in Washington, and, after 1912, Coulter, the native Minnesotan, spent much of his professional life in Washington think tanks and commissions, including what would become the National Raw Materials Council. A Progressive Republican after the Roosevelt fashion, Coulter's rise in Washington corresponded to a period of increased foreign trade, making him a logical choice for U.S. Tariff Commissioner. In that role, he anticipated and planned for America's role in world trade. His training in law, his governmental experience, and his impressive academic pedigree made him attractive to higher education administrations, and he would eventually return to his native Red River Valley to assume the presidency of North Dakota A & M, now North Dakota State University in Fargo.

"The present age," Coulter concludes in *Co-operation Among Farmers*, is an age of organization." The individual, according to Coulter, could not match the economy or efficiency of society. The Commission for Country Life, published three years prior, had likewise challenged the American farmer to grow more efficient through organization with his peers. A farmer's son, Coulter, however, is quick to distance himself from the American corporation—with its exclusivity and uneven shares and dividends—as a potential model. Instead, he favors the agricultural co-op, where each man is a voting member and where profits or losses are shared equally among the group. While the advent of the Farm Bureaus would give isolated farmers the collective voice Coulter believed they needed, the Bureaus themselves quickly drifted towards exactly the exclusivity Coulter warned against.

The Argument for Cooperation[141]

It would be a much pleasanter task to describe the beauties of nature and to tell about country scenes than to pick out the weak spots in the organization of country life, attempt to present some explanation for the presence of these unsatisfactory conditions, and suggest methods of improving the same. It is much pleasanter (but probably less useful), to draw pictures of the beautiful flowers, to roam in the fields and listen to the twittering birds, and to enjoy the blessed sunshine, than it is to turn to the conditions surrounding us which need improvement and worry over the existing social, economic, and political problems. There are many beautiful things about country life. No one will attempt to deny that. But, in spite of their presence, the country is not able to compete with the city as the two exist at the present time. This must mean that the city has advantages over the country, or at least seems to have.

Life in the country is not what it should be; it is not what it can be made to be; and it is not what it must be made to be if it is to be placed on a level with the city. This is clear from the rapid movement of population from the country to the cities. From all of the evidence at hand, it is clear that during the last ten years the movement has been as rapid as during the preceding twenty-five years. And this movement will continue until conditions in the country are brought up to a par with those in the city.

Man is a gregarious animal. He wants meet with others of his kind. He wants to talk with others, to attend political and business meetings with others, to go to social and educational gatherings, to see performances of others, to hear them sing and speak. Men like to eat and drink, and laugh and cry and sing together. It is the exceptional man who is a miser in economics, an anarchist in politics, or a social recluse.

Now, if the farm is to be merely a place where people can make a living, and where they are to be shut out from most or even

---

[141] John Lee Coulter, *Co-operation Among Farmers* (New York: Sturgis & Walton Company, 1911), 3–16.

a considerable number of the other activities which are worthwhile, I say, leave the farm. If that is the situation, it fails to supply some of our highest wants. But *if farming can be made a profitable business, if it is to continue to be the most healthful business known to man, and if at the same time life in the open country can be made worth living in the largest sense of the word*, I say hasten the day when the good people of the country may have their eyes opened to the opportunities which the farm affords, and hasten the day, too, when many who are now cooped up in our large cities making a bare existence may see the light and move into the open country.

I believe that I am a normal human being having much the same likes and dislikes as others. I am going to choose to the best of my ability, and by the aid of the best judgment that I can get, the place to live and carry on my work which will bring the greatest total satisfaction to myself and to those who are dear to me, and which at the same time will be to the greatest benefit to all. Having been born on a farm and lived many years close to the frontier, and having experienced the unpleasant side and the difficulties which surround the farmers, as well as having enjoyed some of the good things of country life, having also lived in large cities and traveled through all parts of our country, and spent much time studying conditions in other industries, I feel ready to pass judgment on the possible advantages and disadvantages of country life as it exists at the present time. It is because I believe that agriculture can be made much more profitable than it has been without causing any hardships to others—indeed, that it can be made as profitable as other business enterprises and that it can be made much more livable than at present, equal in most respects to, and in many ways far surpassing the city—that I hope to spend much of my time in the future on the farm.[142]

The present volume would have had absolutely no place in the library of the farmer of one hundred years ago. At that time, the average

---

[142] The original, concluding words of this sentence, "… and that I present this book to the public" have been omitted for readability.

farmer provided largely for himself; the market was a very secondary consideration. He produced his own poultry and eggs, his own pork and beans, which meant his salt pork, his bacon and smoked hams, his own small and large fruits so far as he had any, his own potatoes and other vegetables, his own milk and cream, butter and cheese, meat from the meat cattle, his own mutton, and his wool from which to make socks and mittens and often all of the clothes for the family. He got out his own wood for fuel and did not patronize either the coal barons or the Standard Oil Company. He built his own house and was little troubled about the price of lumber. He often made his own nails or did not use any. A little grist mill at the falls in the neighboring stream ground his grain and furnished him with flour, if, indeed, he did not have to grind it at home. Let me repeat the statement: *the present volume would have had absolutely no place in a farmer's library one hundred years ago.*

One hundred years ago the man who wanted to farm was very apt to have dealings with the Indians; now it is with bankers and real estate men and absentee landlords. And now, when he produces, he produces for the market, and has the railroads, and the warehouse men and the speculators and commission men to deal with. Things have changed since those times. The farmer of the present, if he is going to succeed, must awaken to a realization of the fact that he is living in a new era. He must become a businessman in every sense of the word, or he will fail.

During the last half century there has been a great move in the direction of education in agriculture, but it has been one-sided; it has only covered one-third of the activities of the modern farmer. It has taught him to produce, but has not taught him anything about either selling the things which he has produced or about spending the money received. In an age of commercial agriculture, it is as necessary to buy well and to sell well as to produce well. Any manufacturer will tell you that.

When the people of the United States first realized that it was necessary to begin a systematic study of agriculture, they began with a study of the production of all farm produce. At that time the chief business of the farmer was to produce things. There was little selling and buying, and there were practically no speculators or other middlemen as we know them today. Experiment stations were established to find out about, and agricultural colleges to teach, soil physics and chemistry, animal and seed breeding, diseases of fruits, grains and vegetables, and diseases in the animal kingdom, ways to make two blades of grass grow where one grew before, how to make the cow give more and richer milk, how to build roads, run machines, et cetera.

All of this was necessary at that time and is as important today as at any time in the history of farming. But we must not befog the issue. In addition to this the farmer must be a businessman if he is to succeed at the present time. Too often the pure scientists studying production overlook this phase of farm activity.

At this point, let me ask one question and answer it. I have asked large numbers of people this question in many states and I have their answers. The question is: why did you leave the farm? The first answer is that which most young people have given me: "I left because I was sure I could get bigger wages in the city," or "Because of the greater opportunities in the city," or "Because it is much more interesting in the city, there are so many people there, close together, and there is so much that one can do."

These three excuses are perfectly legitimate. Each one of them has influenced the present writer. The only way to keep young people on the farm or to draw other young people to the farm, or indeed, to influence the present writer to return to the farm, is to make farming as profitable as other industries—or show that it is now as profitable—to demonstrate that there are as great opportunities in the open country as there are in the city and to make country life as interesting as city life.

The second answer to the question asked above is the one which is generally given by middle-aged or older men and women. Some of these have moved to the city to work. Others have become retired farmers solely on account of the increased value of their lands and not because they had been able to save much money from year to year. Some said, "We want our children to be educated, and the city schools are better than those in the country." I know men on farms today who are considering the advisability of leaving the farm. I know others who have left and who would return if satisfactory schools were there. I know still others who left as boys and girls, and who would return now in middle life with their families if the schools, libraries and other in situations were within reach. Therefore these educational facilities must be supplied.

The third answer is one which I hear ringing in my ears as if it were given in chorus by hundreds. It takes many forms, but runs something like this: "They have paved streets, and sidewalks in town; we have muddy roads in the country. They have streetcars, carriages, and automobiles; we have heavy workhorses and wagons or, at best, heavy buggies, on the farm. They have electric lights or gas in the city; we have the old oil lamp or tallow candle. They have telephones, telegraph, and free mail delivery, and they don't need them half as badly as we; yet few of us who live far apart, a long distance from the doctor and from friends, have these. They have furnaces in their homes, and when they get up in the morning their house is warm and pleasant, but we get up to find everything frozen and the house chilled. They have hot and cold running water and baths; we must go out to the old pump or melt snow and ice, and take our bath in the dishpan or washtub. They have toilet facilities and sewers; we must go out in the cold of winter to the snow-filled privy, and in the summer to a foul-smelling, unsanitary one. They have theaters, concerts, and orchestras; we read about them. They have good dancing halls, with fine music and smooth floors; we wish that we had half as good. They have good stores, where they can telephone for the things they want when they want them; we get what is left over, and we have to go after it and have very poor choice. They have libraries where there are papers, magazines, and books which we cannot get with out buying, and so we seldom see these."

Having heard the cry, what shall we do about it? In answer I will say make farming profitable. We must stop talking about "businessmen and farmers." All must be placed on as nearly the same plane as possible. It must henceforth be "bankers, farmers, merchants, and other businessmen." The first step in every town should be to start a club, a good live businessmen's club, in which farmers are admitted as members on the same basis as other businessmen. There should be a clubroom where all may meet, or where groups may meet at will and be on an equal basis.

Second, whenever a bank is to be started, to be supported by some home capital, fifty or one hundred representative farmers should become stockholders, even if they do not purchase more than one share of stock each. They should have at least one member of the board of directors and thus join in the good work. This will take only enough of their time and money from the farm to make businessmen of them. So, too, when a cracker factory or starch factory, canning factory, or packing plant is to be established, this local club should be called upon to give careful consideration to the opportunities which seem to present themselves. The factory should not start out as if it were some scheme to exploit the other fellows. The farmers supply the corn or tomatoes, the potatoes, beans, peas or animals. At least a few of the farmers should take a share or two of stock. This should be done if for no other reason than to learn the business principles involved and to establish a relationship of confidence between the dwellers in the city and the country. The above suggestions are merely the first which are necessary in order to develop the proper spirit of cooperation between the people in the country and in the towns.

But before farmers can be brought up to a level with other businessmen, they must learn to finance their undertaking more intelligently than most of them do now. They must learn to buy and sell as businessmen; they must learn to be more than producers of crops and livestock. To be sure, the farmer must learn the best way to rotate his crops; he must seek out the best fertilizer; he must study the various breeds of corn; he must know how thick to plant and at what seasons, how often and how deep to plow and how and when to harvest; but he must also learn to keep track of what it costs to produce the crops; he must study the most economic storage systems; he must know how to insure, how to market what he produces, and how to invest the money received. Farmers must organize.

In order to carry out this movement, farmers must emulate other businessmen. Farmers live farther apart, and distance is a handicap. Organization is much more difficult among farmers than among businessmen living in the cities, but it is becoming easier each year with growing intelligence and with denser population. Each year it is easier than the year before in thousands of communities to organize active grain, fruit, seed, and stockbreeders' associations, cow-testing societies, mutual insurance societies, debating clubs, good-government clubs, and other associations and organizations corresponding to those which are found in cities.

We must have *better farming*. Without this, great progress is impossible. This is not enough. We have waited altogether too long to add that we must also have *better business*. Without these we cannot have *better living*, and we must have this better living in the open country, or the movement to the cities will not stop and farming will be looked down upon and scorned.

Mary Huston Gregory

Mary Huston Gregory ranges widely in her comprehensive study *Checking the Waste; A Study in Conservation*, touching on immediate problems of health and safety in American cities but also setting forth principles of the fledgling American nature conservancy movement. What little is known of Gregory's motives for writing the book is contained in the book's preface, which challenges the reader to accept personal responsibility for what she suggests is "the greatest question before the American people." The audience for her practical guide, she describes as "teachers, reading circles, farmers' institutes, women's clubs, the advanced grades in schools, and general library purposes." In addition, Gregory cites a number of sources that contributed to the book, including the Report of the Conference of Governors at the White House and the Report of the National Conservation Committee. Gregory's only previously published book was a work of juvenile fiction, *Once Upon a Time*, likewise published by Bobbs Merrill in Indianapolis, Indiana, where she also wrote *Checking the Waste*.

In her lyric, sometimes empirical chapter "Birds," Huston acts as commentator-advocate more than scientist-technician. And while the chapter devotes itself ostensibly to birds and their feeding habits as misconstrued by the American farmer and public as a whole, her writing does reveal, on further examination, fundamental conservation tenets widely held by the emerging field of conservation studies, principally: 1) Nature maintains its own, organic predator-prey balance and is thus best left undisturbed 2) Nature is self regulating and naturally finds its proper carrying capacity if unadulterated and 3) Careful study, rather than uninformed scapegoating, is needed to appreciate the actual habits of wildlife and their real impact on the land. In sum, Gregory's explanatory and often cautionary work succeeds at reaching its target audience: the broader American public.

Birds[143]

Birds give us pleasure in three ways: by their beauty, by their song, and by their usefulness in destroying animals, insects or plants which are harmful to man.

But although they are among man's best friends, they have been greatly misunderstood, so that to the many natural enemies that are constantly preying on birds we must add the warfare that man himself wages on them, and the cutting down of their forest homes. This work of bird destruction has gone on until all the best species are greatly reduced in numbers and some species have been almost entirely driven out.

To see how serious a matter this is, we must study the food habits of birds, and we shall find that although the different species eat a large variety of food, in almost every case, their natural food is something harmful to man.

The large American birds, the eagles, hawks, owls and similar kinds, are called birds of prey because they feed on small birds and

---

[143] Mary Huston Gregory, *Checking The Waste; A Study in Conservation* (Indianapolis: Bobbs-Merrill, 1911), 236–244.

animals. Some of these are of the greatest benefit to the farmer, while others are altogether harmful. Another large class of birds lives almost entirely on injurious insects and this class is entitled to the fullest care and protection from the farmer.

Still another class lives largely on fruits, wild or cultivated, and on seeds, which may be either the farmer's most valuable grains, or seeds of the weeds that would choke out the grain.

It can not be denied that birds often do serious damage through their food habits, but the great mistake that has been made in man's treatment of birds has been in hastily deciding that, if birds are seen flitting about fields of grain, they are destroying the crop. A better knowledge of their food habits will lead to proper measures for destroying the harmful kinds and protecting the useful ones.

Successful agriculture could hardly be practiced without birds, and the benefit to man, though amounting each year to millions of dollars, can hardly be estimated in dollars and cents, since it affects so closely the size of our crops, the amount of timber saved for use in manufactures, and even the health of the people.

Here again we see the careful balancing that runs through nature, how carefully each thing is adjusted to its work. Naturally, the balance between birds, insects, and plants would remain true, no one increasing beyond its proper amount. But when man begins to destroy certain things and to cultivate others, this balance is seriously disturbed. The birds that destroy weed seeds being killed, weeds flourish in such vast numbers as to drive out the cultivated crops. The birds which destroy mice, moles, gophers, et cetera, being killed, these animals become a nuisance and cause serious losses. If insect-destroying birds are driven out, the farmer will be at the mercy of the insects unless he employs troublesome and expensive methods of getting rid of them. Certain favorable conditions cause large numbers of birds to gather in a small region and they become a pest. Very careful observation has shown that, in nearly every case, the favorite food of the birds is something which is not valued by man, and if this food is provided, the farm grains and fruits will not be seriously molested.

Few birds are altogether good, still fewer are altogether bad; most species are of great benefit, even if at the same time they do some harm. Some birds do serious damage at one season and much good at another. The most notable example of this is the bobolink, which in northern wheatfields is loved no less for his merry song than for the thousands of weed seeds and insects he destroys; while in the South he is known as the reed bird or rice bird, the most dreaded of all foes to the rice crop.

Flying down on the fields by hundreds of thousands, these birds often take almost the entire crop of a district. The yearly loss to rice growers from bobolinks has been estimated at two million dollars.

If crows or blackbirds are seen in large numbers about fields of grain, they are generally accused of robbing the farmer, but more often they are busily engaged in hunting the insects that, without their help, would soon have destroyed his crop, and even if they do considerable damage at one season, they often pay for it many times over.

Whether a bird is helpful or the reverse, in fact, depends entirely on the food it eats, and often even farmers who have been familiar with birds all their lives do not know what food a bird really eats. As an example of the misunderstanding that is often found in regard to birds, when hawks are seen searching the fields and meadows, or owls flying about the orchards in the evening, the farmer always supposes that his poultry is in danger, when, in reality, the birds are quite as likely to be hunting for the animals which destroy grain, produce, young trees, and eggs of birds.

In order to correct such mistaken ideas, the Department of Agriculture has made a most careful and accurate study of the habits of birds, and it is the results of these observations that are recorded here.

Fieldworkers from this department who have observed the habits of the principal birds that live among men have watched them all day and from one day to another as they fed their little ones, and, to be more certain of their facts, they have examined the stomachs of hundreds of birds, both old and young, to learn exactly what each bird had eaten. In this way they have proved absolutely that many species that are supposed to eat chickens, or fruit or grain, in reality never touch them, but are among the farmer's best friends.

Among other things they have learned that while they are feeding their young, birds are especially valuable on a farm. Baby birds require food with a large amount of nourishment in it that can be easily digested. Almost all young birds have soft, tender stomachs, and must be fed on insects; as they grow older, the stomach or gizzard hardens and is capable of grinding hard grain or seeds. The amount of food required by the baby birds is astonishing. At certain stages of their growth they require more than their own weight in insects. And the young birds are to be fed just at the season that insects do the most injury to growing crops of grain and young fruit and vegetables.

Birds vary so much in the kind of food eaten, not only by different varieties of the same species but by the same birds at different seasons, that it is necessary to make a careful study of each bird to know whether, if he is sometimes caught eating cultivated fruit and grains, he helps in other ways enough to pay for it.

When insects are unusually abundant, birds eat more than at other times and confine themselves more strictly to an insect diet, so that at such times the good they do is particularly valuable.

Birds of prey may do harm in a particular place, because in that region mice, rabbits, and other natural food are scarce, and they are driven to feed on things that are useful to man, while, in places where their natural food is plentiful, the same birds are altogether helpful.

In the same way, birds which naturally eat weed seeds frequently find these almost altogether lacking where the farms are most carefully cultivated, but in their place are fields of grain whose seed also furnishes them desirable food. Is it any wonder, then, that, their natural food being taken from them, they turn to the cultivated crops? The fruit-eating birds seem always to choose the wild fruits, but, where these are not to be had, they enter the orchards and soon become known as enemies of the farmer.

A careful examination of the harm done by birds leads to the belief that the damage is usually caused by a very number of one species of birds living is a small area. In such cases so great is the demand for food of a particular kind that the supply is soon exhausted, and the birds turn to the products of the field or orchard. The best conditions exist when there are many varieties of birds in a region, but no one variety in great numbers, for then they eat many kinds of insects and weeds, and do not exhaust all the food supply of one kind. Under such circumstances, too, the insect-eating birds would find plenty of insects without preying on useful products, and the insects would be held in check so that the damage to crops would be slight.

The following are examples of the food eaten by birds and the good that they thus accomplish to man:

During the outbreak of Rocky Mountain locusts in Nebraska, a scientific observer watched a long-billed marsh wren carry thirty locusts to her young in an hour and the same number was kept up regularly. At this rate, for seven hours a day, a nestful of young wrens would eat two hundred and ten locusts a day. From this, he calculated that the birds of eastern Nebraska would destroy daily nearly 163,000 locust.

A locust eats its own weight in grain a day. The locusts eaten by the baby birds would therefore be able to destroy 175 tons of crops, worth at least $10 a ton, or $1,750.

So we see that birds have an actual cash value on the farm. The value of the hay crop saved by meadowlarks in destroying grasshoppers has been estimated at $356 on every township thirty-six miles square.

An article contributed to the *New York Tribune* by an official in the Department of Agriculture estimated the amount of weed seeds annually destroyed by the tree sparrow in the state of Iowa on the basis of one-fourth of an ounce of seed eaten daily by each bird. Supposing there were ten birds to each mile, in the two hundred days that they remain in the region, we should have a total of 1,750,000 pounds, or 875 tons, of weed seed consumed in a single season by this one species in the one state. In a thicket near Washington, DC was a large patch of weeds where sparrows fed during the winter. The ground was literally black with the seeds in the spring but on examining them it was found that nearly all had been cracked and the kernels eaten.

A search was made for seeds of various weeds but not more than half a dozen could be found, while many thousands of empty seedpods showed how the birds had lived during the winter.

In no place are birds more important than in the forests, where they save hundreds of thousands of dollars worth of valuable timber each year. In forests there can be no rotation of crops and no cultivation, and spraying, which keeps down the insect pests in the orchard, is impossible here because of the expense. It would not pay to spray two or three times a year a crop of timber that requires a lifetime to grow. So, in the forests, the owner must depend entirely on birds for his protection. How great the destruction of our forests would be is shown by the fact that the damage at present is estimated at $100,000,000, in spite of the fact that a vast army of birds is working tirelessly, summer and winter, to devour the insects! The debt of the forester to the birds can hardly be estimated.

ELLEN H. RICHARDS

Ellen H. Richards' many firsts include the first woman admitted to the Massachusetts Institute of Technology, the first woman permitted to teach at MIT as an instructor, the first woman in America accepted to a school of science and technology, and the first American woman to earn a degree in chemistry. Called a "pragmatic feminist" by scholar Barbara Richardson for Richards, acknowledgment of the economic value of women's work within the home, Richards once wrote to her parents, "Perhaps the fact that I am not a Radical and that I do not scorn womanly duties but claim it as a privilege to clean up and sort and supervise ... is winning me stronger allies than anything else."

Caroline Hunt's 1958 biography, *The Life of Ellen H. Richards, 1842–1911*, details how Richards essentially pioneered the field of home economics in the 1890s, bringing together Richards' various interests in chemistry, sanitation, education, culinary arts, physical fitness, and nutrition, under the heading of *ecology*, a term she is credited for having introduced into English from a German word meaning "household of nature." Richards passion for home economics and

Progressive reforms resulted in her opening of public kitchens throughout New England as well as her design and demonstration of model kitchens throughout the region, an interest which lead to the Massachusetts exhibit of the Rumford Kitchen at the Chicago World's Fair of 1893, where, incidentally, Frederick Jackson Turner would deliver "The Significance of the Frontier in American History." Taken as a whole, Ellen Richards' unprecedented expertise positioned her well for the Progressive Era women's reform movement, which encouraged women's direct involvement in the betterment of their communities via outreach that became known as "municipal housekeeping."

"Protection of Water Supplies as a Conservation of Natural Resources" exemplifies Richards' socially-minded research as well as the hands-on experience she gained analyzing, on behalf of MIT and the Massachusetts State Board of Health, tens of thousands of water samples. Her work on water pollution, beginning as early as 1873, resulted in the first state experimental station for water pollution research, according to historian Bill Kovarik, as well as the first water quality standards in America and first modern sewage treatment facility. Richards would author a number of books leading up to her conclusions in "Protection of Water Supplies …," with such titles as *The Cost of Cleanness*, *The Cost of Shelter*, *The Art of Right Living* and *Air, Water and Food From a Sanitary Standpoint*. Collectively, her research published post 1900 describes a second term she coined, *euthenics*—signifying the improvement of the environment inside and outside the home—as contrasted with *eugenics*, a concept she disapproved of along with Social Darwinism. At her most indignant— "Man has made himself much extra trouble by reckless waste of nature's provision"—Richards is unequivocal, declaring no "pure" water remains in America. Particularly suspicious of commercial interests fouling water and food (Richards' research also demonstrated the need for factory and food inspection laws) her work anticipates the creation of the Environmental Protection Agency and other regulatory bodies. Richards addresses the question of individual water usage, too, arguing for common-sense, in-home conservation. While primarily addressed to the sanitation engineer of her era, Richards' writing was sufficiently plainspoken, and alarmist, to merit the attention of the American public, including the farmer, whose wells were likewise impacted by the pervasive watershed pollution.

Protection of Water Supplies as a Conservation of Natural Resources[145]

By 1877 the preliminary survey of the drainage basins of Massachusetts was concluded and the opinion reached that while there were some things to be remedied, "as a whole throughout the state the evil from the pollution of streams is small compared with that arising from the accumulation of filth in cesspools and accumulations near dwellings. ... In order to encourage towns ... it will be necessary to regulate rather than wholly prohibit the contamination by filth of our waters."

In 1887 there were 123 sources of public water supply, furnishing 82 percent of the population, 50 group waters, 73 surface waters, and only 5 streams.

The oversight of these given to the State Board of Health instead of to a separate Rivers Commission, and it was enjoined especially to prevent further pollution and to advise towns as to means of prevention.

Protection of water supplies as a conservation of natural resources means (1) Clean soil and prevention of fouling; (2) Husbanding rainfall by storage; (3) Legal protection of the storage basins.

The carrying further of the idea of prevention as both a sanitary and an economic measure involves the long look ahead in a close study of watersheds both for quantity to be available in years to come and for quality maintainable according to the standards already discussed.

Water supply problems have changed in the last one hundred years from the securing of a gallon or two of "pure" water for drinking and cooking to thirty gallons for cleanliness and one hundred to two hundred gallons for manufacturing or transportation purposes. At present it is all drawn from the same source and returned as dirty water for

---

[144] Ellen H. Richards, *Conservation by Sanitation; Air and Water Supply; Disposal of Waste* (New York: J. Wiley & Sons, 1911), 85–88.

the use of another community. Just how dirty the supply may be allowed to become is one of the burning questions of the day. When commercial interests are 98 percent of the whole, the 2 percent interest in the health of the people is apt to be disregarded, and only the high commercial value being now set upon human brain and energy has brought the question to a business point. It is slowly being recognized that safety of human life is to be considered in the advance of mechanical and manufacturing processes. Accidents on the one side and conditions of living on the other are robbing the nation of thousands of valuable citizens.

Air and water are two of the conditions now most before the world. It is being shown that in terms of human life it will pay to care for the water supply. Just how much that means we shall see later at the point of intensive interest—the great cities. The question of "pure" water may be dismissed. There is no such available. All abundant sources, rain, lakes, streams, wells, have been contaminated. The deep unpolluted sources from rocks and sands, so-called artesian wells, contain, for the most part, large amounts of mineral salts, and in most regions such sources are not sufficient, so that impounded rain is the chief supply of most large cities.

Water is the universal solvent and the common carrier; hence anything is liable to be found on or in it—animal, vegetable, mineral.

There is held to be less daughter from mineral substances, unless near arsenic works or lead mines. Vegetable decay used to be held responsible for malaria, and even now waters carrying much dead organic matter or those highly colored are looked upon as suspicious, but animal matter is disrepute the world over, both on account of the disease germs which may accompany it and on account of the solubility of its more or less alkaloidal compounds which may prove depressants of vitality if not direct poisons.

Certain things should be borne in mind: First, that at times of low water surface supplies are subject to vegetable growths of which they may show no trace at high water.

Ground supplies are subject to contamination in low-water times from sewage which may be held back in high water.

If it is true good water is becoming scarce and that water pure and undefined is as nearly gone as the coal supply, it behooves the nation not to think of a substitute as in the case of coal, but of a more provident manner in its use.

However, unused land is rare; even public domains have travelers, and careless travelers, as fire ravages show. It is no longer safe to assume that mountain streams are clean, and irrigation is making unused water scarce.

Wastes must be disposed of somehow, and soil, air, and water are all unclean.

In man's hurry to get through with his inheritance, he cannot wait for nature to filter, so he adopts nature's methods with a hurry attachment.

The impurities are either coarse or fine—either mineral, vegetable, solid or in solution, harmless or poisonous, but impurities all the same.

Some will settle out if left in quiet; some may be strained out. Some must be caught and clotted and some must be actually filtered through the finest net. It is a mistake to say that water once soiled can ever be made "pure" again. It may be made clear if turbid, palatable and colorless if disagreeable and brown, safe if suspicious or dangerous, but that does not mean pure. However, accepting it as a conventional term, since neither rainwater nor the best spring water is really pure, "purification," as used, means renovating spoiled water—almost as good as new.

Man has made himself much extra trouble by reckless waste of nature's provision. The rain has been allowed to flow away to the sea without doing its full work. Ten gallons have been used and fouled where one would have served. Pipes have leaked and water run to waste. Used water has been allowed to soil many times its volume in good water. Crops have thirsted for water which might have refreshed them and brought a portion of food as well.

The future is to bring more care for both quantity and quality as bearing on the food supply as well as the health of the people. Even wash waters may be turned to increasing national wealth.

Because, through ignorance of biological principles, the first attempts at sewage farming failed, it is not wise to neglect so positive a source of income, but it is largely the water that is of great value, and while the plant food it carries is of minor consequence, yet it is in a most available form. When more is known of the office of mineral matters in plant growth, these "farms" may be better carried on.

Just as insurance companies balance facts and probabilities and count on a law of chance, so the sanitary engineer is to be called upon to balance the two risks, damage to business and damage to health.

Men have refused to insure and come through life safely. Also others have lost their all in a month.

It is the province of the future sanitary engineer to make good his promises to protect both business and health, till that time when both interests will become identical.

WARREN H. WILSON

Warren Hugh Wilson published *The Church of the Open Country*, just three years after earning a Ph.D. in sociology from Columbia University in 1908 and three years into his work as one of two superintendents in the Department of Church and Labor for the Home Missions of Presbyterian Church. Wilson's dissertation director at Columbia, Frank H. Giddings, reportedly held the first chair in sociology in any university in the world, and Wilson's dissertation (later a book) on a small church in Quaker Hill, New York, "Quaker Hill: A Sociological Study," is considered by many the first, in-depth study of rural sociology. Drawn from *The Church of the Open Country*, "The Community Church" is the heady work of a newly minted scholar in a newly minted field (rural studies) in a newly created discipline (sociology) in an era that, thanks to Roosevelt's Country Life Commission, showed an unprecedented interest in rural religion.

A pastor at Arlington Avenue Presbyterian Church in Brooklyn, New York, until 1908, Wilson left his post to focus on the research, training, and policy recommendations associated with the Department of Country Life, where he made a name for himself by suggesting "innovations and additions" to the programs in rural churches including soil conservation and rural recreation. A devotee of both the country and the church, Warren H. Wilson, like Gifford Pinchot, found himself in the middle of what often seemed conflicting concerns. *The Church of the Open Country* was not exactly a work of independent scholarship, having been commissioned by the editorial committee of the Missionary Education Movement in the hope of adding it to a growing list of home-grown texts written by leading scholars and prepared especially for study by church youth. But Wilson manages to walk a fine line in "Church and Community," challenging the church to become more relevant by attuning itself to the farmer's needs for socialization and celebration, while at the same time admitting that the truly isolationist farmer was a man not likely swayed by novel programs or gimmicks. Arguing against those who would consolidate rural schools and churches for the sake of convenience, Wilson goads, "Others there are who hold, as a matter of theory, that all rural institutions should be assembled at the population centers, presumably at the railway stations, and that ultimately the farmers will follow their stores, schools, and churches, and will live in congested villages.... But the course of American history indicates no such future peopling of the land. It becomes the Church to serve the farmer where he lives." In sum, Wilson posits as foundational the last line of the poem "The Country Church" by Liberty Hyde Bailey: "to love and to work is to pray." In honor of Wilson's lifelong research and advocacy, which resulted in twelve books and pamphlets, the Presbyterian Church opened the Warren H. Wilson Vocational Junior College in North Carolina in 1942, an institution that would later evolve into Warren Wilson College.

Church and Community[145]

The tilling of the soil is an occupation which cannot be carried on except by sober men, and it has never been maintained in a

---

[145] Warren Wilson, *The Church of the Open Country: A Study of the Church for the Working Farmer* (New York: Missionary Education Movement of the United States and Canada, 1911), 23–29.

population through succeeding generations except among religious people.[146] The relation between work and worship is more evident in the country than in the city, for rural life is simple and its social texture is clear to the observer. The working of cause upon effect is plain in the country. There are not so many disturbing or extraneous forces, and there are but few interferences between cause and effect. The bearing of economic life upon religious institutions is much more evident than in the city because country life is highly organized, but simple.

What is a community? A man or woman in the country lives his whole life within the radius of a team haul from his home. However much he visits without this circle, his knowledge of the community is one hundred times more intense and personal than his knowledge of any other community. In this small republic to which he is limited by the common means of transportation, he visits, he buys and sells, he worships, he marries, and, within this radius, he buries his dead.

This enables one to define the community in popular terms, as a child might define it, as "the place where we live." This includes locality, personal and social relations, and vital experience. The community is the larger whole in which the members of a we-group find satisfaction of their vital needs. This means that the community is the virgin soil of the Kingdom. The church when it pervades the whole life of the community constitutes with it a little republic of God, because the whole life of the people in that community is lived within this range.

Another definition of a community is given by University of Missouri professor Charles A. Elwood in the *American Journal of Sociology*: "A society is, therefore, a group of people living together by means of interstimulation and response."[147] What its total life is depends very largely upon the attitude of its members toward one another. How

---

[146] This sentence marks the seventh sentence of Chapter Two, "Church and Community," as it appears in the 1911 edition and has been selected as an opening for the sake of brevity and readability.

[147] Wilson's original footnote read "Charles A. Elwood, of the University of Missouri, in *American Journal of Sociology*, March 11, 1911, p. 835."

they cooperate depends, therefore, upon common will, belief, and opinion, and the agencies by which common will, belief, and opinion effect social control. These agencies are chiefly religion, government and law, and education.

The church, because of its relation to the community, should be a community center. It can not exist and prosper unless it is the focus of the life of the community. Now communities are of infinitely varied size and form. They are not perfect circles or squares or ellipses. They cannot be brought down to any geometrical terms. One uses the word "center" as a suggestive term, because there is no better. If the church, being vitally connected with the community, does not make itself central to the life of the community, it will not continue to exist. There may be several such centers in a community, but their relation to the whole people must in some way be vital, or they will pine and die.

The churches of individualist communities have served their people by ministering to persons alone. Their method is the preaching of sermons. They have no other. The typical man in pioneer and settler days was an individualist. He was made such by his work and by the lonely struggle of his life. He could be no other than what God, by means of the forest, the vast open prairie, and the lonely work at the furrow, made him. His wife spent her solitary hours over the varied occupations of the primitive household, and she too was a strong, resourceful individual. She had but few social traits as compared with modern women. The church that ministered to such people could preach a gospel of individual salvation alone. It would have been false to its duty if had preached any other. It needed no methods such as our later churches use. It had the one method: the periodic revival of religion for individual souls.

For the individualist is a man of warm heart, of passionate interest, of devoted friendships, and resolute loyalty. To him life means nothing but persons. Therefore the church of settler folk and the church which has pioneer individuals in any numbers in its membership must use emotional methods. Of course if you do not care to win and to hold this kind of people, you can omit the measures by which they are to be won, but these measures are always emotional, because the emotional, individual type of Christian is produced by his occupation and by his inheritance. He can be only what he is. He must be dealt with on his own terms. Country people are very many of them pioneers. The number of pioneers is greater among the older people than among the younger, but for some time to come the country church will have to deal with the proud, solitary, passionate personalities of men who can be won and can be served by their feelings alone.

In some parts of Alabama, conspicuously in the sandy stretches of the lower Appalachian Mountains, the people have a custom known as "all-day sings." Certain singing masters make a business of going through the country and collecting the folk on Sundays for singing. These men have a rough and effective power in song. They use generally no instrument, and therefore the rhythm and the swing of the music are its most notable elements. For hours the people sit under the leadership of the singing master and sing simple, popular, religious songs. Little by little the master selects those whom he calls his school. They are the best singers and he keeps them in permanent connection with himself, coming back again to that community for later performances.

The churches through all this country have learned that people will not go to church within a radius of several miles of an all-day sing. Sunday schools have to be closed and church services are but little attended. Yet families will drive ten miles to the all-day sing and spend the whole Sunday, eating their meals in the intervals between sessions and driving home at night, every way content. These gatherings are the expression of the overflowing emotionalism of the people of that country. It would seem that the churches there could use the method which has grown out of the life of the people and make effective for their own use what is a matter of private profit to the singing masters.

The church in settler days, when everyone is highly individualized, is scarcely to be called an institution. Its building is a mere roof over a pulpit. Its work consists of preaching and no more. It has no societies or organizations. The settler and the pioneer would believe it wicked to organize the societies which an ordinary village church of our day thinks necessary. Among the southern mountains, where the pioneer type, strongly individualized, remains, where every man is an independent person and every woman is a strong character, the churches have but one method of religion, namely, the periodic revival. They think the methods of the people in the valley to be wicked, unreligious, and expressive of unregenerate minds. To them Ladies' Aid societies are unspiritual, Sunday schools are sinful, and boys' clubs are extremely worldly.

In every church of modern times there remain some pioneers. The settler has come down through later generations. The individualist is a factor and he must be dealt with in his own terms. It is not fair for him to tyrannize over others any more than it is right for others to exact of him what he cannot furnish. Teach him the gospel of personal salvation, for religion means personal things to him, and these alone. Arouse him with emotion. Attach him to persons. Teach him to command and to obey those whom he loves, but do not expect in his own life to change him into another type, for that is impossible. Ministers who are serving in communities of settlers, of pioneers and mountaineers, have a great duty of evangelism. If they cannot preach a gospel that moves the heart and holds the affections, they can accomplish nothing among a people dominated by emotion.

WILLIAM A. MCKEEVER

William A. McKeever, a professor of philosophy at Kansas State University (then Kansas State Agricultural College) dedicates *Farm Boys and Girls* to the 10 million boys and girls then enrolled in country schools. McKeever, who was the juvenile director of the National Presbyterian Temperance Board, is perhaps better remembered for his moral crusades against the film industry, famously dubbing theaters "schools for criminals" and reportedly calling for martial law on the streets of Lawrence, Kansas as a group of youths protested the censorship of a Fatty Arbuckle picture with what McKeever called a "mad mob spirit." Beyond his public grandstanding and temperance zealotry, however, McKeever was an influential and respected scholar and commentator. In addition to publishing a handful of books on child-rearing and psychology, McKeever penned *The Pioneer; A Story of the Making of Kansas*, which he described in the introduction as a "metrical narrative" telling the stories of his pioneering parents. The work has been somewhat inaccurately classified as "cowboy poetry," as, like Warren Wilson's study of Nebraska and Roosevelt's study of the Badlands, it melds environmental history with sociological thought.

Collaborating with superintendents of schools throughout the Great Plains and Middle West in researching *Farm Boys and Girls*, McKeever's work is intended for rural parents as well as experts in child development. *Farm Boys and Girls*, which covers every aspect of the development of rural youths, from camping and hiking to sexual deviancy, is intended as a thorough manual for the rearing of responsible, responsive farm youth. In particular, Professor McKeever emphasizes parental vigilance, such that parents are alongside their children at the time of what the professor calls "awakenings"—natural, developmental stages otherwise termed "instincts." The firm, educated hand could guide the farm boy or girl through these awakenings, as could structured participation in the County YMCA, Boy Scouts, or social and economic clubs. In keeping with the Commission on Country Life report, the selections that follow pay little attention to the condition of the farming father. McKeever's other titles, including, among others, *Training the Girl* (1911) and *Training the Boy* (1913), demonstrate the public's hunger for child-rearing manuals in the late Progressive period.

"Selections From Farm Boys and Girls"

The Farm Boy's Choice of Vocation. Turn which way you will upon the great broad highway of life and there you will always be able to find the wrecks and broken forms of humankind, men and women who have failed in their life purposes.[148] Strange to say that particular aspect of the science of character building which has to do with the substantial preparation for vocational life has been very

---

[148] William A. McKeever, *Farm Boys and Girls* (New York: Macmillan, 1912), 41–52, 275–286.

much neglected. By what rule do men succeed in their callings and by what different rule do other men fail? Are some foreordained to success and others to failure? Is there an inherent strength in some and a native weakness in others? Is there a type of education and training which specifically fits and prepares for each of the native callings? None of these questions has been thoroughly gone into with a view to finding out what were best to be done and what best to leave undone. So, we blunder away, hit or miss, in the vocational training of our boys and girls.

Should the farmer's son farm? In attempting to give helpful suggestions to farm parents relative to their boy's vocation, perhaps this question will first demand an answer. The tentative reply to it is this: the farmer's son, or any other man's son, should follow that calling for which he is best suited by nature and in which he will thereby have the greatest amount of native interest, provided it be practicable to prepare him for such calling. Some farm boys are destined by nature for mechanical pursuits, others for social or clerical work, others for captains of industry, and so on. Likewise, the city boys may reveal in their natures a great variety of instinctive tendencies and interests which will be found of great worth in guiding them into a successful life occupation.

Yes, the farmer's son should by all means take up his father's business, provided that at maturity he may have both native and acquired interest in the same and that to a degree predominating any other native or acquired interest.

It can be proved that the country boy matures more slowly than the city boy. For example, at the age of sixteen, he is behind the latter in height, weight, school training, and sociability. But while the city boy matures more rapidly, the country boy makes up for the loss by a longer period of development. It is the author's firm belief that this fact of slow growth proves a tremendous advantage to the country youth in that it allows for greater stability of character, and especially for a greater amount of courage and aggressiveness in form of permanent life habits.

But one might well wish that all rural parents could realize the evil consequences of being impatient with the son in respect to his choice of a life work. Many a good boy yet in his teens is hounded and driven about by the continuous nagging of his parents, who ignorantly believe that he should have his future destiny all planned and ready for its realization. As a result, this same good boy is often driven to desperation and to the point of leaving the home place or breaking away from the affectionate ties that bind him to parents and of seeking the position wherein he might earn a living. As a matter of fact, few young men have any very clear or reliable vision of their future life at the age of eighteen, or even twenty. Many of the best men in the world are faltering and uncertain even as late as twenty-five. However, if the relatives and friends would only exercise all due patience, offering only such helps and suggestions as can be given, and trusting the future finally to throw upon the problem a light from within the youth himself then, we may be assured, practically every man will finally come to some line of effort that will bring him a comfortable living.

What of predestination? The old-fashioned idea of a boy's being marked by the hand of destiny, cut out for some particular calling in life, still has a place in the minds of the masses. The kindred belief that some men are "natural-born failures" has also wide currency. A third superstition is the very common opinion that others are just naturally lucky. All these traditional opinions are the outgrowth of ignorance of human nature such as may be dispelled by means of a course of instruction—or a carefully arranged course of home reading—in modern psychology.

None of the foregoing superstitions would be worthy of our attention were it not for the gross injustice which they entail upon children. Parents everywhere in both city and country are dealing with their children upon the assumption that one and all of these fallacies are true. "My oldest boy just naturally has no luck," said the father of three sons and two daughters. "He changes around from one thing to another and fails every time." But what of this particular boy's early training? Was it the same as that of the others? Did he enjoy equal advantages? Did his parents, when married, really know anything about rearing children? Or did they really mistreat their firstborn through ignorance and use him as a sort of practice material from which they learned how to do better by the succeeding ones?

Until the foregoing inquiries about the unlucky son's boyhood life be fully answered, we cannot reasonably permit ourselves to condemn him. There is nothing more in predestination than this; namely, it can be shown that the child is born with not a few latent abilities—aptitudes for doing and learning this and that—and that one of these aptitudes is likely to have correlated with it more than the average amount of nerve development in the corresponding brain center. As a result, that particular aptitude will require less training than the others and will tend to predominate over them as maturity is approached.

The reply of the psychologist to the statement that some men are "natural-born failures" is this: few if any of those possessed of ordinary physical and mental qualities at birth are necessarily so. Excepting the feebleminded and the like, whose marks of degeneracy are usually apparent to all, it may be asserted on the highest authority that none are natural-born failures to any greater extent than they are natural-born successes, but that they have within the inherited nerve mechanisms many possibilities of both success and failure.

We should be willing to overlook almost any other interest in this discussion for the sake of inducing in the farm father the belief that his young boy is a potential success—the belief that this boy is furnished by nature with the latent ability to shine somewhere in the broad field of human endeavor—provided he be rightly trained and disciplined during his growing years. Here, then, is probably the greatest of all the human training problems, namely, the vocational one.

Roughly speaking, there have been three methods of vocational training.

First, historically, there has been the apprentice method, the youth being bound out to learn a trade. The chief faults of this traditional way of teaching the boy to be self-supporting were these: it made no allowance for intellectual development, and it gave the father too much authority to choose the calling for the boy.

A modern offshoot of the old-time apprentice course is the trade school which flourishes in many of the big cities today. This new institution has one great advantage over its prototype. It offers such a great variety of forms of training that the youth may exercise much free choice. But it preserves one of the serious defects of apprenticeship in its neglect of the intellect of the learner. The modern trade school can never hope to do more than prepare young men and women to make a good living. It is a get-ready-quick institution and can never be expected to give the student breadth of view and depth of insight into the great problems of human life.

The second oldest method of preparing men for a vocation is what has been called the cultural method. It has aimed at high advancement in book learning with the thought of finally enabling the student to enter a professional class comparatively few in numbers and supposed to possess a superior advantage over the great mass of human kind. One fault of this method has been to emphasize learning for its own sake and to defer too long the training of the individual in the material and practical side of his calling.

But the chief fault of this cultural method has been its contempt for common labor and ordinary industry, its theory being that true education prepares one to avoid such practices. If the young man wished to prepare for law or medicine or teaching or the ministry—one of the learned professions—then the old classical school was at his service. But, if he would become a mere artisan or industrial worker, there was no advanced course of schooling available.

The third and newest method of preparing the young person for his vocational life is, in reality, a compromise between the first and second. It provides that the learner shall have book instruction and industrial training at the same time and that both of these are to be regarded as cultural, since taken together they prepare for independence of thought and action, and for the vocation as well. This new method of preparing young people for their life work would call nothing mean or low. It aims to serve all. Its motto is the development of head and hand together. It seeks to produce cultured handicraftsmen as well as cultured artists and professional men.

Our justification for the foregoing, somewhat lengthy discussion of the different theories of education is that of wishing to be certain of bespeaking the father's patience and forbearance in the preparation of his son for the vocational life. The farmer is most fortunate in having ready at hand a large amount and variety of industrial practice to supplement the boy's book lessons. In this respect he probably has a superior advantage over all other classes.

But in guiding his boy gradually toward the vocational life, the farm father can easily mistake what is merely a passing interest on the former's part for a permanent one. The carefully kept records of farm boys show that they take up many different lines of work with great enthusiasm, and yet soon tire of them and drop them. These serial and transitory interests are usually mere juvenile responses to the awakening of some new nerve centers. They are not much different in nature from the brief passing interest which the child has in his various playthings.

Now, the chief function of these transitory interests in special forms of work and learning as shown by the young growing boy is this: to furnish the occasions for a great variety of activities and practices for trying him out on all the possible sides of his nature. Not one of these intense, boyish interests is necessarily very directly preparatory to his final choice of a vocation, while all are indirectly so. Therefore, if the fifteen-year-old son chances to win in a corn-raising contest, or at a livestock exhibition, or if he manifests unusual interest in arithmetic, declamation, or nature study, do not regard any of these as necessarily pointing to his best possible vocational work. Presumably, at such an undeveloped age, he is still in possession of some latent interests and aptitudes, one of which may far outweigh any such thing hitherto awakened in his life. Give him time to mature and, if at all practicable, send him on to college.

It is the opinion of the author that the state agricultural college, as now situated and organized, is the ideal institution of higher learning for the country-bred youth. It offers him every reasonable incentive and opportunity for continuing in the calling of his father, if he be so inclined, while, at the same time, it gives instruction in many other departments of learning. Whether the state institution be a separate one, or merely a college within the organization of the state university, matters little. In either case the young man will be brought within reach of a course in scientific farming, stock raising, horticulture, and the like, either to choose or let alone—and the so-called cultural work will still be there for the taking.

Many rural parents, weighted down with the overwork of the farm, cherish and express a very earnest desire that their sons may find some easier form of earning a living. So they deliberately plan with the boy the easy course to be pursued. Said one such farmer, "Wife and I decided that there would not be much in it for Henry except hard work if he settled down on the home place, so we decided to send him to college and educate him for something that offered less work and more pay." So they shielded the son from the heavier duties of the farm and encouraged, in every way, the boy's thought of an easy way to success.

But one thing these well-meaning parents failed to foresee. That is, when the boy entered college, he began to look for that same sort of royal road to learning. The assigned lessons and tasks soon took the appearance of drudgery, and he dodged and avoided them wherever possible. In less than a year, the youth had failed at college and was back home. "The confinement of the college did not agree with his health." More than three years have passed since, and the boy has spent the time drifting from one job to another and all the while growing weaker in character and integrity.

Here we have but another instance of the old, old story with its tragic aspects. Yet, nearly all the faltering, vacillating men now drifting about the country might have been saved through careful training in the performance of work. The boy who would be insured success in his coming vocation must be required to buckle down to solid work of a kind and amount to suit his years and strength. He must learn through the character-building experience of toil, not only what it means to stay by an assigned duty till it is performed, but he must also experience the unfailing joy of work well-done. He will thus have the advantage of the spur of successful effort and acquire the beginnings of that splendid self-reliance which is a distinguishing mark of all successful men.

But there is a sort of drudgery and of ugliness against which the boy's nature instinctively rebels, and it ought to. By this we mean to refer to the actual conditions of overwork and the accompanying rundown appearance that characterizes so many farm homes today. No wonder the boys hasten away to the city to find a "job." Why not clean up the place by cutting away the underbrush and weeds, by planting shade trees and repairing fences and outbuildings, by painting and renovating the house and barn? And all this as an investment in behalf of the children and their possible future interest in the farm home as the best place on earth in which to dwell. All this and more might be urged as means of guiding the thoughts of the farm boy towards the possibilities of his taking up the calling of his father. And while all these material advantages may not serve to overcome the natural tendency of the young man to seek a radically different type of occupation, they will at least make it more certain that his natural abilities for an agricultural pursuit were not left unawakened.

The Country Mother and the Children. Greater attention needs to be given to the conservation of the farmer's wife.[149] Although there are many other justifications for giving more thought to the care and the comfort of the country mother, the single fact of her very close relation to the children growing up in the home, and of her peculiar responsibilities as center of life there, warrant greater attention to her interests. Recently, while passing upon a country highway, I met a funeral procession. A little inquiry revealed a pathetic situation, one that has been repeated thousands of times throughout the length and breadth of this fair country. The deceased was the wife of a young farmer, both of them under thirty-five years of age, hard-working and ambitious for success, but thoughtless of their own health and comfort. Their farm was somewhat new and unimproved; there were hundreds of things to do other than the routine affairs of homekeeping and

---

[149] This sentence marks the beginning of Chapter XVIII, "The Country Mother and the Children" on page 41 of the 1912 edition. The chapter title is represented here as a subheading of the same name to suggest the closely related subject matter of these two thematically-grouped chapters.

crop raising. Worst of all, there was a mortgage to be lifted. After all reasonable improvements were made and the mortgage paid off, then, according to their plans, they were going to take matters easy. But the delicate cord of life suddenly broke in the case of the wife, and left the young husband as overseer of the farm and home and sole caretaker of three little children.

How can parents hope to produce a better crop of boys and girls in the farm communities so long as the typical farm wife is crushed into the earth with the overweight of the burdens placed upon her? A few minutes enumeration in this same rural neighborhood brought out the startling fact that in fully half of the homes a scene similar to the one just described had been enacted during the last score of years. That is to say, during the twenty years, fully one half of the farm mothers living in that particular neighborhood had died before their time from one cause or another. In most instances the death occurred during what we usually speak of as the prime years of life and at a time when the rose bloom should naturally be fresh upon the cheek. Fortunately, this serious condition, still present in some communities, is being gradually improved by the improved methods.

The report of the Country Life Commission makes the following suggestions:

> The relief to farm women must come through a general elevation of country living. The women must have more help. In particular these matters may be mentioned: development of a cooperative spirit in the home, simplification of the diet in many cases, the building of convenient and sanitary houses, providing running water in the house and also more mechanical help, good and convenient gardens, a less exclusive ideal of money-getting on the part of the farmer, providing better means of communication, as telephones, roads, and reading circles, and developing of women's organizations. These and other agencies should relieve the woman of many of her manual burdens on the one hand and interest her in outside activities on the other. The farm woman should have sufficient free time

and strength so that she may serve the community by participating in its vital affairs.

In discussing this same matter, Henry Wallace, a member of the Commission, says in his paper, *Wallaces' Farmer*:

> They have been saying that the mother is the hardest worked member of the family, which is often and we believe generally true. They have been saying that in the anxiety of the farmer to get more land, he not only works himself too hard, but his wife too hard, and the boys and girls so hard that the boys get disgusted and leave the farm, and the girls marry town fellows and go to town.
>
> Now, the farmer's wife is really the most important and essential person on the farm. As such she needs the most care and consideration. You are careful, very careful, not to overwork your horses. How much more careful you should be not to overwork the mother of your children. You rein back the free member of the team. You take special care of the brood mare, and the cow that gives three hundred pounds of butter. Have you always kept the freest of all workers, your wife, from doing too much? How about this?

We shall attempt to show a number of specific conditions that may be sought as tending to conserve the strength and the life of the rural mother, with a view to her continuing to be, in every best sense of the word, a caretaker and conserver of the lives of her own children.

However it may be achieved, the first thing to work for in this connection is a surplus of nerve energy. If the child training is to go on in a satisfactory manner, the mother especially, and if possible both parents, must have stated times and occasions for looking after such training and for inculcating a series of important fundamental lessons. The first and best test of this child-rearing situation may be made at evening. If, after the work of the ordinary day, the mother is still fresh enough to take a real interest in the children's affairs, to read to them briefly and perhaps tell them a story or two, or to read for further preparations of her work with them, then it may be said that her life energies are being conserved in a fairly satisfactory manner. The children will most certainly reap the benefits. But if the close of the ordinary day's work finds the farm mother suffering from physical and nervous exhaustion, cross and impatient with the other members of the family, depressed in spirit and gloomy as to the future, these are signs which should give alarm to the head of the household and arouse him to the point of looking into such distressful conditions and setting them right.

How would it do to plan for the mother a daily period of rest and relaxation? Would not such a program furnish something of a guarantee of length of life in her own case and of peace and contentment in the home and of improved wellbeing in respect to the children? How shall we state this question? Must the very lives of the rural mother and her children be run through the mill of overwork as a grist for the improvement and upbuilding of the farm animals and the farm crops? Or should all of these material things be valued only in proportion as they contribute to the happiness and contentment and the long life of the members of the family? Too many farmers seem to say, as expressed by their conduct: *I must lift that mortgage this year. I must market so many bushels of corn and so many head of livestock. So here goes my wife, and here go my children into the hopper. Perhaps they will have to give up their lives. At any cost I must make this thing pay!*

Then, how would it be to set apart an hour or more each day, regularly, for the rest and relaxation of the mother, and call it Mother's hour? During that time let it be the policy of the entire family to require no work, no assistance, no favors of her, unless it be in case of illness. During such a time of recuperation, the ordinary woman would regain her poise. The nerve energy would be more or less restored, while she would tend to view the better things of life more nearly from their right angle. Best of all, she would re-gather during the hour not a little strength to be used later in the caretaking of her children. Try it for a week.

This is not the place for a detailed discussion of what might or ought to be put into the house for the sake of the convenience of the homemaker. But if such materials be thoughtfully arranged, they may be made most effective, even though they be small and inexpensive. A little inquiry among the ordinary homes will show what is meant here, by either the presence or the lack of the things indicated. It is not so much a question of expense as it is one of thoughtful provision. The guiding principle of the home convenience is that of saving and conserving the strength of the housekeeper.

There is especially one day in the week which might be appropriately called the "mother-killing day." That is the occasion of her doing the washing and ironing for the family. Not infrequently two or three days thereafter are required for the restoration of her normal strength and health. Now, it is clearly the specific duty of the farmer to take hold of just such matters as this and attempt seriously to put them right. Doing the washing for four or five, and that with the use of the washtub, is a man's work so far as required muscular energy is concerned, and very few women are able to do it regularly and live out their allotted lives. Therefore, let the conscientious farmer see to it first of all that some kind of machinery be installed for lightening such wife-killing tasks as that just named. Let him provide such household helps and conveniences *first*, and for the sake of the housemother and her children. And then, if there be other means available, let him provide the man-saving machinery about the barn and the fields.

The farmer who is seriously interested in providing for the care and comfort of his family, and for the instruction and intelligent direction of his children, will see to it that his life companion be allowed her share of outings. This matter must be just as much on his mind as that of marketing the produce. The usual habit of the farmer's wife is to give up willingly her rights and opportunities of this sort. But she cannot well continue to be spiritually strong and mentally well-disposed toward the world unless she be permitted to get out among her friends and acquaintances at frequent intervals.

So, arrange carefully a series of outings for the country mother. The beginning of such a program is to provide that there be available for her use, and at her command, a horse and carriage. This equipment need not be of the finest quality, and it may be used for other purposes, but when her needs appear, it should be given up to her purposes. At least one afternoon a week she should go away from the place and be free as much as possible temporarily from the cares of the household while she finds congenial company among some of the neighboring women, or at the library or elsewhere.

The unending problem of the home life throughout much of the civilized world is that of obtaining adequate assistance in the performance of the household work. Much of the time such assistance from outside sources is practically unavailable. And yet something must be done to meet the situation. If there be young girls growing up in the home, the solution of the problem may, and should, be met by means of requiring the daughters to assist with the home duties. But in case there be no daughters, it is seriously recommended that either the father or the boys do certain parts of the heavier housework.

It is not necessarily beneath the dignity of the best and most brilliant man of this country for him to get down on his hands and knees in his own home and help perform the menial work there which threatens to break the health of his life companion. If there be growing sons in the family, there is every justification for training them to assist in the housework in a case where such assistance is needed to shield the health and strength of the mother. It prepares for better manhood and for more sympathetic protection of his own wife-to-be if the boy be required to do such things, and thus to become intimately acquainted with what it means to perform the many burdensome tasks that tend to wear away the lives of so many good women.

There will be no better occasion than this to remind parents of the necessity of carefully training the growing children to perform such deeds as will shield the mother in the home, and show a sympathetic interest in her welfare. These matters will not naturally be acquired by children. The country today is full of grown men whose mothers and wives have worked themselves to death, and yet these men did not detect the seriousness of the situation until it was too late. There are many men of this same general class who are willing and even anxious to protect the women of the home from the crush of overwork, but who know not how to do it. Such faults as we have just named might easily have been avoided had these men, during very early boyhood, been brought into an intimate acquaintance with the burdensome tasks of the household. Especially should they have been drilled time after time in the performance of deeds of love and sympathy in respect to their mother. It may seem a little thing for a younger child to rush to the table, call for and partake of the best the table provides and, inattentive to the wants of any other members of the family, hurry off to his play full fed and happy. And yet this very thing may be indicative of a serious lack of attention to the rights and requirements of others, such as may be carried over into his future home life and there amount to serious abuse. Again, it must be insisted that deeds of sympathy and altruism are acquired through the actual and continued practice of the performance of such deeds.

Among the other splendid results of the conservation of the nerve energy and the vital interests of the house mother may be mentioned that of her ability to plan thoughtfully for the instruction of the boys and girls. It is not an easy task to select appropriate stories and readings for the young. It is neither an easy nor a trifling matter for the parent to be able to read suitable stories to them and to interpret helpfully such stories. It is not a trifling matter for the parents to converse together an hour at evening and there plan as to the future home instruction of their young. When should this be introduced into the boy's life and when into the girl's life? What is a fair allowance for the boy for what he does and for his spending money for the Fourth of July, Christmas, and the like? What is a fair allowance for the girl with which to purchase her clothes and for her pin money? When should each of them be told this and that about the secrets of life, and where may helpful literature thereon be obtained? Just when and how much should the boy and girl be allowed to go among the young people of the community? When we consider the far-reaching results which their solution may mean for the developing young lives, these and many other such questions become exceedingly important.

In many a farm home today there is a secret compact which goes far to shape the destiny of a great number of lives. Go if you will to the farm home where the life of the mother is being gradually crushed out by the overwork and the lack of sympathetic protection on the part of the husband, and you will almost invariably find a secret understanding between the mother and the growing children in reference to the future careers of the latter. It is implied by these words put into the mouth of the mother:

> Your father is too ambitious about the work and in his desire for accumulating wealth about the farm. He is overworking me, is thoughtless of me, and indifferent to your present needs and your future welfare. Work on as you must, driven by him, but do as little as you can and grow up to manhood and womanhood. Study your books, get through with your schooling, and in time find something easier for your own life work. Perhaps we can persuade him to give it up after a while and move to town, where you can go out more, dress better, and get more enjoyment out of life."

Thus, the children grow up to mistrust and dislike their father, and to despise the vocation in which he is engaged. Such a state of affairs will precipitate their flight from the home nest. This will take place at the earliest possible moment and will often be in the nature of a leap into the dark, anything to get away from the drudgery of the farm.

Mark you this situation well, you farm fathers, and attack it in all possible haste with the best available relief. A happy, contented, well-protected farm mother almost certainly means the same sort of farm children, while the converse situations will also run in the same unvarying parallel. Do not satiate your desire for more hogs and more land with the sacrifice of the peace and happiness and the very lifeblood of your wife and children.

JOHN MUIR

John Muir somehow managed to be a reclusive yet ubiquitous presence during the Golden Age of conservation. Muir penned the "The Hetch Hetchy Valley," excerpted here from Chapter 16 of Muir's book-length ode to the park, *The Yosemite*, at a time when the first efforts to dam the Valley had long since been rebuffed. But with the elections of 1912 ushering in a new cast of political players and new president Woodrow Wilson, Muir clearly realized the Park's ongoing vulnerabilities. His strike here, therefore, might be considered preemptive. All of the stock characters of Progressive Age America find a place in Muir's passion play, including the "gainseekers," the "mischiefmakers," "Satan," and "Senators." The author's overriding religious metaphor—equating Hetch Hetchy with Eden and its commercializers with temple destroyer and infidels—reflects the righteous indignation Muir felt. Nearing the end of his life at the time *The Yosemite* was published, a whisper of Muir's growing cynicism and fatigue is implicit in the sentence: "Experience shows that there are people good enough and bad enough for anything." By 1912, John Muir and been fighting on behalf of the Sierras for over thirty-five years.

Elsewhere, Muir lovingly catalogues Yosemite's natural wonders, arguing by accretion for the Valley's preservation in its sublime natural state. In the second half of the chapter, where this excerpt begins, Muir concludes his wildlife survey and launches into the thicket that is the early history of San Francisco's application for the use of waters from Hetch Hetchy and the ambivalent governmental action since. It is worth noting that one year earlier Muir had included in-progress portions of Chapter 16 in his pamphlet *Let Everyone Help Save the Famous Hetch-Hetchy Valley and Stop the Commercial Destruction Which Threatens Our Parks*. For the pamphlet, Muir collected fragments of his own earlier writing about the Valley along with the quotations of others in support of its preservation. The pamphlet is also a practical manual for the cause, as it includes a two-page spread telling how supporters could lobby Congress.

Hetch Hetchy Valley[150]

It appears, therefore, that Hetch Hetchy Valley, far from being a plain, common, rock-bound meadow, as many who have not seen it seem to

---

[150] John Muir, *The Yosemite* (New York: The Century Co., 1912), 254–257.

suppose, is a grand landscape garden, one of Nature's rarest and most precious mountain temples. As in Yosemite, the sublime rocks of its walls seem to glow with life, whether leaning back in repose or standing erect in thoughtful attitudes, giving welcome to storms and calms alike, their brows in the sky, their feet set in the groves and gay flowery meadows, while birds, bees, and butterflies help the river and waterfalls to stir all the air into music—things frail and fleeting and types of permanence meeting here and blending, just as they do in Yosemite, to draw her lovers into close and confiding communion with her.

Sad to say, this most precious and sublime feature of the Yosemite National Park, one of the greatest of all our natural resources for the uplifting joy and peace and health of the people, is in danger of being dammed and made into a reservoir to help supply San Francisco with water and light, thus flooding it from wall to wall and burying its gardens and groves one or two hundred feet deep. This grossly destructive commercial scheme has long been planned and urged (though water as pure and abundant can be got from outside of the people's park, in a dozen different places), because of the comparative cheapness of the dam and of the territory which it is sought to divert from the great uses to which it was dedicated in the Act of 1890 establishing the Yosemite National Park.

The making of gardens and parks goes on with civilization all over the world, and they increase both in size and number as their value is recognized. Everybody needs beauty as well as bread, places to play in and pray in, where Nature may heal and cheer and give strength to body and soul alike. This natural beauty-hunger is made manifest in the little windowsill gardens of the poor, though perhaps only a geranium slip in a broken cup, as well as in the carefully tended rose and lily gardens of the rich, the thousands of spacious city parks and botanical gardens, and in our magnificent national parks—the Yellowstone, Yosemite, Sequoia, et cetera—Nature's sublime wonderlands, the admiration and joy of the world. Nevertheless, like anything else worthwhile, from the very beginning, however well-guarded, they have always been subject to attack by despoiling gainseekers and mischief-makers of every degree from Satan to Senators, eagerly trying to make everything immediately and selfishly commercial, with schemes disguised in smug-smiling philanthropy, industriously, famously crying, "Conservation, conservation, panutilization," that man and beast may be fed and the dear nation made great. Thus, long ago a few enterprising merchants utilized the Jerusalem temple as a place of business instead of a place of prayer, changing money, buying and selling cattle and sheep and doves, and earlier still, the first forest reservation, including only one tree, was likewise despoiled. Ever since the establishment of the Yosemite National Park, strife has been going on around its borders, and I suppose this will go on as part of the universal battle between right and wrong, however much its boundaries may be shorn, or its wild beauty destroyed.

The first application to the government by the San Francisco Supervisors for the commercial use of Lake Eleanor and the Hetch Hetchy Valley was made in 1903, and on December 22nd of that year it was denied by the Secretary of the Interior, Mr. Hitchcock, who truthfully said:

> Presumably the Yosemite National Park was created such by law because within its boundaries, inclusive alike of its beautiful small lakes, like Eleanor, and its majestic wonders, like Hetch Hetchy and Yosemite Valley. It is the aggregation of such natural, scenic features that makes the Yosemite Park a wonderland which the Congress of the United States sought by law to reserve for all coming time as nearly as practicable in the condition fashioned by the hand of the Creator—a worthy object of national pride and a source of healthful pleasure and rest for the thousands of people who may annually sojourn there during the heated months.

In 1907, when Mr. Garfield became Secretary of the Interior, the application was renewed and granted, but under his successor, Mr. Fisher, the matter has been referred to a commission, which as this volume goes to press, still has it under consideration.

The most delightful and wonderful campgrounds in the Park are its three great valleys—Yosemite, Hetch Hetchy, and Upper Tuolumne—and they are also the most important places with reference to their positions relative to the other great features—the Merced and Tuolumne Canyons and the High Sierra peaks and glaciers, et cetera, at the head of the rivers. The main part of the Tuolumne Valley is a spacious flowery lawn four or five miles long, surrounded by magnificent snowy mountains, slightly separated from other beautiful meadows, which together make a series about twelve miles in length, the highest reaching to the feet of Mount Dana, Mount Gibbs, Mount Lyell and Mount McClure. It is about 8500 feet above the sea, and forms the grand central High Sierra campground from which excursions are made to the noble mountains, domes, glaciers, et cetera, across the range to the Mono Lake and volcanoes and down the Tuolumne Canyon to Hetch Hetchy. Should Hetch Hetchy be submerged for a reservoir, as proposed, not only would it be utterly destroyed, but the sublime canyon way to the heart of the High Sierra would be hopelessly blocked and the great camping ground, as the watershed of a city drinking system, virtually would be closed to the public. So far as I have learned, few of all the thousands who have seen the park and seek rest and peace in it are in favor of this outrageous scheme.

One of my later visits to the Valley was made in the autumn of 1907 with the late William Keith, the artist. The leaf colors were then ripe, and the great, godlike rocks in repose seemed to glow with life. The artist, under their spell, wandered day after day along the river and through the groves and gardens, studying the wonderful scenery, and, after making about forty sketches, declared with enthusiasm that although its walls were less sublime in height, in picturesque beauty and charm, Hetch Hetchy surpassed even Yosemite.

That any one would try to destroy such a place seems incredible, but sad experience shows that there are people good enough and bad enough for anything. The proponents of the dam scheme bring forward a lot of bad arguments to prove that the only righteous thing to do with the people's parks is to destroy them bit by bit as they are able. Their arguments are curiously like those of the devil, devised for the destruction of the first garden—so much of the very best Eden fruit going to waste, so much of the best Tuolumne water and Tuolumne scenery going to waste. Few of their statements are even partly true, and all are misleading.

Thus, Hetch Hetchy, they say, is a "low-lying meadow." On the contrary, it is a high-lying natural landscape garden, as the photographic illustrations show.

"It is a common minor feature, like thousands of others." On the contrary, it is a very uncommon feature—after Yosemite, the rarest and, in many ways, the most important in the national park.

"Damming and submerging it 175-feet deep would enhance its beauty by forming a crystal-clear lake." Landscape gardens, places of recreation and worship, are never made beautiful by destroying and burying them. The beautiful sham lake, forsooth, should be only an eyesore, a dismal blot on the landscape, like many others to be seen in the Sierra. For, instead of keeping it at the same level all the year, allowing Nature centuries of time to make new shores, it would, of course, be full only a month or two in the spring, when the snow is melting fast; then it would be gradually drained, exposing the slimy sides of the basin and shallower parts of the bottom, with the gathered drift and waste, death and decay of the upper basins, caught here instead of being swept on to decent natural burial along the banks of the river or in the sea. Thus the Hetch Hetchy dam lake would be only a rough imitation of a natural lake for a few of the spring months, an open sepulcher for the others.

"Hetch Hetchy water is the purest of all to be found in the Sierra, unpolluted, and forever unpollutable." On the contrary, excepting that of the Merced below Yosemite, it is less pure than that of most of the other Sierra streams, because of the sewerage of campgrounds draining into it, especially of the Big Tuolumne Meadows campground, occupied by hundreds of tourists and mountaineers, with their animals, for months every summer, soon to be followed by thousands from all the world.

These temple destroyers, devotees of ravaging commercialism, seem to have a perfect contempt for Nature, and, instead of lifting their eyes to the God of the mountains, lift them to the Almighty Dollar.

Dam Hetch Hetchy! As well dam for watertanks the people's cathedrals and churches, for no holier temple has ever been consecrated by the heart of man.

WILLIAM TEMPLE HORNADAY

William Temple Hornaday, director of the New York Zoological Society, was, by most accounts, the era's leading advocate for wildlife protection. A past president of the American Bison Society and a former, self-professed "sportsman," Hornaday chose to abandon, circa his book *Wildlife Conservation in Theory and Practice*, that once unsullied appellation, conceding "the conscientious and dutydoing sportsmen of the world are now so hopelessly mixed up with the motley array of gamehogs and gunners-at-large as to be almost unrecognizable." In the preface to *Wildlife Conservation in Theory and Practice*, Dr. Hornaday suggests the importance of Yale University as the site for a presentation on wildlife conservation, writing, "What is needed and now demanded of professors and teachers in all our universities, colleges, normal schools, and high schools is vigorous and persistent teaching of the ways and means that can successfully be employed in the wholesale manufacture of public sentiment in behalf of the rational and effective protection of wildlife."

The often preacherly tone of Hornaday's call to arms is echoed in the remarkably charismatic chapter below—drawn from *Our Vanishing Wild Life; Its Extermination and Preservation* and peppered with exclamation points—which he wrote to accompany his launching of the Permanent Wildlife Protection Fund. Unsparing in its criticism of negligent farmers, irreverent market-gamers, injudicious scientists, and gratuitous fashion mavens sporting fur and feather, "The Former Abundance of Wildlife" sounds a particularly persuasive chord in the burgeoning conservation movement: the necessity of conservation for the next generation. The highly quotable, always influential Hornaday influenced Aldo Leopold, among others. "The Former

Abundance of Wildlife" is a dynamic read that occasionally gives way to zealotry and bigotry. Notable is Hornaday's disdain for predators and the absence of a call for their preservation. Indeed, Hornaday uses the "the cruel wolf and the criminal dog" as metaphors for the dastardly deeds of market gunner.

The Former Abundance of Wildlife[151]

> *By my labors my vineyard flourished. But Ahab came. Alas! for Naboth.*

In order that the American people may correctly understand and judge the question of the extinction or preservation of our wildlife, it is necessary to recall the near past. It is not necessary, however, to go far into the details of history, for a few quick glances at a few high points will be quite sufficient for the purpose in view.

Any man who reads the books which best tell the story of the development of the American colonies of 1712 into the American nation of 1912, and takes due note of the wildlife features of the tale, will say without hesitation that when the American people received this land from the bountiful hand of Nature, it was endowed with a magnificent and all-pervading supply of valuable wild creatures. The pioneers and the early settlers were too busy even to take due note of that fact, or to comment upon it, save in very fragmentary ways.

Nevertheless, the wildlife abundance of early American days survived down to so late a period that it touched the lives of millions of people now living. Any man fifty-five years of age who, when a boy, had a taste for "hunting"—for at that time there were no "sportsmen" in America—will remember the flocks and herds of wild creatures that he saw and which made upon his mind many indelible impressions.

"Abundance" is the word with which to describe the original animal life that stocked our country, and all North America, only a short

---

[151] William Temple Hornaday, *Our Vanishing Wild Life; Its Extermination and Preservation* (New York: C. Scribner's Sons, 1913) 1–10.

The Golden Age, Readings 1900–1920 411

half-century ago. Throughout every state, on every shoreline, in all the millions of freshwater lakes, ponds and rivers, on every mountain range, in every forest, and even on every desert, the wild flocks and herds held away. It was impossible to go beyond the settled haunts of civilized man and escape them.

It was a full century after the complete settlement of New England and the Virginia colonies that the wonderful big-game fauna of the Great Plains and Rocky Mountains was really discovered, but the bison millions, the antelope millions, the mule deer, the mountain sheep and mountain goat were there, all the time. In the early days, the millions of pinnated grouse and quail of the central states attracted no serious attention from the American people at large, but they lived and flourished just the same, far down in the seventies, when the greedy market gunners systematically slaughtered them, and barreled them up for "the market," while the foolish farmers calmly permitted them to do it.

We obtain the best of our history of the former abundance of North American wildlife first from the pages of Audubon and Wilson, next from the records left by such pioneers as Lewis and Clark, and last from the testimony of living men. To all this we can, many of us, add observations of our own.

To me the most striking fact that stands forth in the story of American wildlife one hundred years ago is the wide extent and thoroughness of its distribution. Wide as our country is, and marvelous as it is in the diversity of its climates, its soils, its topography, its flora, its riches and its poverty, Nature gave to each square mile and to each acre a generous quota of wild creatures, according to its ability to maintain living things. No pioneer ever pushed so far, or into regions so difficult or so remote, that he did not find awaiting him a host of birds and beasts. Sometimes the pioneer was not a good hunter—usually he was a stupid fisherman—but the "game" was there, nevertheless. The time was when every farm had its quota.

The part that the wildlife of America played in the settlement and development of this continent was so far-reaching in extent, and so enormous in potential value, that it fairly staggers the imagination. From the landing of the Pilgrims down to the present hour, the wild game has been the mainstay and the resource against starvation of the pathfinder, the settler, the prospector, and at times even the railroad builder. In view of what the bison millions did for the Dakotas, Montana, Wyoming, Kansas and Texas, it is only right and square that those states should now do something for the perpetual preservation of the bison species and all other big game that needs help.

For years and years, the antelope millions of the Montana and Wyoming grasslands fed the scout and Indian fighter, freighter, cowboy and surveyor, ranchman and sheepherder, but thus far I have yet to hear of one western state that has ever spent one penny directly for the preservation of the antelope! And today we are in a hand-to-hand fight in Congress, and in Montana, with the Woolgrowers Association, which maintains in Washington a keen lobbyist to keep aloft the tariff on wool, and prevent Congress from taking fifteen square miles of grasslands on Snow Creek, Montana, for a National Antelope Preserve. All that the woolgrowers want is the entire earth, all to themselves. Mr. McClure, the secretary of the association says: "The proper place in which to preserve the big game of the West is in city parks, where it can be protected."

To the colonist of the East and pioneer of the West, the white-tailed deer was an ever-present help in time of trouble. Without this omnipresent animal, and the supply of good meat that each white flag represented, the commissariat difficulties of the settlers who won the country as far westward as Indiana would have been many times greater than they were. The backwoods Pilgrim's progress was like this: Trail, deer; cabin, deer; clearing; bear, corn, deer; hogs, deer; cattle, wheat, independence.

And yet, how many men are there today, out of our ninety millions of Americans and pseudo-Americans, who remember with any feeling of gratitude the part played in American history by the white-tailed deer? Very few. How many Americans are there in our land who now preserve that deer for sentimental reasons, and because his forbears were nation-builders? As a matter of fact, are there any?

On every eastern pioneer's monument, the white-tailed deer should figure, and on those of the great West, the bison and the antelope should be cast in enduring bronze, *Lest we forget!*

The game birds of America played as different part from that of the deer, antelope, and bison. In the early days, shotguns were few, and shot was scarce and dear. The wild turkey and goose were the smallest birds on which a rifleman could afford to expend a bullet and a whole charge of powder. It was for this reason that the deer, bear, bison, and elk disappeared from the eastern United States while the game birds yet remained abundant. With the disappearance of the big game, came the fat steer, hog and hominy, the wheatfield, fruit orchard, and poultry galore.

The game birds of America, as a class and a mass, have not been swept away to ward off starvation or to rescue the perishing. Even back in the sixties and seventies, very, very few men of the North thought of killing prairie chickens, ducks and quail, snipe and woodcock, in order to keep the hunger wolf from the door. The process was too slow and uncertain, and besides, the really poor man rarely had the gun and ammunition. Instead of attempting to live on birds, he hustled for the staple food products that the soil of his own farm could produce.

First, last, and nearly all the time, the game birds of the United States as a whole have been sacrificed on the altar of rank luxury, to tempt appetites that were tired of fried chicken and other farm delicacies. Today, even the average poor man hunts birds for the joy of the outing, and the pampered epicures of the hotels and restaurants buy game birds, and eat small portions of them, solely to tempt jaded appetites. If there is such a thing as "class" legislation, it is that which permits a few sordid market shooters to slaughter the birds of the whole people in order to sell them to a few epicures.

The game of a state belongs to the whole people of the state. The Supreme Court of the United States has so decided (Geer vs. Connecticut). If it is abundant, it is a valuable asset. The great value of the game birds of America lies not in their meat pounds as they lie upon the table, but in the temptation they annually put before million of field-weary farmers and desk-weary clerks and merchants to get into their beloved hunting togs, stalk out into the lap of Nature, and say *Begone, dull care!*

And the man who has had a fine day in the painted woods, on the bright waters of a duck-haunted bay, or in the golden stubble of September, can fill his day and his soul with six good birds just as well as with sixty. The idea that in order to enjoy a fine day in the open a man must kill a wheelbarrow load of birds is a mistaken idea, and if obstinately adhered to, it becomes vicious. The outing in the open is the thing—not the blood-stained feathers, nasty viscera, and death in the gamebag. One quail on a fence is worth more to the world than ten in a bag.

The farmers of America have, by their own supineness and lack of foresight, permitted the slaughter of a stock of game birds which, had it been properly and wisely conserved, would have furnished a good annual shoot to every farming man and boy of sporting instincts through the past, right down to the present, and far beyond. They have allowed millions of dollars worth of their birds to be coolly snatched away from them by the greedy market shooters.

There is one state in America, and so far as I know only one, in which there is at this moment an old-time abundance of game bird life. That is the state of Louisiana. The reason is not so very far to seek. For the birds that do not migrate—quail, wild turkey and doves—the cover is yet abundant. For the migratory game birds of the Mississippi Valley, Louisiana is a grand central depot, with terminal facilities that are unsurpassed. Her reedy shores, her vast marshes, her long coastline, and abundance of food furnish what should be not only a haven but a heaven for ducks and geese. After running the gauntlet of guns all the way from Manitoba and Ontario to the Sunk Lands of Arkansas, the shores of the Gulf must seem like heaven itself.

The great forests of Louisiana shelter deer, turkeys, and fur-bearing animals galore, and rabbits and squirrels abound.

Naturally, this abundance of game has given rise to an extensive industry in shooting for the market. The "big interests" outside the state send their agents into the best game districts, often bringing in their own force of shooters. They comb out the game in enormous quantities, without leaving to the people of Louisiana any decent and fair quid-pro-quo for having despoiled them of their game and shipped a vast annual product outside to create wealth elsewhere.

At present, however, we are but incidentally interested in the short-sightedness of the people of the Pelican State. As a state of old-time abundance in killable game, the killing records that were kept in the year 1909–10 possess for us very great interest. They throw a startling searchlight on the subject of this chapter—the former abundance of wildlife.

From the records that with great pains and labor were gathered by the State Game Commission, and which were furnished me for use here by President Frank M. Miller, we set forth this remarkable exhibit of old-fashioned abundance in game, *Official Record of Game Killed in Louisiana during the Season 1909–1910*; "*Birds:* wild ducks, sea and river 3,176,000; coots 280,740; geese and brant 202,210; snipe, sandpiper and plover 606,635; quail (bobwhite) 1,140,750; doves 310,660; wild turkeys 2,219. Total number of game birds killed: 5,719,214. *Mammals:* deer 5,470; squirrels and rabbits 690,270. Total of game mammals: 695,740; Fur-bearing mammals: 1,971,922. Total of mammals: 2,667,662. *Grand total of birds and mammals*: 8,386,876."

Of the thousands of slaughtered robins, it would seem that no records exist. It is to be understood that the annual slaughter of wildlife in Louisiana never before reached such a pitch as now. Without drastic measures, what will be the inevitable result? Does any man suppose that even the wild millions of Louisiana can long withstand such slaughter as that shown by the official figures given above? It is wildly impossible.

But the darkest hour is just before the dawn. At the session of the Louisiana legislature that was held in the spring of 1912, great improvements were made in the game laws of that state. The most important feature was the suppression of wholesale market hunting by persons who are not residents of the state. A very limited amount of game may be sold and served as food in public places, but the restrictions placed upon this traffic are so effective that they will vastly reduce the annual slaughter. In other respects, also, the cause of wildlife protection gained much, for which great credit is due to Mr. Edward A. McIlhenny.

It is the way of Americans to feel that because game is abundant in a given place at a given time, it always will be abundant, and may therefore be slaughtered without limit. That was the case last winter in California during the awful slaughter of band-tailed pigeons, as will be noted elsewhere.

It is time for all men to be told in the plainest terms that there never has existed, anywhere in historic times, a volume of wildlife so great that civilized man could not quickly exterminate it by his methods of destruction.

Lift the veil and look at the stories of the bison, the passenger pigeon, the wild ducks and shore birds of the Atlantic coast, and the fur-seal.

As reasoning beings, it is our duty to heed the lessons of history, and not rush blindly on until we perpetrate a continent destitute of wildlife.

For educated, civilized man to exterminate a valuable wild species of living things is a crime.[152] It is a crime against his own children, and posterity.

No man has a right, either moral or legal, to destroy or squander an inheritance of his children that he holds for them in trust. And man, the wasteful and greedy spendthrift that he is, has not created even the humblest of the species of birds, mammals and fishes that adorn and enrich this earth. "The earth is the lord's, and the fullness thereof!" With all his wisdom, man has not evolved and placed here so much as a ground squirrel, a sparrow, or a clam. It is true that he has juggled with the wild horse and sheep, the goats and the swine, and produced some hardy breeds that can withstand his abuse without going down before it, but, as for species, he has not yet created and placed here even so much as a protozoan.

The wild things of this earth are not ours to do with as we please. They have been given to us in trust, and we must account for them to the generations which will come after us and audit our accounts.

---

[152] Chapter II, "Extinct Species of North America," begins with this sentence in the 1913 edition. It is run-in here for the sake of readability and continuity.

But man, the shameless destroyer of Nature's gifts, blithely and persistently exterminates one species after another. Fully 10 percent of the human race consists of people who will lie, steal, throw rubbish in parks, and destroy forest and wildlife whenever and wherever they can do so without stopped by a policemen and a club. These are hard words, but they are absolutely true. From 10 percent (or more) of the human race, the high moral instinct which is honest without compulsion is absent. The things that seemingly decent citizens—men posing gentlemen—will do to wild game when they secure great chances to slaughter are appalling. I could fill a book of this size with cases in point.

Today the women of England, Europe, and elsewhere are directly promoting the extermination of score of beautiful species of wild birds the devilish persistence with which they buy and wear feather ornaments made of their plumage. They are just as mean and cruel as the truck driver who drives a horse with a sore shoulder and beats him on the street. But they do it! And appeals to them to do otherwise they laugh to scorn, saying, "I will wear what is fashionable, when I please and where I please!" As a famous bird protector of England has just written me, "The women of the smart set are beyond the reach of appeal or protest."

Today, the thing that stares me in the face every waking hour, like a grisly specter with bloody fang and claw, is the extermination of species. To me, that is a horrible thing. It is wholesale murder, no less. It is capital crime and a black disgrace to the races of civilized mankind. I say "civilized mankind," because savages don't do it!

There are three kinds of extermination:

The practical extermination of a species means the destruction of its member to an extent so thorough and widespread that the species disappears from view, and living specimens of it can not be found by seeking for them. In North America this is today the status of the whooping crane, upland plover, and several other species. If any individuals are living, they will be met with only by accident.

The absolute extermination of a species means that not one individual of it remains alive. Judgment to this effect is based upon the lapse of time since the last living specimen was observed or killed. When five years have passed without a living "record" of a wild specimen, it is time to place a specimen in the class of the totally extinct.

Extermination in a wild state means that the only living representatives are in captivity or otherwise under protection. This is the case of the heath hen and David's deer, of China. The American bison is saved from being wholly extinct as a wild animal by the remnant of about three hundred head in northern Athabasca, and forty-nine head in the Yellowstone Park.

It is a serious thing to exterminate a species of any of the vertebrate animals. There are probably millions of people who do not realize that civilized man is the most persistently and wickedly wasteful of all the predatory animals. The lions, the tigers, the bears, the eagles and hawks, serpents, and the fish-eating fishes, all live by destroying life, but they kill only what they think they can consume. If something is by chance left over, it goes to satisfy the hunger of the humbler creatures of prey. In a state of nature, where wild creatures prey upon wild creatures, such a thing as wanton, wholesale, and utterly wasteful slaughter is almost unknown!

When the wild mink, weasel, and skunk suddenly finds himself in the midst of scores of man's confined and helpless domestic fowls, or his caged gulls in a zoological park, an unusual criminal passion to murder for the joy of killing sometimes seizes the wild animal, and great slaughter is the result.

From the earliest historic times, it has been the way of savage man, red, black, brown and yellow, to kill as the wild animals do—only what he can use, or thinks he can use. The Cree Indian impounded small herds of bison and sometimes killed from one hundred to two hundred at one time, but it was to make sure of having enough meat and hides and because he expected to use the product. I think that even the worst enemies of the Plains Indians hardly will accuse them of killing large numbers of bison, elk, or deer merely for the pleasure of seeing them fall or taking only their teeth.

It has remained for the wolf, the sheep-killing dog, and civilized man to make records of wanton slaughter which put them in a class together, and quite apart from other predatory animals. When a man can kill bison for their tongues alone, bull elk for their "tusks" alone, and shoot a whole colony of hippopotami, actually damming a river with their bloated and putrid carcasses, all untouched by the knife. The men who do such things must be classed with the cruel wolf and the criminal dog.

It is now desirable that we should pause in our career of destruction long enough to look back upon what we have recently accomplished in the total extinction of species, and also note what we have blocked out for the immediate future. Here let us erect a monument to the dead species of our own times.

It is to be doubted whether, up to this hour, any man has made a list of the species of North American birds that have become extinct during the past sixty years. The specialist have no time to spare from their compound differential microscopes, and the bird-killers are too busy with shooting, netting, and clubbing to waste any time on such trifles as exterminated species. What does a market shooter care about birds that can not be killed a second time? As for the farmers, they are so busy raising hogs and prices that their friends, the birds, get scant attention from them—until a hen hawk takes a chicken!

Down South, the Negroes and poor whites may slaughter robins for food by the ten thousand, but does the northern farmer bother his head about a trifle of that kind? No, indeed. Will he contribute any real money to help put a stop to it? Ask him yourself.

Let us pause long enough to reckon up some of our expenditures in species and in millions of individuals. Let us set down here, in cold blood, a list of the species of our own North American birds that have been totally exterminated in our own times. After that we will have something to say about other species that soon will be exterminated, and the second task is much greater than the first.

LIBERTY HYDE BAILEY

By comparison with his titles as professor of horticulture at Cornell University and chair of Roosevelt's Country Life Commission, Liberty Hyde Bailey's editorship of the important periodical *Country*

Life in America is often overlooked, as is his championing of nature study in lieu of instruction in vocational agriculture and home economics as favored by many of his colleagues. Bailey's conservationist ideology is arguably the most comprehensive of his day, as it applies equally to land use on farms, in suburbs, and in the wilderness. Though Aldo Leopold is often considered heir to Bailey's ecological beliefs, Wendell Berry, in his role as traditional family farmer and conscientious environmental steward, may be a superior analog.

In fact, Bailey's and Berry's conservationist views are both overtly steeped in agrarianism and Christianity, a fact which some readers find endearing and others alienating. Firm in his belief that the earth ought to be considered religiously, Bailey was concerned not only with the common good—aptly expressed in his statement "We begin to foresee the vast religion of a better social order"—but in a equitable division of the earth's resources. Sounding an almost socialist note, Bailey's writing in *The Holy Earth* is replete with aphorisms both pithy and profound, including such definitive statement as "Dominion does not carry personal ownership." In "It is Kindly," excepted from *The Holy Earth*, the author references a popularly held belief concerning the nearness of end times and environmental apocalypse—fears born in part of the ecological disasters caused by the rapid settling and capitalization of the West.

It is Kindly[153]

The contest with nature is wholesome, particularly when pursued in sympathy and for mastery.[154] It is worthy a being created in God's image. The earth is perhaps a stern earth, but it is a kindly earth.

Most of our difficulty with the earth lies in the effort to do what perhaps ought not to be done. Not even all the land is fit to be farmed.

---

[153] Liberty Hyde Bailey, *The Holy Earth* (New York: C. Scribner's Sons, 1915) 11–16.

[154] The excerpt here begins, for the sake of concision and readability, with the fifth paragraph in the chapter "It Is Kindly" as it appeared in the 1915 edition.

A good part of agriculture is to learn how to adapt one's work to nature, to fit the crop scheme to the climate and to the soil and the facilities. To live in right relation with his natural conditions is one of the first lessons that a wise farmer or any other wise man learns. We are at pains to stress the importance of conduct; very well: conduct toward the earth is an essential part of it.

Nor need we be afraid of any fact that makes one fact more or less in the sum of contacts between the earth and the earth-born children. All "higher criticism" adds to the faith rather than subtracts from it, and strengthens the bond between. The earth and its products are very real.

Our outlook has been drawn very largely from the abstract. Not being yet prepared to understand the condition of nature, man considered the earth to be inhospitable, and he looked to the supernatural for relief; and relief was heaven. Our pictures of heaven are of the opposites of daily experience—of release, or peace, of joy uninterrupted. The hunting grounds are happy and the satisfaction has no end. The habit of thought has been set by this conception, and it colors our dealings with the human questions and, to much extent, it controls our practice.

But we begin to understand that the best dealing with problems on earth is to found it on the facts of earth. This is the contribution of natural science, however abstract, to human welfare. Heaven is to be a real consequence of life on earth, and we do not lessen the hope of heaven by increasing our affection for the earth, but rather do we strengthen it. Men now forget the old images of heaven, that they are mere sojourners and wanderers lingering for deliverance, pilgrims in a strange land. Waiting for this rescue, with posture and formula and phrase, we have overlooked the essential goodness and quickness of the earth and the immanence of God.

This feeling that we are pilgrims in a vale of tears has been enhanced by the widespread belief in the sudden ending of the world, by collision or some other impending disaster, and in the common apprehension of doom, and lately by speculations as to the aridation and death of the planet, to which all of us have given more or less credence. But most of these notions are now considered to be fantastic, and we are increasingly confident that the earth is not growing old in human sense, that its atmosphere and its water are held by the attraction of its mass, and that the sphere is at all events so permanent as to make little difference in our philosophy and no difference in our good behavior.

I am again impressed with the first record in *Genesis* in which some mighty prophet-poet began his account with the creation of the physical universe.

So do we forget the old-time importance given to mere personal salvation, which was permission to live in heaven, and we think more of our present situation, which is the situation of obligation and of service, and he who loses his life shall save it.

We begin to foresee the vast religion of a better social order.

Verily, then, the earth is divine, because man did not make it. We are here, part in the creation. We cannot escape. We are under obligation to take part and to do our best, living with each other and with all the creatures. We may not know the full plan, but that does not alter the relation. When once we set ourselves to the pleasure of our dominion, reverently and hopefully, and assume all its responsibilities, we shall have a new hold on life.

We shall put our dominion into the realm of morals. It is now in the realm of trade. This will be very personal morals, but it will also be national and racial morals. More iniquity follows the improper and greedy division of the resources and privileges of the earth than any other form of sinfulness.

If God created the earth, so is the earth hallowed, and if it is hallowed, so must we deal with it devotedly and with care that we do not despoil it, and mindful of our relations to all beings that live on it. We are to consider it religiously: put off thy shoes from off thy feet, for the place whereon thou standest is holy ground.

The sacredness to us of the earth is intrinsic and inherent. It lies in our necessary relationship and in the duty imposed upon us to have dominion and to exercise ourselves even against our own interest. We may not waste that which is not ours. To live in sincere relations with the company of created things and with conscious regard for the support of all men now and yet to come, must be of the essence of righteousness.

This is a larger and more original relation than the modern attitude of appreciation and admiration of nature. In the days of the patriarchs and prophets, nature and man shared in the condemnation and likewise in the redemption. The ground was cursed for Adam's sin. Paul wrote that the whole creation groaneth and travaileth in pain, and that it waiteth for the revealing. Isaiah proclaimed the redemption of the wilderness and the solitary place with the redemption of man, when they shall rejoice and blossom as the rose, and when the glowing sand shall become a pool and the thirsty ground springs of water.

The usual objects have their moral significance. An oak tree is to us a moral object because it lives its life regularly and fulfils its destiny. In the wind and in the stars, in forest and by the shore, there is spiritual refreshment: and the spirit of God moved upon the face of the waters.

I do not mean all this, for our modern world, in any vague of abstract way. If the earth is holy, then the things that grow out of the earth are also holy. They do not belong to man to do with them as he will. Dominion does not carry personal ownership. There are many generations of folk yet to come after us who will have equal right with us to the products of the globe. It would seem that a divine obligation rests on every soul. Are we to make righteous use of the vast accumulation of knowledge of the planet? If so, we must have a new formulation. The partition of earth among the millions who live on it is necessarily a question of morals, and a society that is founded on an unmoral partition and use cannot itself be righteous and whole.

MARTHA FOOTE CROW

Martha Foote Crow brings a woman's sensibility to *The American Country Girl*, which she dedicates to the six and a half million girls living in the country and small towns of the era. Crow, better known for her literary studies *Elizabethan Sonnet Cycles* (1896), and *Christ in the Poetry of Today; An Anthology of American Poets* (1923), corresponded with many of the most influential women of her era, including poets Sara Teasdale and Elinor Wylie. Indeed, Crow dedicated the substance of her life and scholarship to young women, including service in the capacity of Dean of Women at Northwestern University—the same position earlier occupied by Frances Willard. Like Willard, Crow was active in the National Women's Temperance Union, though perhaps her most enduring institutional affiliation was with the sorority Alpha Phi, for which she served as the first national president and administrator of education.

In the chapter that follows, "The Country Girl: Where Is She," Crow elaborates on the themes of her 1915 *The Nation* article "Seven Million Country Girls," writing elegantly about the most overlooked member of the farm family. Demonstrating her poetic acumen and insight, Crow writes, "Since the days of Eve, the woman young and old has been adapting herself and readapting herself, until, after all these centuries of constant practice, she has become a past master in the art of adaptation." While Crow praises the work of the Country Life Commission and the effects of the Country Life movement in general, she also categorically dismisses its generalizations, stating "There is no rural mind in America." Moreover, she argues that the country girl is not substantively different from her city brethren, except perhaps in the burden she shoulders. A native New Yorker and a lifelong social activist, Crow published *Harriet Beecher Stowe: A Biography for Girls* in 1913 as part of her mission to highlight role models for America's young women.

The Country Girl: Where is She?[155]

The clarion of the Country Life movement has by this time been blown with such loudness and insistence that no hearing ear in our land can have escaped its announcement. The distant echoes of brutal warfare have not drowned it: above all possible rude and cruel sounds, this peaceful piping still makes itself heard.

---

[155] Martha Foote Crow, *The American Country Girl* (New York: Frederick A. Stokes Company, 1915), 3–11.

The Golden Age, Readings 1900–1920 425

It has reached the ears of the farmer and has stirred his mind and heart to look his problems in the face, to realize their gigantic implications, and to shoulder the responsibility of their solution. It has penetrated to the thoughts of teachers and educators everywhere and awakened them to the necessities of the minute, so that they have declared that the countryside must have educational schemes adapted to the needs of the countryside people, and that they must have teachers whose heads are not in the clouds. It has aroused easygoing preachers in the midst of their comfortable dreams and has caused here and there one among them to bestir himself and to make hitherto unheard of claims as to what the church might do, if it would, for the betterment of country life.

And all of these have given hints to philanthropists and reformers, and these to organizations and societies; these again have suggested theories and projects to legislators, senators, and presidents; the snowball has been rolled larger and larger; commissions have sat, investigations have been made, documents have been attested, reports handed in, bills drafted and, what is better, passed by courageous legislation, so that now great schemes are being not only dreamed of but put into actual fulfillment. Moreover, lecturers have talked, and writers have issued bulletins and books, until there has accumulated a library of vast proportions on the many phases of duty, activity, and outlook that may be included under the title, "A Country Life Movement."

In all this stirring field of new interest, the farmer and his business hold the center of attention. Beside him, however, stands a dim little figure hitherto kept much in the background, the farmer's wife, who at last seems to be on the point of finding a voice also. For a chapter is now assigned to her in every book on rural conditions and a little corner under a scroll work design is given to her tatting and her chickens in the weekly farm paper. Cuddled about her are the children, and they, the little farm boys and girls, have now a book that has been written just about them alone—their psychology and their needs. Also, the tall, strong youth, her grown-up son, has his own paper as an acknowledged citizen of the rural commonwealth. But where is the tall, young daughter, and where are the papers for her and the books about her needs? It seems that she has not as yet found a voice. She has failed to impress the makers of books as a subject for description and investigation. In the nationwide effort to find a solution to the great rural problems, the farmer is working heroically; the son is putting his shoulder to the wheel; the wife and mother is in sympathy with their efforts. Is the daughter not doing her share? Where is the country girl and what is happening in her department?

It is easier on the whole to discover the rural young man than to find the typical country girl. Since the days of Eve, the woman young and old has been adapting herself and readapting herself, until, after all these centuries of constant practice, she has become a past master in the art of adaptation. Like the cat in the story of Alice, she disappears in the intricacy of the wilderness about her and nothing remains of her but a smile.

There are some perfectly sound reasons why American country girls as a class cannot be distinguished from other girls. Chief among these is the fact that no group of people in this country is to be distinguished as a class from any other group. It is one of the charms of life in this country that you never can place anybody. No one can distinguish between a shopgirl and a lady of fashion; nor is any schoolteacher known by her poise, primness, or imperative gesture. The fashion paper, penetrating to the remotest dugout, and the railway engine indulging us in our national passion for travel, see to these things. Moreover, the pioneering period is still with us and the western nephews must visit the cousins in the old home in New Hampshire, while the aunts and uncles left behind must go out to see the new Nebraska or Wyoming lands on which the young folks have settled. We do not stay still long enough anywhere in the republic for a class of any sort to harden into recognizable form.

New inhabitants may come here already hardened into the mold of some class, but they or their children usually soften soon into the quicksilver-like consistency of their surroundings. There is also no subdividing of notions on the basis of residence, whether as townsman or as rural citizen. The wind bloweth where it listeth in this land. It whispers its free secrets into the ears of the city dweller in the flat and of the rural worker of the cornfield or the vine-screened kitchen. The rain also falls on the just and the unjust whether "suburbanated" or countrified. There is no rural mind in America. There has indeed been a great deal of pother of late over the virtue and temper of "rural-minded people." This debate has been conscientiously made in the effort to discern reasons why commissions should sit on a rural problem. Reasons enough are discernible why commissions should sit, but they lie rather in the unrural mind of the rural people, as the words are generally understood, than in some supposed qualities imposed or produced in the life of sun and rain, in that vocation that is nearest to the creative activities of the divine.

And, if there is no rural mind, there is no distinctive rural personality. If the man that ought to exemplify it is found walking up Fifth Avenue or on Halstead Street or along El Camino Real, he cannot be discovered as a farmer. He may be discovered as an ignorant person, or he may be found to be a college-educated man, but in neither case would the fact be logically inclusive or uninclusive of his function as farmer.

The same is almost as exactly true for his wife and his daughter. If one should ask in any group of average people whether the farmer's daughter, as they have known her, is a poor, little, undeveloped child, silent and shy, or a hearty, buxom lass, healthy and strong and up to date, some in the group would say the latter and some the former. Both varieties exist and can by searching be found along the countryside.

To be convinced of this, one who knows this country well has but to read a book like *Folk of the Furrow* by Christopher Holdenby, a picture of rural life in England. In such a book as that one realizes the full meaning of the phrase, "the rural mind," and one sees how far the men and women that live on the farms in the United States have yet to go, how much they will have to coagulate, how many centuries they will have to sit still in their places with wax in their ears and weights on their eyelids, before they will have acquired psychological features such as Mr. Holdenby gives to the folk of the English furrow.

A traveler in the Old World frequently sees illustrations of this. For instance, in passing through some European picture gallery, he may meet a woman of extraordinary strength and beauty, dressed in a style representing the rural life in that vicinity. She will wear the peasant skirt and bodice and will be without gloves or hat. A second look will reveal that the skirt is made of satin so stiff that it could stand alone; the velvet bodice will be covered with rich embroidery, and heavy chains of silver of quaint workmanship will be suspended around the neck.

On inquiry, one may learn that this stately woman was of what would be called in this country a farmer family that had now become very wealthy, that she did not consider herself above her "class"—so they would describe it—no, that she gloried in it instead. It was from preference only that she dressed in the fashion of that "class."

Now, whether desirable or not, such a thing as this would never be seen in America. No woman (unless it were a deaconess or a Salvation Army lassie or a nun) would pass through the general crowd showing her rank or profession in life by her style of dress. And that is how it happens that neither by hat nor by hatlessness would the country woman here make known her pride in the possession of acres or in her relation to that profession that forms the real basis of national prosperity. Hence no country girl counts such a pride among her inheritances. Therefore it is not easy to find and understand the country girl as a type, it is not because she is consciously or unconsciously hiding herself away from us; she is not even sufficiently conscious of herself as a member of a social group to pose in the attitude of an interesting mystery. She is just a human being happening to live in the country (not always finding it the best place for her proper welfare), just a single one in the great shifting mass.

Although it may be difficult to find what we may think are typical examples of the country girl as a social group, yet certain it is that she exists. Of young women between the ages of fifteen and twenty-nine, there are in the United States six and a half million (6,694,184 to exact) who reside in the open country or in small villages. This we are assured is so by the latest census report.

By starting a little further down in the scale of girlhood and advancing a trifle further into maturity, this number could be doubled. It would be quite justifiable to do this because some farmers' daughters become responsible for a considerable amount of labor value well before the age of fifteen—and on the other hand the energy of these young rural women is abundantly extended beyond the gateway of womanhood, far indeed into the period that used to be called "old maidism," but which is to be so designated no more—the breezy, executive, freehanded period when the country girl is of greatest use as a labor unit and gives herself without stint (and often without pay) to the welfare of the whole farmstead. The American country girl is not by any means behind her city sister in her ability to make the bounds of her youth elastic, though the girl on the farm may go at it in a somewhat different way. Then, perhaps, too, the word "youth" may, alas, have another connotation in the mind of one from what it has in the dreams of the other.

If we should, however, thus enlarge the scope of our inquiry, we should increase but not clarify our problems. Moreover, it is the country girl that interests us, the promise and hope of her dawn, the delicate, swiftly changing years of her growth, the miracle of her blossoming. There is something about the kaleidoscope of her moods and the inconsistencies of her biography that fascinates us. The moment when she awakes, when the sparkle begins to show in her eyes, when we know that a conception of her mission and of her supreme value to life is beginning to glow before her imagination—that is the crisis to work for and to be happy over when it comes. As for us, we ask no greater happiness than once or twice to catch a glimpse of that.

That great host of six million country girls is scattered far and wide; they are everywhere present. A certain number of millions of them are working industriously in myriads of unabandoned farms all over the Appalachian plateau, and on the wide prairies to the Rockies, and beyond. In thousands of farmsteads they are helping their mothers wash dishes three times a day three hundred and sixty-five days in the year, not counting the steps as they go back and forth between dining room and kitchen. They are carrying heavy pails of spring water into the house and throwing out big dishpanfuls of waste water, regardless of the strain in the small of the back. They are picking berries and canning them for the home table in the winter; they are raising tomatoes and canning them for the market; they are managing the younger children; they are baking and sewing and reading and singing; they are caring for chickens and for bees and for orphan lambs; they ride the rake and the disc plow and sometimes join the roundup on the range. Moreover they go to church and they go to town and they look forward to an ideal future just as other girls do. The country girl is a human being also.

It has been intimated that young women living on remote, secluded farms have not, with all their singing, been always able to dispel the monotony of a thousand inevitable dishwashings a year; they are said nowadays to have opened their ear to the lure of the town and to have started out, keeping step with their brothers, to join what someone has called "the funeral procession of the nation" cityward. If we could, in fact, get them to confide in us, we should find that they have longings and aspirations, many of which are unsatisfied, and that is the reason why it seems to be high time for their voice to be heard.

Some of the younger farm women are showing themselves equal to the larger burdens in the business of agriculture. They are running their own farms in Michigan and their own automobiles in Kansas. They are taking up claims. They are developing them and proving-up in the Dakotas and through Montana and Wyoming. From four to six in the morning they till an acre; then they ride twenty miles to the school and teach from nine to four; after that they ride back and work in their cornfields till the stars twinkle out. They stay alone in their shack and are happy and fearless and safe.

Moreover, some thousands of the girls are laboriously teaching schools in thousands of one-room schoolhouses, where they provide almost 100 percent of the common instruction for 50 percent of the population.

Besides this, there is no one of all the gainful occupations in which young women of this country engage which has not drawn upon the reservoir of country strength for supplies. Among those women blacksmiths and engineers, those clerks, secretaries, librarians and administrators, those lawyers, doctors, professors, writers, those nurses, settlement workers, investigators, and other servants of the people in widely diverse fields, there are many whose clearness of eye and reserve of force have been developed in the wholesome conditions of the open country. The country girl has no reason to be ashamed of the part she has borne in the nonrural world. It has been said that about 80 percent of the names found in *Who's Who in America* represent an upbringing in the rural atmosphere. The proportion of women in this number or the special proportion of grownup farm girls to be found among those women can not be stated, but the number must be large enough to justify a belief that to spend a childhood in the open country or in the rural village will not, in the case of women any more than in the case of men, form an impassable barrier to eminence.

From this great rural reserve of initiating force, sane judgment, and spiritual drive have come, in fact, some of the most valued names in philanthropy and literature. Among them we find the leader of a great reform, Frances Willard, the inaugurator of a worldwide work of mercy, Clara Barton, the president of a great college, Alice E. Freeman, the wise helper of all who suffer under unjust conditions in city life, Jane Addams, and the writer of a book that has had a national and worldwide influence, Harriet Beecher Stowe. It heartens us up a bit to name over examples like these. They give us a vista and a hope. But now and then there is a country girl who would rather have, say, a better pair of stilts over the morass or a stronger rope thrown to her across the quicksand, than a volume of *Who's Who* tossed carelessly to her in her difficulties. For all the country girls on their farms do not sing at their work. They are not idle, heaven knows, but their work does not invariably inspire the appreciation it deserves.

HENRY WALLACE

It was Henry Wallace, more than his son Henry Cantwell Wallace or his still more famous grandson, Henry Agard Wallace, who would enjoy the greatest popular audience in the Gilded and Golden Ages of agriculture and conservation writing. Born in Pennsylvania, Wallace came to Iowa in 1863 as a home missionary of the United Presbyterian Church and stayed, turning to farming and, after a provocative speech given on the Fourth of July in Winterset, Iowa, began writing a weekly column for the Winterset *Madisonian*, a newspaper he would eventually come to edit before moving on to a post at the Des Moines newspaper *Iowa Homestead*. After being relieved of his duties at *IH* for writing editorials attacking railroad monopolies and shipping rates, the sixty-year-old Wallace joined forces with his son Henry Cantrell to form what would soon become an industry standard, the periodical *Wallaces' Farmer*.

In the remarkable epistle that follows, "The Health of Farm Folk," Henry Wallace brings to bear his trademark talents for plainspoken speech and down-home homilies that would earn him the nickname "Uncle Henry" and would endear him to Teddy Roosevelt, who appointed him to serve on the prestigious Country Life Commission in 1907. Never interested in public office, Wallace spent his most influential years championing the causes of others—including Iowan James "Tama Jim" Wilson's successful bid to became the U.S. Secretary of Agriculture in 1897—and the cause of scientific and conservation-minded farming, passions that resulted in his tenure as president of the Third National Conservation Congress in 1910. In the piece below, taken from one of two books of letters to farm boys and farm families, Wallace is at his most natural, assuming the bully pulpit, a la his friend and fellow Republican Teddy Roosevelt, to elucidate many of the on-the-farm conservation themes that would be taken up as the life's work of his great-grandson, Henry A. Wallace. The younger Wallace would serve as secretary of agriculture and vice president under Franklin D. Roosevelt and would run for the presidency as a Progressive in 1948.

The Health of Farm Folk[156]

Our efficiency, our comfort, our joy in life, depend more than on any one other thing on our good health.[157] There is joy in living,

---

[156] Henry Wallace, *Letters to the Farm Folk*, 2nd ed. (Des Moines: Wallace Publishing Company 1915), 61–66.
[157] Wallace's salutation, "Dear Farm Folks," has been removed for organizational clarity.

a joy in work, a joy even in what seems at first sad drudgery, if we were in abounding health and every organ in the body and every department of the brain working smoothly, easily, efficiently. Without good health, there is usually little joy in life.

Theoretically, the farm folk ought to be the healthiest people in the world. Wise men tell us that what folks need for health is pure water, pure air, pure and wholesome food and healthy outdoor exercise, and a cheerful spirit. There ought to be all of this on the farm. These conditions of good health are within easier reach of the farmers than any other class of people on the face of the earth. And yet, if the reports of the American Medical Association, of investigations made in cooperation with the National Educational Association can be given half credence, the health of the children of the rural schools the United States over is not nearly so good as that of children of the cities, even including the slums. Doctor T. W. Wood, chairman of the committee, reports that the country schoolchildren are from 15 to 20 percent more defective than the city children. By "defective" he means "in some respect inferior physically."

Much of this may be accounted for by the fact that out in the country the children can not so readily secure the services of dentists, oculists, and aurists. We apprehend, however, that the great trouble lies in the fact that most country houses are either drafty or overheated, or both; hence there is more danger of infection, and also more chance of contracting disease than in the better ventilated houses of the city. There is not comparison between the heating and ventilation of the country schoolhouse and the modern city school. These difficulties and disadvantages may be overcome in time, if the attention of the country people is called to these facts.

Certainly, it is possible for any farmer, be he landlord or tenant, to secure good water. We are about three-fourths water anyhow, and if we don't get good water, we can't expect to have good health. Water from a well or spring is nearly always pure. The soil itself is a great filter, and furnishes us the beverage of life in purity. If it is impure, it is because we have allowed impurities to get into it, in the country, for example, in the shape of surface water, the drainage of the barnyard, or the stable or open closet. Typhoid fever comes from impure water. It is a filth disease. It is practically unknown in the cities that have good water systems, except in the fall of the year, when the people come back from vacations in the country. It is now recognized as a country disease. No landlord should permit a tenant to go into a farm that does not have pure water. No tenant should ever rent a farm that does not have pure water. He can't afford to, for his family is more to him than anything that grows out of the land.

Certainly, the farmer can have an abundant supply of pure air. There is plenty of that in the country. It is always ready to come into his house if he will give it a chance, and yet we suspect there is more disease in the country through lack of pure air than from lack of pure water. When our grandfathers built their houses out on the prairie, there was no lack of ventilation. There were enough cracks to let in all the pure air necessary. And the open fireplaces carried impure air. When they came to build better houses, the good man aimed to keep out the cold and forgot to make provision for letting in pure air and letting out the bad.

Many a farmer allows his children to sleep in an unventilated bedroom. He sleeps in one of that kind himself. While the men folks of the family do not suffer so much, because they lead an open-air life during the day, the women folks do not stand it so well. Many of them become pale, anemic, "go into decline," and contract consumption, simply because they breathe their own breath over and over again all night long. They are fearful of "night air" as if there could possibly be any other air at night than "night air." When the third house is built, let us hope that architects will have learned how to build houses and provide for the admission of pure air and the removal of the impure, as successfully as the really wise and progressive farmer now ventilates his cow barn or pig pen.

Why is it that people attend church in an unventilated building, or go to the hall in town and breathe air which is a combination of stale tobacco and the odor of bad breath and onions, and colds and sneezes, until the speaker is dull and the audience drowsy? I seldom attend a country church or a public meeting in which I do not long to get out and hire a boy to throw stones at a window and let in some of God's fresh air. There are two people in the world who are stupid beyond measure: one is the average farmhouse architect and the other is the average church janitor.

If any class of people can have pure food, it is the farmer. A large part of his food comes from his own farm. He has milk from the cow almost as directly as the calf itself, and if he is careful, it is as clean as milk can be. If the farmer milks a clean and healthy cow, in a clean stable, and his own hands are clean, there will be no harmful germs in it. The eggs on the farm are always fresh laid. The hen can't lay any other kind. The roots and vegetables and fruits come from the hand of nature in absolutely good condition. Farm folks should always cure their own bacon. If they eat anything that conveys disease, it is their own fault, and theirs alone. When it comes to that small portion of the food that he has to purchase, the farmer must take his chances with others. If he buys the cheap stuff from a cheap and dirty grocery, he must expect inferior and often unwholesome goods.

I believe that much of the weakness and lack of physical development of farm folks grows out of the fact that farmers have not yet learned to apply to their own families the principles which the best of them apply so successfully in the care of their livestock. They now understand quite generally that for the proper development of livestock, there must be a balanced ration, plenty of protein for the young and growing things, and for those that are giving milk, carbohydrates and fats for those that are being fattened for market. They should learn that in a family of growing children, and for the nursing mother, there must be flesh- and bone-forming food.

Some of the best cooking I have ever tasted has been on the farm, and, I must admit, some of the very worst. The farmer's table should have the very greatest variety, because a great variety of fruits and vegetables can be grown on the farm. If he will, he can grow about every fruit and vegetable adapted to his locality, and thus give that variety which is necessary to the health of the living, growing human beings.

The farm certainly affords exercise in abundance, exercise in the open air, exercise adapted to every stage, and to both sexes. Yet it must be admitted that there is not a sufficient variety of exercise on most farms for the young folks, nor for the older folks either. A riding implement of any kind tends to bend the form, and this is particularly injurious to the boys. This lack of variety can be remedied by having a croquet ground and a tennis court on the farm. These cost but little. There is always room for them on the farm, and these are games not confined to the young. If we are to have robust health on the farm, there must be games or recreation for the girls and women, as well as for the men and boys.

Given a sufficient variety of food which is really wholesome and palatable, given sports and recreation, given pure air and pure water, and the farm folks should be the healthiest folks on the face of the earth, which we fear they are not. The farm ought to be the place in which can be developed the finest type of human beings on this planet, and it will be so when we realize what it is possible for us to do in the way of growing men and women. There is no good reason whey we can not develop fine human beings, as healthy at least as the livestock on the farm. We ought to develop in every township in the United States a class of young men who can take the lead in sports and games, and in argument. In the matter of health and vigor, they ought to be superior to anything that can be grown in the city or in any other country on the face of the earth.

We can not expect to realize that dream of the psalmist—"When our sons shall be as plants grown up in their youth. And our daughters as cornerstones hewn after the fashion of a palace"—if our women follow the fashions of the day. Do you suppose the Lord did not know how to make feet on girls? Do you suppose He meant them to put the heel three inches higher than the toes, because Paris decreed it, and New York, and Chicago, and St. Louis and Los Angeles followed fashion? Do you suppose He ever intended they should wear hobbles in the way of skirts, so that they could not have free exercise of their limbs—or let us say legs—that He saw fit to put on them? Do you suppose He ever intended they should ruin their eyes by wearing veils? Do you suppose He ever intended them to twist their spines—which is really the main thing in man or beast—into awkward in order to follow the dictates of fashion? When will we have an independent spirit among farm folks, so that they will refuse to bow to the decrees of fashion in any part of the world?

The farm folk might be the most healthy and vigorous and efficient and energetic people on the face of the earth because they have within easy reach the fundamental conditions of health and vigor and energy and efficiency. They can be, if they will, and they will be, after they get the ideal clearly fixed in their minds.

CHARLES JOSIAH GALPIN

Associate professor of agricultural economics at the University of Wisconsin's College of Agriculture at the time *Rural Life* was published, Charles Josiah Galpin establishes in its pages a methodology strangely divergent from agricultural scholarship of the time. In fact, sociology as a field did not exist when, in the 1880s, Galpin studied at Colgate University in New York, where the young Galpin naturally assumed he would gravitate towards law. The grandson of farmers on both sides of his family, Galpin's interest in rural sociology seemed fated, as he grew up in Van Buren County, Michigan along with Liberty Hyde Bailey and, slightly further afield, Kenyon L. Butterfield, both of whom would end up on Roosevelt's famous Country Life Commission. In his 1938 memoir *My Drift Into Rural Sociology*, Galpin recalls, "While a rural milieu was my native habitat, religious constraint was my daily companion—religion interpreted in the home as the active obligation to others." Indeed religion, as for so many of the agrarians of the Golden Age, would remain a constant in Galpin's life, even after he left his professorial job at the University of Wisconsin to pursue a career as the head of the Farm Life Studies division in the newly developed Office of Farm Management within the U.S. Department of Agriculture. The position began in the latter days of the Woodrow Wilson administration in 1919, just a year after *Rural Life* was published. It was while firmly ensconced in his role in the USDA in 1924 that Galpin wrote the following jeremiad in the periodical *The Country Gentleman*: "Thus of the fifteen million farm children under twenty-one years of age, more than four million are virtual pagans, children without the knowledge of God. If perchance they know the words to curse with, they do not know the Word to live by."

Despite Galpin's moralizing, *Rural Life* examines rather than perpetuates the decade-old prescriptions offered rural folks under the umbrella of the Country Life movement and the Commission on Country Life. While Galpin concedes the need for change in the countryside, he warns that corrective measures will fail that are not embraced by rural residents themselves. "The new rural life," Galpin states, "cannot be achieved by epigrams from the lips and pens of exceptional persons"—by which he meant professors and their ilk. Deliberately eliminating bibliographic references, Galpin's work seems ethnographic, even folkloric, a style that marked the very beginning of rural sociology. Harkening back to his altruistic upbringing, he describes his unconventional purpose in the book as "to instigate observation of local conditions, study of one's community, and action, confident, self-reliant action." In its sometimes celebratory tone, "Country Fetes" reads as a breath of fresh air, or, as Galpin himself described it—"one voice calling the rural mind to use its own powers of discernment upon its own social problems." Apropos to agrarian and conservation land ethic, the "fetes" Galpin documents—husking bees, paring bees, chopping bees, "roundups," "boiling downs," and "sugaring offs"—all begin with the land. Introducing Galpin's memoirs in 1938, T. Lynn Smith wrote, "More than anyone else he commands the admiration of all people interested in the sociology of rural life."

Country Fetes[158]

Urban life has created the vacation idea and organized it quite generally into the various kinds of industrial employment. A period of a week or two or even a month, free from work duties, often with pay, is counted upon by many a government employee, salaried salesman,

---

[158] Charles Josiah Galpin, *Rural Life* (New York: The Century Company, 1918), 261–276.

clerk, and salaried professional man or woman as a part of regular earnings. Among the basic city workers, rest time has been wrought into half holidays, and some additional whole holidays, but especially into reduced hours of daily labor, and so into a longer rest period each day; while Sunday, with some exceptions, is a day of complete cessation from urban wage labor. Employers and proprietors in city enterprises have not been slow to incorporate the vacation into their own scheme of living. Rest time, play time, release from business and routine, have seeped into the plans of the ordinary urban family, and even the youngsters look forward to camps, tours, house parties, mountain climbings, or sea coast pleasures. Transportation companies recognize the vacation of the urbanist. Commerce is ready with pre-vacation hints, and, when vacation begins, business becomes comatose until the rest period is over. Urban vacation, holidays, half holidays, and daily leisure, moreover, are characterized by jauntiness and the gala costume of a good-time personality.

Farm life in America is in sober contrast with urban life in respect to its organization of leisure, rest, and play. Immersed in the atmosphere of solemn labor, the farmstead during work seasons is a stranger to the jaunty air of holiday. The carefree manner and dressed-up costume of leisure lack support in rural opinion even when the daily task is over and relaxation is the rule, perhaps because the virtue of work is afraid of the trappings of indolence.

When the country school lets out, the children go to work on the farm, instead of planning for amusement. Outings of pleasure apparently must be draped in a camouflage of business, errand, exchange of work, tangible advantage, in order to pass muster among the farmers: Farmers may attend the national stock shows for a few days in the winter, "to buy some better stock." Farm women go to a "sewing circle" or a "canning demonstration." Young men go to town Saturdays "to get the team shod."

Reconstruction of rural life must unquestionably face a reorganization of leisure, bringing about some release from the strain of continuous labor, and a general acceptance of the legitimacy of play and its costume and its dramatic manner. However, with these aspects of working hours and the utilization of time free from labor, we are not concerned just now. It is our task to point out the opportunity to help solve the rural social problem through organization and use of the country gala spirit generally present at the completion of a season's toil and the harvest of the annual crops.

The exulting radiance of rural satisfaction that comes from garnered wheat, corn in the ear, cotton in the bale, potatoes in the bin, and fruit in the barrel is like the glow on sunset skies. Optimism in the home is generally at its height as the social spirit waxes in the presence of the fruits of long continued toil. There is, moreover, a satisfaction, similar in character but reduced in degree perhaps, present in the rural heart when the spring plowing is over and the spring crops are all in the ground, just before the hay harvest, and before the struggle of one great stage in the annual agricultural process is plainly accompanied by a sense of satisfaction which presumably can be socialized. These two recurring periods of joy are already recognized by rural populations to a greater or less degree by means of festal occasions and events. Rural social organization can drive two great wedges into rural asceticism and aloofness at these points of periodic social joy and gladness. The festal spirit can be utilized in neighborhood and community events: celebrations, sociables and entertainments, picnics, fairs, and tournaments wherein the tide of good feeling shall float man, woman, and child into a larger sea of acquaintance, tolerance, and participation in life.

Any description of rural festivals would be incomplete without some reference at least to the work parties, called "bees," which, while especially characteristic of earlier decades, are still prevalent in America. The thoroughgoing ruralist would recall and revive the husking bees, paring bees, chopping bees, quilting bees, barn raisings, roundups, boiling-downs, and sugaring-offs.

The production or work element in the bee is distinctly rural, and may very well have served the young folks as an excuse for getting together and having a good time. The social instincts, including altruism, the virtue of the work habit, the joy of harvest time, all found common ground in the bee. "Shoveling out" in winter, roadmaking, and road dragging in spring and fall are practically "bees." The old time country parson's donation, when the products of the farm—oats for the parson's horse, potatoes and apples for the winter cellar, wheat for the mill, bushels of biscuits and cakes—were brought to the parsonage and showered upon a smiling family, was a rural screen for a joyous occasion, compatible with the stern requirements of religion. Even the country auction, appealing strongly to the economic and altruistic instincts of farmers over many a mile, is a screen in front of a good time—the fine "feed," the comedy, as the auctioneer plays his part of bargainer and jokemaster. With the entrance of the machine into agriculture, the bee will probably decrease. Threshing time, corn-shredding, and silo-filling, are pieces of continuous social labor affording small opportunity for the gala spirit. Serious work and play in one and the same costume belong to the days of handicraft. As the rural bee declines, a larger place is made for pure play on a community scale.

Rural organizations employ a few simple forms of social gatherings wholly without work features, with which to express their common life. The *sociable* is such a form, combining the basic features of the urban *reception*, along with indoor games and usually some refreshments. The journey to a country function is so considerable a part of the occasion that it may be made the center of attention, and there results a *straw ride* or *sleigh ride*; or the feast may be emphasized, and it becomes a *supper*, a *box-social*, an *oyster supper*.

Entertainment features of a somewhat more formal character—literary, musical, esthetic, even religious—often surpass the merely conversational features, and the result is a lecture or lyceum course, a concert, a home talent play or exhibition. The old-fashioned singing school was a musical sociable. The country band or orchestra is a selective social group, related directly to *entertainments*.

The country dance, at its best, is an esthetic sociable, of exceedingly complex character. As a conventional medium of getting the two sexes acquainted with one another and mutually interested, while surrounded by the atmosphere of romance, the country dance competes strongly with the sociable and the tamer forms of gathering.

When the country dance is commercialized highly and located in proximity to the tavern bar, it will not be strange if we find the dance idea exploited, in the interest of liquor sales and raffling schemes and undesirable devices on the side. As a problem in rural morality, the country dance presents the same difficulties found in the commercialized city dance.

When the religious interest is central, in the season of some leisure, a *camp meeting* or *revival* or even a *funeral* becomes a rural social medium entirely outside the sphere of work and work ideas. The social spirit expands in certain directions under the intense emotionalism of these occasions.

The rural *Chautauqua*, although under commercial stress of late years, is an event of potential social character, bringing together on a community scale the features of the ordinary country sociable and entertainment.

The foregoing brief analysis of one general form of rural rest and recreation faces us with the question, "What reorganization is possible in the interest of life, more life, and still more life in the country?" The answer to this question is being given in communities all over America: rural organizations possessing the larger vision are staging their sociables and entertainments on the platform of community service, community upbuilding, community goodwill, community acquaintance, and not in the arena of community exploitation and partisan organization-profiteering. A "community supper" or "community social," advertised by a club, school, or church, can make good its claim only by a generous, wholehearted community spirit in the management of the event and the use of the proceeds. Profits are not to be "split" feebly with the community, but are to be lavished upon the community at large. Such a faith in the community is of the essence of religion; and notable instances in the United States prove that where a club, a school, or a church has had this faith, and actually planned its good times on a community scale instead of a partisan scale, the "windows of heaven" open and pour out "unexpected blessings."

The picnic is perhaps the simplest form of a distinctive country fete. The picnic spirit, full of joy of anticipation for the children, is nearly delivered from the reminders and the vexations of work, especially if we do not press too hard upon the fact that every picnic depends upon some extra labors of the housewife. Family picnics, school picnics—are all founded upon some adventure, for a day, some home-leaving and some surrender to the drift of circumstance and emergency. There is the romantic wood and lake or river, luring everyone out of his habitual personality. But when the feast is ready and the goodies spread, the people flock together into familiar groups and the idea of the clan persists.

Only where the consciousness of neighborhood and community breaks up the hereditary shyness of families, does one find the basket picnic clan disappearing, merging into one great festal group. If a farmers' club, which has achieved the larger spirit of neighborhood, establishes a great annual community picnic, after plowing and seeding time, the social momentum of this one spectacular occasion will carry clear through the period of harvest to the next great community event.

A county federation of rural clubs finds in its county picnic a strong emotional auxiliary to its general program of technical cooperation. Memory clings to the kindly circumstances of the county picnic and enforces during the year every rational argument for rural consolidation.

The "county poor farm" affords a place for a county picnic, where demonstrations of various sorts may be combined with the more informal and general features of the picnic.

Managers of country picnics, when alive to the open heart and open mind of the picnic occasion, will tactfully force apart the units of the clan and bring them together in one common group.

The dramatic interest in rural life has appropriated the idea of the celebration as an opportunity for repeating in miniature or in symbol or in costume some notable event. The housewarming celebrates the completion of the new farmhouse. As the fire crackles on the hearth, tales are told—some of early love, some of the plans, some of the difficulties. Friends wish the house well on its mission of shelter. Good cheer, hospitality, and deeper acquaintance with people are the result.

The country church, built anew on the sacrifices of a parish, is dedicated with befitting ceremonies which recount all the steps of the cherished enterprise and give credit to the faithful workers. The last day of school is a joyful event for the country youngster. He must end the exacting period of mental discipline by a celebration which shall express his pent-up joy.

In lighter vein, the old-time "horning," or *charivari*, will pester the newly wedded couple and remind them that the worst is yet to come. Surprise parties, on birthdays and anniversaries of weddings, are packed with good feeling. Halloween is seldom passed without its country pranks.

The national feast days and holidays, Memorial Day, Thanksgiving, Christmas, Watch Night, the Fourth of July, are uniformly celebrated, Arbor Day, May Day, Old Settlers' Day, and latterly Alfalfa Day, Good Roads Day, Cheese Day, Homecoming—all have their drama and play actors. The barbecue, the watermelon feast, and the like give a rough and primitive character necessary to the art setting of these occasions.

Out of these ancient customs modern rural life can choose at will. Each particular form of celebration can be suffused with the wider spirit of our time. Feuds, prejudices, old scores, can all be "called off" for the great game of "community." Any rural institution in any locality can serve the social spirit by bringing back once in a while to the community all those still living who have ever lived there. The sentiment of youth will respond to an invitation from the little school, or the literary society, or the church, and people will leave their cities and their honors and come to celebrate in the country. They will wish to bring some gift to the home of their youth, and if the local spirit of the occasion is large enough, many a substantial contribution in the way of rural art, architecture, and landscaping, may be looked for. The reaction upon the little community will be remarkable. Community self-respect and ambition will displace the disintegrated, negligible sense of country isolation. Many a successful city man and woman having a country lineage would welcome an invitation to a "homecoming celebration" at the country school of his youth in preference to an invitation to attend college or university week at his greater alma mater.

Every country community has its historical mind that carries names and facts from one generation to another. A more systematic use of historical material in connection with the celebrations of rural community life will tend not only to bind the past to the present, but to give significance to the present in view of the future.

Trials of strength and skill in the presence of spectators are occasions, which, while they may be of educational value, are distinctly social in character. All contests of a public nature of rural children and youth should be inspected with reference to their social effects. If public contests engender strife leading to community disintegration, they should be modified or rejected, for community building, social cohesion is more precious than the training of a few local star performers.

Athletic field days, baseball matches, and basketball or volleyball team contests have great drawing power when they come at the breathing spells between the great country labors. County athletic meets, where winners from the country schools in township contests compete, have proved popular and useful as socializers. Spelling, speaking, arithmetic contests, either alone or linked with athletic contests in a county, stir up considerable school loyalty and enthusiasm. Ski tourneys are popular winter spectacles among our Scandinavian populations.

Neighborhood hunts, wolf drives, rabbit drives, squirrel hunts, have the features of a community sport. Target practice and fancy shooting are prevalent in some quarters. When hunting takes the aspect of ridding the country of a pest, a tournament becomes significant. A rat-killing contest was once known to excite so much dormant community action and enthusiasm that, as the two hundred banqueters viewed the thousands of rattails hung as trophies on the wall, they asked the question, "Can we not do more important things in cooperation than join in killing rats at one and the same time?"

Acre and five-acre contests in corn-raising and potato-raising have some of the characteristics of a spectacular match. There is no reason why such contests in a community cannot be so managed that not only each field during the growing season may be advertised to the public by a sign as an entrant in the contest, but so that on a set day for inspection a grand parade of inspection may pass from one farm to the next till all have been surveyed.

Of all the possible rural tournaments, the plowing match seems to have the largest possibilities of social and economic value. Plowing as a fundamental agricultural process contains the central work interest common to all. On a large field, where all the contestants are in plain view, there is an appealing social spectacle: horses and men in action—representatives of various farms, families, neighborhoods—excite loyalty and a degree of heated local hope. Breeds of horses and makes of plows are also matched against one another. Plow manufacturers will be present. The artistic side of farming comes out in the straight furrow, the neatly laid and well-covered field. The plowing match in the Scottish communities of America has proved a persistent medium of social value, drawing great numbers of people together. Not only is plowing in this way much improved by constant standardization, but a champion plower rallies the same romanticism about him as a champion quarterback on a college football team. Publicity given to farm skill adds interest to farm life. A plowing tournament can easily be added to other festal occasions after harvest. Scorecards will be furnished by colleges of agriculture.

Boards of commerce, in search of methods of cementing town and farm, have employed the contest method, especially in farm boys' and girls' project work. Some have attempted to stimulate rural social contests by offering prizes for the largest number in attendance of community club meetings. A county seat town commercial club, may offer prizes to all farmers' clubs in the county for the largest number of people in attendance at the greatest number of meetings during one year. The theory of the board of commerce is this: the more farmers attend farmers' club meetings the more they will wear their best clothes, the more they will wear these out, and purchase new; the oftener they meet and discuss farm life, the more rapid will be the spread of ideas of betterment, and the more will they be in the town market for lumber, furniture, plumbing, and all manner of professional service. It would hardly be legitimate for a ruralist to spread this doctrine, if it were not also true that a greater degree of consumption of these town-bought goods and services will result in "life, more life, and still more life." Well-staged country tournaments—well-organized, well-advertised, well-arranged in all details for the comfort and pleasure of spectators—always draw people from the cities to share the occasion. This adds to the self-respect of the countryman and tends both to hold population to the land and to annex the restless city dweller.

KENYON L. BUTTERFIELD

Arguably the foremost rural social scientist of his day, Kenyon L. Butterfield got his start editing the *Grange Visitor* in his home state of Michigan from 1892–1896 before serving as the superintendent to the Michigan Farmers' Institute and as a field agent for the Michigan Agricultural College. In 1902 he earned an advanced degree from the University of Michigan and was rewarded with an appointment as Instructor of Rural Sociology. One year after earning his graduate degree at Michigan, Butterfield accepted a position as president of the Rhode Island College of Agriculture and Mechanic Arts, a meteoric rise as much attributable to the desperate need for qualified, credentialed administrators at the new agricultural colleges as to Butterfield's considerable leadership potential. After three years of disappointing enrollment at the Rhode Island College, Butterfield assumed the presidency at nearby Massachusetts Agricultural College in Amherst, where he remained until 1924 and from whence he wrote *The Farmer and the New Day*.

A favorite of Theodore Roosevelt and, later, of Woodrow Wilson, Butterfield embodied the Progressivist ideal of the educated intellectual devoted to solving the most pressing problems of the average citizen. His quick rise in academe was perfectly timed to merit his appointment to Roosevelt's Country Life Commission in 1907 and to Woodrow Wilson's Commission for the Study of Agricultural Credit and Cooperation in Europe in 1913. Such was Butterfield's abiding belief in democracy that, when the U.S. entered the War in 1917, he took a leave of absence from his presidential duties to serve as a member of the Army Educational Commission of the Young Men's Christian Association with the American Expeditionary Force in France. In fact, as Butterfield put the finishing touches on the preface of the *Farmer and the New Day,* he wrote that he was finishing "under pressure of a demand to join at once colleagues overseas in educational work among our soldiers." Further lamenting his inability to find time to, as he put it, "chew the cuds of reflection" ("Executives," he wrote, "find it difficult to secure these requisites"), Butterfield resigned himself to the fact that his new book would be a statement of the "larger problems farmers face" rather than an "attempt to furnish solutions."

By Butterfield's exceedingly high standards, he had only scratched the surface of the new opportunities awaiting farmers in a postwar economy. In the passage that follows, provocatively titled, "The Farming That Is Not Farming," Butterfield touches on nearly all of the emerging agricultural and conservation trends of the early twentieth century in virtuoso fashion. Referencing in an epigraph Wilson's 1918 declaration of "the birth of a new day" of social justice and economic opportunity, Butterfield postulates an agrarian cultural moment in suspension between the question posed by *Ecclesiasticus*—"How shall he become wise that holdeth the plow"—and the Roman veneration of husbandman. Into this equipoise, what Butterfield calls a "twilight zone," steps the new farmer of the book's title, and that new farmer—be he a returning serviceman, retiring professional, or abiding victory gardener—faced limitless possibilities on the land. In "The Farming That Is Not Farming" the author spells out a dizzying array of choices—hobby farm, urban garden, community supported agriculture, prison farm, factory farm, conservation forest, backyard woodlot, landscaped park, nature school, and pharmaceutical farm, among others. The chapter aptly summarizes the varied, on-the-farm breakthroughs of the Golden Age while looking forward to an anything-goes postwar mélange of agricultural and conservation opportunities. In the spirit of the age, Butterfield would himself pull up roots after directing the expeditionary forces through 1919, implementing his newly globalized vision through the formation of the World Agricultural Council and the American Country Life Association. In later life, Butterfield returned home to serve as the president of the Michigan Agricultural College and traveled the world as an international missionary.

The Farming that is Not Farming[159]

There has been developed in America, gradually but very steadily, an interest in the soil that is not farming in the older or ordinary sense of the word. It might be called the twilight zone between farm and city. It has to do with the food production in some measure, but its greatest significance arises from quite other aspects and influences. Heretofore this twilight zone has not been of very much interest to the farmer. Indeed, he has been inclined to treat it as something of a joke. He has enjoyed the thought of the city dweller fussing with a few vegetables and calling it farming. In a few cases where it has become a factor in production, the farmer has perhaps been moved to oppose it. But the war has brought out in a stronger light this new interest. The war gardens have grown apace. There have been millions of them. Now that the war is over, most of them will be discontinued, but many will persist, and some aspects of these war ventures will become important. In fact, we must recognize that in this twilight zone there is a very important field of effort in which the soil plays a large part. The farmer ought to be sympathetic toward it. He can afford on the whole to ignore the question of its effect on the prices of his products, because its influence is not likely to be very detrimental to him, while its development means so much for humanity that it ought to enlist his sympathy. At any rate, it is probably inevitable, and even if the farmer's business is affected, he will probably have to adjust his mind and plans thereto.

---

[159] Kenyon L. Butterfield, *The Farmer and the New Day* (New York: Macmillan, 1919), 70–83.

We must not confuse this new field with what has been called the back-to-the-farm movement. There are still some who believe that our agricultural problem is to be solved by a return migration from city to farm. This twilight zone of farming does not at all solve the farm problem; perhaps it complicates it. It may help mightily to solve the city problem, for looking at it in the large way, it promises not so much an economic gain for humanity as the evolution of a great welfare movement. It is likely to become a real asset in improved methods of living. It consists of a rather miscellaneous group of activities. At present, it is more in evidence in the East where the population is crowded, but it arises wherever there are large cities with huge factories and crowded living conditions. Let us recite quite briefly some of the items in this twilight zone of farming that is not farming.

The Five-Acre Farm. The acreage suggested is a rough measure for what might be called a farmlet. It ranges perhaps from three to ten acres. It has to be carried on as a rule near a large market, under a system of intensive cultivation and chiefly with vegetables, fruit, poultry, or some combination of these, although it is quite common in some irrigated valleys in the West, especially where fruit is grown. There are cases of a more general type of farming practiced by the owner of the little farm well-tilled, but these are exceptional. This small farm can support a family only where the market is good, the soil fertile, either naturally or under commercial fertilization, and where the family can do the work without hiring extra labor, except possibly for harvesting. In some cases such a place will be occupied by a family which has partial support from other sources, but desires the country life and work for the sake of health or the better education of children, or just for sheer love of the country itself. There is evidence that the number of these little farms is increasing quite rapidly, particularly near the Atlantic seaboard, north and south. Negro farmers in the South and recent negro immigrants to the North seem to seek these small places rather than continue as wage workers. There is every reason to suppose that with the growth of cities and the resultant better markets and the increase in the price of land, very small farms will become a characteristic feature of

American agriculture and will have a considerable influence upon certain types of production.

The Working Man's Homestead. This is primarily not a matter of growing food but a chance to get a house. It is an expression of the desire to leave the crowded tenement and to find a separate house with land enough about it to insure good health, sunshine, and privacy. These little plots of one-tenth or perhaps not over one-twentieth of an acre, worked night and morning by the head of the house with more or less help from other members of the family, will grow a considerable quantity of fresh vegetables and fruits, accomplish quite a substantial saving in money, induce a larger consumption of fresh fruit and vegetables of much better quality than has heretofore been the case. This is by no means all the good that may come. In such a home, family life can be better developed than in the tenement. Children are educated by contact with growing things and get a little at least of the same advantage that comes to the farm boy who learns early in life to deal with things practical. Not the least of its advantages is that it creates respect for the farmer. This movement had gained quite a headway in Europe prior to the war. It had shown itself chiefly in what are called the garden cities of England and to some extent in this country. No working men in the world are housed so well or, on the whole, live so well as those grouped in separate houses, not over eight families to an acre. Do farmers realize the difference between a housing plan that takes care of perhaps forty people on an acre and a housing plan, or lack of plan, that purports to care for four thousand people on an acre? This arithmetic preaches its own sermon on behalf of humanity. In some cases more ambitious workmen will undertake larger areas—perhaps the one-acre or two-acre plot, in which case more of the work will be done by the women and children in the family, or by the man himself if employed chiefly in the winter, with light summer work. More and more frequently the working man who can get enough land will seek to retire from wage-earning before he reaches the deadline, because when his children are grown it may be possible for him and his wife to earn very comfortably the larger share of their living from this small plot. Before the war Belgium was perhaps the best instance of the development, on a large scale, of the working man's homestead insofar as numbers are concerned. Thousands upon thousands of Belgian working men living on farms of an acre or one-half an acre went many miles every day to and from their work. This was only possible where rapid transit at very low fares was common. In Belgium the government-owned railways provided these requirements. It is clear also that this movement involves the cooperation of large employers of labor, not only in the location of factories but by helping to provide plans, credit, supervision, and education. The provision of working men's homesteads promises to be one of the great social movements of the new day.

The Factory Garden. This is, in America, purely a war development. The manufacturer sets apart or rents a considerable area, perhaps ten to forty acres, organizes it as a unit of management, and allots parcels to individual employees to till. This plan requires expert supervision, as well as the preparation of the ground, the purchase and application of fertilizers, and probably the purchase of seed, by the employer. This scheme has proved substantially helpful in increasing the food supply of the working man, but it is likely to be rather temporary as a large movement. It may, however, play quite a part after the war for those working men who are for any reason barred from garden cities and yet who wish to work parcels of ground.

Use of Vacant Land in Cities. Twenty years ago, the mayor of Detroit, Michigan, caused a national smile by advocating the "Pingree potato patch"; but it was a good idea. An enormous amount of absolutely idle land within the confines of every city is worse than useless because it is usually unsightly; it spreads weed seeds, and in a day when thrift is again coming to be a virtue, one rebels at the thought of waste anywhere. Again we may learn from Europe where, to a much greater extent than with us, these idle lands have been put to use. Generally speaking, this plan should be handled by municipalities. It cannot be very successful or widespread without invoking a compulsory law to bring such land into use under terms that are fair to the owner. The use of these plots needs organization and superintendence because most of the workers are not experienced. They especially need protection from vandalism. To thousands of dwellers of the tenements, the vacant-land garden would be a great boon.

The Community Garden. The English government, during the war, has made a multitude of allotments of land to working men by which they can grow a portion of their own food. It is understood that food production in England has increased fourfold during the past two or three years, and that this increase is largely due to the small allotments to thousands of people who had never before grown any part of their food supply. Allotments may be handled by cities as just suggested, utilizing the vacant land. Another development may be the provision by the community, small or large, for its own fruits and vegetables. This may be either by arrangement with individual growers or by municipal management of the enterprise.

Several successful community gardens have been conducted in Massachusetts during the past season. The city of Worcester furnished a tract of land, plowed and fertilized it, and divided it into plots of one-eighth of an acre. Any one might secure one of these, pay for the plowing and the fertilizer and plant what he wished. A garden supervisor was provided by the county farm bureau. The gardens have been counted as very successful. The town of Newton adopted a somewhat different method. The town furnished the ground, plowed and fertilized it and supplied seed potatoes for planting it. Any citizen of the town was allowed to work upon this tract, and according to the amount of work each had done, the crop was divided at the end of the season, after the expenses of plowing and fertilizer were deducted. This method likewise has been very satisfactory.

The Home Garden. The home garden in the village or in the suburban town has long been characteristic of America. It does not need much attention from outsiders. It is an individual matter. To many people it constitutes one of the great attractions of the life in the smaller group. While it is to be encouraged, it can hardly be organized.

The Farmer's Garden. There are thousands of farmers who regard a garden as a nuisance. They wont "fuss" with it. They have no time for it. They have "bigger things to do." So the garden is neglected and the result is an astonishing lack of variety on the farmer's table, where one would naturally expect the greatest variety. The luxury of fresh fruits and vegetables is missed by those who could have them most easily.

In some whole regions of our country, canned fruits and vegetables are bought on store credit and used in lieu of homegrown products. One of the big educational campaigns of recent years has been conducted by the Department of Agriculture to try to meet this astonishing situation. Some farmers have proposed that in a community of farmers whose chief interest is in stock growing or general farming, there might be either a community garden or an arrangement with certain individuals for the growing in the community itself of fresh fruits and vegetables, the other farmers furnishing the market. In other words, it would be perfectly feasible for an organized community either to arrange with one of its number to grow "garden sauce" for the neighborhood, or to hire a specialist to manage the community garden.

Boys' and Girls' Gardens. The development of boys' and girls' gardens and of boys' and girls' agricultural and canning clubs has been one of the great educational movements of our time. Indeed, its educational value is its chief value, although if we could have an accurate census of the value of the products grown by these hundreds of thousands of American boys and girls both in city and country, they would receive great praise for their practical contribution to our food supply. For the farm boy and girl, this movement has awakened new interest in the science of farming, new interest in farm processes, new knowledge of scientific methods, and a new love for growing things. It has given the zest of responsibility and possession. It is also astonishing to discover the extent to which city and village boys and girls have participated in this movement. It is estimated, for example, that in the state of Massachusetts this past season not less than 75,000 boys and girls who are not living on farms carried on gardens or even larger enterprises. There is much testimony to the awakening that has come to many a farmer and a farmer's wife through the successes of the boy and girl in trying new methods. It is difficult to overstate the importance to agriculture and country life of the boys' and girls' gardens. It is a selective process. It is foolish to try to keep all farm-bred boys and girls on the farm. It is equally foolish to seek a great migration of city people to the country. But we do want something that will tend to keep the farm-minded boy and girl on the farm and to send the farm-minded city boy and girl to the farm. This the boys' and girls' gardens tend to do. It also will help greatly in making future consumers appreciative of good food, what it costs to grow it, and how it is to be cared for in the home.

The Estate. By this is meant the country home of the man of means whose business is not farming at all. The practice of living in the country for at least half the year is rapidly growing. It is a healthy, normal, educative movement. It leads to outdoor life, to a new understanding of country things, and occasionally helps to educate a community to better farming. However, the growth of estates in this country is likely in the near future to become a real problem. One can see what the possibilities are, if unrestricted, by studying the situation in Great Britain and Ireland up to very recent times. Of course with our abundance of land it will be a long while before the problem of the estate is a national concern. But already in some of the smaller states of the East, land that ought to be producing crops for nearby markets is monopolized for mere pleasure. Of course, if the farming of these estates were really made to pay, the estate would simply become a large-scale farm and would be judged on its own merits. It is said that in one county in the East nearly one-half of the land, some of it the very best farming land of the county, has gone into estates that probably will produce one-fourth of what the land would produce if farmed by small farmers growing truck crops for the nearby large markets. We have seen not only in Great Britain and Ireland but in Germany the government itself stepping in to break up the large estates. It is a question that may need our attention in America.

Forestry. Theoretically forestry is a branch of agriculture. When we are fully alive to its importance, we shall treat trees as crops. It will require, however, a very great stretch of the imagination to think of forestry as a branch of farming. The farm woodlot, however, is much more worthwhile than it seems to most farmers. The time will come when it will be worthwhile really to conserve our coal, and a not unimportant item in this conservation program will be the prevalence of wood-burning furnaces in the farm homes, the wood being obtained either from woodlots or the home farm. In the aggregate, this practice would result in an immense saving of coal. Forests are the only crop that can be grown in all rough or mountainous regions. Great areas in both the East and West are useless for anything else, and they are now producing only intermittently and fitfully their full capacity of forest products. It is almost impossible to expect individual owners to change this situation. Possibly trust companies can be encouraged to invest in and develop forest areas on a scientific basis as a means of utilizing funds in their charge. But in general it may be said that the only possible way of establishing and maintaining an adequate forest policy is for the government to do it. Not only the federal government but each state should be moving in this direction as rapidly as possible. It means the best use of the land, better conditions for farming due to the effect on conservation of water, cheaper lumber, and so on. Not the least of the possibilities of forestry consists in the fact that a scientific forestry policy carried on by a state over a series of years could be made to yield a substantial income for the support of some permanent interest of the state, such as the public schools. Variations of the effort to grow trees for wood products, are the growing of nut-bearing trees, sugar- and oil-bearing trees, and of ornamental trees and shrubs.

City Forestry. The water supply of a city is a vital concern. As a region becomes exceedingly populous, the difficulty in keeping a supply both adequate and pure increases. There is no doubt that for larger cities, at least, the question of adequate forestation of the areas which supply the reservoirs will become a regular part of municipal policy. The matter of parks and playgrounds is of course of little direct interest to the farmer but of most intense importance to the city, and these, in connection with forests as a part of the park system of a city, have to do intimately with soil and the soil treatment. It is not impossible to conceive of cities ultimately gaining some substantial revenue from their forests. This is already done in some places in Europe. The use of trees and shrubs for decorative purposes in streets and parks and in fact the whole question of what is called city forestry, including planting, care, protection from the ravages of insects, and disease, constitutes a large factor in city planning.

The Landscape. When we say landscape, we are likely to think of parks or estates of the wealthy, but the farmer has a landscape with him every day. Perhaps he too seldom uses his opportunities to make his surroundings beautiful at small expense. The beautification of country highways, the establishment of village parks and playgrounds, the landscape adornment of public buildings, schools, churches and Grange halls is too much neglected. But there is the landscape as nature has it. The farmer has access to beautiful views. Does he not sometimes need education in landscape appreciation?

Soil Specialties. There are other uses of soils than the growing of food, feeds and fibers. Floriculture is a large industry. The systematic production of the medicinal plants is increasing. Specialties, such as mushrooms, rhubarb, et cetera sometimes make quite profitable returns. Seed farms and nurseries are common. These specialties will increase in number and in the aggregate they will eventually comprise quite a substantial business. More than that, they represent a very intelligent use of the soil and a highly skilled utilization of plants.

The Soil and Social Amelioration. It has recently been stated in France that the victims of shell shock recover much more rapidly if they can be put to work in the fields. Better than medicine, better than nursing, better than the hospital is the soil in the open country. Now this statement is only a new illustration of the fact that both physicians and social experts discovered some time ago and to an increasing extent are putting into operation. It is common knowledge that schools for delinquents, the old-fashioned "reform" schools, have usually been placed in the country. The most enlightened prison policies now provide for farms in connection with the prisons and the habitual use of prisoners in producing their own food. Not long ago, one of the most prominent experts connected with the treatment of the feeble-minded asserted that in the future these institutions must consist of colonies so located that the inmates could not only have open air but farm work. It has been found that even the insane can be used to a very large extent in many farm operations. The outdoor work tends to health; steady employment makes for mental poise and sanity. In all these institutions there is some considerable saving of expense to the state. If properly managed, these farms could be also demonstrations in good farming. It should be more generally understood that the increase in delinquents of various sorts, physical and mental and moral, is becoming a serious menace to our civilization, both in country and in city. If, therefore, the use of the soil as a means of amelioration and possibly of curing is practicable, it has very far-reaching consequences. We shall also need to provide easy facilities by which those who are partially disabled, either physically or nervously, can be placed upon the land. It will not do to arrange for these people under the expectation that invalids can do farm work. But men who need to be outdoors, and can do fairly active work, or men only partly disabled can farm small pieces of land in many cases to advantage. This is destined to be a part of a national policy for taking care of the considerable current of men and women who would seek the country if they knew the way.

Game Farming. Just as there are soil specialties, so there are animal specialties, growing of pets, of fur-bearing animals, of game. In general, the state itself or large land-holding concerns can carry on these types of farming to best advantage. In some portions of the country, naturally wooded, and in connection with the forestry policy, game farming can be made a considerable factor both in the production of meat and in the increase in value of animal products. A variation of game farming is fish farming, that is, the use of fresh lakes, ponds and streams for the production of fish. We are just beginning the development of this field.

The Soil as a Machine. The average American thinks of the soil merely as a storehouse of plant food. But in all older settled regions, farmers discover that it is desirable to make a highly intensive use of the soil, not so much a reservoir of fertility as a container of fertility. Commercial fertilizers are added to the soil and furnish the major part of the plant food. Glass farming is dependent upon this use of the soil, as are also crops that are grown out of their normal season. When an effort is made to get unusual yields of special quality, the same principle is brought into operation. It is a principle that has the utmost significance in all countries where population presses upon available farming areas. This is not farming in the ordinary sense of the word, but again it is a highly intelligent and skilled use of the soil for growing things that man wants. We have here a powerful social appeal to people to tie themselves up with a bit of the land for the sake of health and sanity and good influences.

All this field of farming that is not farming is therefore sure to broaden. It ought to have the sympathetic understanding of the farmers. It is really big with importance for humanity. These things also mean a gradual change of attitude on the part of consumers. When they have their own gardens, they will come to know that cabbages come from the land instead of from the grocery. They will know something of the toil and sweat and disappointments of the producer and of the real costs of production. They will themselves develop more discriminating tastes and will increasingly call for higher quality, and of course the demand for quality in the long run spurs the farmer to his best effort and best profit. On the whole, it will make for a freer consumption, especially of fruits, vegetables and poultry products, which can usually be grown in areas near the market, and a reduction of costs and wastes of transportation, storage, and distribution.

This twilight zone also has a tremendous significance in an educational way. It is working itself gradually into the system of public education, and calls for trained administrators. It promises to send students in largely increasing numbers to the agricultural schools and colleges. It even means something in the way of quantity production and ought thereby to assist in solving the problem of food supply.

MARION FLORENCE LANSING

The daughter of Jenny Stickney Lansing, who compiled the well known Stickney Readers, Marion Florence Lansing seemed destined to carry on the familial literary tradition. A graduate of Mount Holyoke College, Lansing first made her literary fame between 1905 and 1910 by editing volumes of fairy tales she discovered while browsing the Harvard Library in her hometown of Cambridge, Massachusetts and publishing them with her mother's publisher: Ginn and Company of Boston.

Lansing's essay "At a World Table" comes from the ultimate chapter of her 1920 monograph *Food and Life*, a timely volume released during the aftermath of World War I and roughly coincident with Iowan Herbert Hoover's efforts, on behalf of the Wilson administration, to coordinate food shipments first to the Allies and, later, to the defeated. Lansing's writing demonstrates the complete sea change resulting from the war, which fueled American food production and once more elevated the status of the farmer. Lansing's encouragement of American youth towards the field of farming—in order to prevent famine and, subsequently, to increase political stability—suggests a newly globalized American agrarianism, as does her central metaphor of the world food table. Important here is Lansing's application of the conservationist modus operandi to food production, food consumption, and food aid to less prosperous nations. Also implicit in the world food table metaphor is the notion of food as an agent of foreign policy and national supremacy.

A believer in science—she would later publish such titles as *Great Moments in Science* and *Moments in Exploration*—Lansing's emphasis on agrarianism and Christian brotherhood reminds of Liberty Hyde Bailey, as does her emphasis on young adults as vehicles for social change. In her preface to *Food and Life*, Lansing summarizes the country's shifting priorities thusly: "War did a real service in bringing people back from the conventionally remote attitude of modern civilization to a vivid realization of the interest in and importance of the universal human need."

At a World Table[160]

Even as we buy at a world market and eat from a table spread with foods from all over the world, so we sit at a table at which is seated with us all the rest of the hungry world.[161]

---

[160] Marion Florence Lansing, *Food and Life* (Boston: Ginn and Co., 1920), 153–159.
[161] This sentence marks the second sentence of the second paragraph of the original chapter. The excerpt begins here for contextual clarity.

War has brought the picture of a world table freshly to our minds. The United States gave its splendid example of voluntary self-rationing because of the appeal of other members of the world family who were rising from the table hungry because there was not food enough to satisfy them. The United States had food in plenty for itself. It deliberately set aside a part of that food for the needs of the warring nations. It sent out of the country food which it might have eaten, because Americans would not stuff themselves with plenty while others starved. This was a beautiful thing to do, but it was the only thing to do. No man or woman or child could have enjoyed food if he had been actually sitting at table with hungry Belgians or Serbians or Poles or Armenians who were not being fed. The danger was that we should forget these other members of the world family because they were out of sight. That is the danger always. We need to stir our imaginations to picture this world table. When we are tempted to leave good food on our plates or to throw away a piece of bread, we must train ourselves to see some hungry child reaching out for bread and not getting it because there is not enough to go around. The chief lesson of sitting at a world table is not to waste. If we eat what is set before us, we release for sending overseas other foods which are needed there.

The sharing of food is the sign of a new world brotherhood for which men everywhere are hoping and working. Science has made it possible for the world to become one, sitting at one table. A man can speak from Wales to Australia by wireless message in a fifteenth of a second. Surely no nation need go hungry without other nations' knowing of its need. Our land and ocean systems of transportation make it possible to send food quickly. Our new scientific farming makes it possible to raise food to feed adequately the nations of the earth. The land has never been worked as it can be worked. Two blades of grass can be made to grow where one grew before. A small plot of ground properly enriched and tended and protected from pests will yield far more than it ever yielded under old farming methods. Yet without the vision of a world table and the desire for a world brotherhood, science alone would be slow in saving the world from famine.

Today we are all summoned to take part in a new crusade—to drive famine from the earth. Famine is a dreadful specter that has always stood just behind the poor man and the nation whose food supply was barely equal to its needs. It has been ready to pounce on its victims the moment wages ceased or a crop failed, bringing with it attendant woes of disease and anarchy. It is the enemy of law and order, the foe of prosperity and contentment.

To drive famine from the earth more food must be raised. Even before the war the world was in danger of going hungry. The population of the globe is increasing. Its food supply must therefore increase. Every human being must eat; he must be a consumer of food. The more need there is of food, the more producers there must be. "Everyone who creates or cultivates a garden helps, and helps greatly, to solve the problem of the feeding of the nations," said President Wilson. To solve this problem there must be more than gardeners; there must be farmers. The farmer is the leader in the world crusade against famine. To be a food producer is to be an active partner in the world's food business, and active partners were never more needed. Boys and girls should think, when they are choosing what they will be, whether they can choose this for their vocation. If they can, they will be doing a splendid service.

To drive famine from the earth there must be less waste. Here everyone can take a part. "This is the time for America to correct her unpardonable fault of wastefulness and extravagance. Let every man and woman assume the duty of careful, provident use and expenditure as a public duty."

To drive famine from the earth there must be world brotherhood. Here boys and girls can help. When boys and girls do anything, they do it with all their might. They do it joyfully as an adventure. They do it all together as they would play a game. They do it in the spirit of King Arthur's knights, who "rode abroad redressing human wrong." For them "every morning brought a noble chance / And every chance brought out a noble knight." They gloried in a vision of a world protected and purified by their valor, and in that vision wrought "All kind of service with a noble ease / That graced the lowliest act in doing it."

To raise food or save food without a vision, as many worthy folks are doing, is good service, but it is not the kind of service that makes of the world one brotherhood. For that we must have the vision. Boys and girls are the ones who can catch the vision and work for it. They can keep before themselves and others the vision of all the world seated at one table, repeating together the familiar prayer: *Give us this day our daily bread.* They can help to answer that prayer for the world, and so become, like Arthur's knights, "the fair beginners of a nobler time."

About the Editor

Zachary Michael Jack, fourth generation Iowa farmer's son and great-grandson of the celebrated farmer and conservation writer Walter Thomas Jack, is the editor of two previous collections on rural life, both nominated for the Theodore Salutous Award for the year's best book on agricultural history: *Black Earth and Ivory Tower: New American Essays from Farm and Classroom* and *The Furrow and Us: Essays on Soil and Sentiment*. Jack's love of nature originates in his family's 152-year-old Iowa Heritage Farm and timberland and in his years spent teaching at colleges and universities at the foothills of the Smoky and Organ Mountains.

A poet, essayist, literary journalist, community arts activist, and environmental scholar, Jack is the founding director of the Iowa-based School of Lost Arts for children, advisory board member for the Humanities-Net forum for place studies (H-Place) and author of two book of poems, *The Inanity of Music and Wings* and *Perfectly Against the Sun*. An assistant professor of English, he teaches courses in writing and rural studies at North Central College.

Printed in the United States
89405LV00007B/2/A